NO
TURNING
BACK

NO TURNING BACK

The Life and Death of Animal Species

WRITTEN AND ILLUSTRATED BY

RICHARD ELLIS

HarperCollins *Publishers*

HarperCollins books may be purchased for educational, business, or sales promotional use. For information, please write: Special Markets Department, HarperCollins Publishers Inc., 10 East 53rd Street, New York, NY 10022.

FIRST EDITION

Designed by Joseph Rutt

ISBN 0-06-055803-2

To-morrow and to-morrow and to-morrow
Creeps in this petty pace from day to day,
To the last syllable of recorded time;
And all our yesterdays have lighted fools
The way to dusty death. Out, out, brief candle!
Life's but a walking shadow, a poor player
That struts and frets his hour upon the stage,
And then is heard no more; it is a tale
Told by an idiot, full of sound and fury,
Signifying nothing.

Contents

III. FINALE

If you're booked for extinction, there's nothing much you can do about it.

—Will Cuppy,
How to Become Extinct, 1941

Let us divide up extinction into more manageable pieces. First, there are the great extinctions caused by climatic shifts. It's simply too hot or too cold for something to survive. This kind of extinction we can understand, since the change of climate attacks food supplies and all aspects of habitat at the same time.

Then there are the extinctions caused by hunting—the dodo and the solitaire supposedly come into this category.

Then come the other kinds of extinction, which are altogether more mysterious. Some species are too rigidly specialized. Their lack of adaptability to even small changes, changes so minute that they do not register as such to us, causes them to die out.

As humans we rightly value the ability to adapt to new surroundings. Those who don't, we cheerfully label *dinosaurs*. Companies that can't adapt deserve to go bust.

And don't we feel, deep down, that those species that have become extinct somehow "deserved it" too, that their very inability to survive was a moral black mark against them? By a curious alogic, not surviving means you didn't deserve to survive in the first place.

—Robert Twigger,
The Extinction Club, 2001

Preface

I think I have been writing one book all my adult life. Well, maybe two. The one book, largely about life on earth, the other was about Atlantis. The one book, largely about life in the sea, which I began almost thirty years ago, was conveniently chopped into book-sized pieces by various editors. I began with sharks, moved on to whales and dolphins, then on to whaling history, followed by the great white shark, sea monsters, creatures of the depths, the giant squid, and the origin and evolution of life in the sea. I then chronicled the modern-day depletion of the oceans' resources—the disappearance of many of the creatures I had written about earlier. I recently took a paleontological detour to investigate the lives of the long-extinct marine reptiles—the ichthyosaurs, plesiosaurs, and mosasaurs, and this excursion into the past inevitably brought up the topic of extinction, the subject of this book.

As told in *The Empty Ocean,* extinction is no stranger to the sea. In historical times, some of the most noteworthy species losses have taken place in the water; more important, there are others to come. Steller's sea cow, discovered in 1741 by Vitus Bering's shipwrecked crew on the island that would later bear his name, was hunted into extinction a mere 28 years later. There are no more great auks, Labrador ducks, or Caribbean monk seals, all eliminated in the past two centuries. Overfishing has driven many species to the brink, and our children's children may never see a barndoor stake, a northern right whale, or a broadbill swordfish. An important study, published in 2003 by Ransom Myers and Boris Worm just as *The Empty Ocean* was coming off the presses, noted that industrialized fishing has reduced the numbers of all large predatory fishes—tuna, swordfish, marlins, groupers, codfish, sharks, skates and rays—by 90 percent. Failure to apply the brakes to longlining and gillnetting, the most destructive fishing methods ever devised, will surely result in mass marine extinctions in the near future.

As tempting as it is for conservationists like me to couple "overfishing" with "extinction," the latter process has been part of the history of life on earth since the first cells subdivided billions of years ago, long before *Homo destructivus* made his appearance on stage. Even though many—probably most—of the pieces of the puzzle are missing, almost everyone acknowledges the long, complex process of evolution, where early species gave rise to later ones, and so ad infinitum. But what happened to all those earlier species? They went extinct, of course, and *H. sapiens* had absolutely nothing to do with it, since we've been around only for about 100,000 years.

Many of the earlier extinctions were caused by forces that exceed anything that humans could devise, and in some cases, may even exceed their powers of comprehension. Try to imagine volcanic eruptions lasting for a million years (while you're at it, try to imagine a million years), releasing forces that destroy 90 percent of all life on earth. The great Permian extinction, which occurred 250 million years ago, was so catastrophic that life was almost brought to a close. A hundred and eighty-five million years later, at the close of the Cretaceous, an asteroid the size of Mt. Everest slammed into the Earth at 60,000 miles an hour, and created such environmental havoc that the oceans boiled, skies darkened, acid rain fell for years, and tsunamis half a mile high drowned everything in their path. The Chicxulub impact, named for the Mexican village where the impact crater was discovered, took out various protozoans and algae, the ammonites, the pterosaurs, the pterosaurs, the mosasaurs, and the last of the nonavian dinosaurs, including *Tyrannosaurus rex,* the most formidable land predator that ever lived.

One might expect that such an enormous cataclysm would destroy virtually every living thing, but many creatures somehow survived the impact and the chaos that followed, including ancestral birds, early mammals, and ancestral crocodiles and turtles. The "Age of Reptiles" ended with a big bang 65 million years ago, but some reptile lineages made it through, and the warm-blooded vertebrates began the journey that would bring one of them (us) to a position of such global dominance that we can now contemplate our awesome power to generate destruction and modifications that rival the magnitude of the end-Cretaceous extraterrestrial impact.

The sheltering sea offers no protection from the relentless power of extinction. Indeed, while life is believed to have originated in primitive oceans, most of the early life forms are long gone, including trilobites, ammonites, jawless fishes, and weird sharks and fishes that left no descendants—even the amphibians that are believed to have taken the first hesitant steps out of the water, which eventually led to the terrestrial vertebrates (like you): gone, long gone. At one time, gigantic marine reptiles dominated the world's oceans: 65-foot-long ichthyosaurs chased down their Jurassic prey; four-flippered, long-necked elasmosaurs somehow caught small fishes and cephalopods; and gigantic mosasaurs, with teeth larger than those of *T. rex*, terrified anything and everything that swam within range. Gone.

Eventually, the sea was conquered by starfishes, sea cucumbers, cephalopods, sharks, fishes, and whales, but the ones we see today are not the same as their ancestors. There were sharks that looked as if they had antlers, others that had teeth on top of their heads, still others that had a whorl of teeth like the blade of a circular saw. The giant *Megalodon*, an oversized version of today's great white shark, had serrated teeth the size of your hand. There were early whales with feet, and one species that had a single, six-foot long, backward-pointing tusk. Their existence is known only from fossils.

We all know that the great dinosaurs are long gone, despite the cinematic fantasies that they might somehow be cloned. The Chicxulub impact eradicated the last of them, but long before the impact, other dinosaur species became extinct without benefit of asteroid: *Iguanadon, Protoceratops, Velociraptor, Hadrosaurus, Apatosaurus* (neé *Brontosaurus*), *Brachiosaurus,* and hundreds of others with less familiar names. (Most of the dinosaurs of *Jurassic Park* actually lived during the Cretaceous period, not the Jurassic.) We know where and when they lived, but we don't know why they disappeared.

Perhaps it was to make room for their successors. *Tyrannosaurus* is extinct, but thanks to one of the most astonishing revelations in all of paleontological history, you can recognize its lineal descendants today, at your bird feeder, in the trees and bushes outside your window, flying over meadows and moors. The dinosaurs are not extinct at all. A lineage can be traced from certain smaller Cretaceous dinosaurs directly to living birds, indeed, *birds* are *dinosaurs*. It has long been known that the primitive bird *Archaeopteryx* could fly, but ever since the first fossils were found in 1860, *Archaeopteryx* has posed more questions about the descent of birds than it answered. But new amazing answers have come from the fossil beds of China, where numerous specimens have been unearthed since 1990, showing dinosaurs with feathers, and some that could actually fly. One species even had four wings. Of all the events discussed in this book, probably the most surprising is the *nonextinction* of the dinosaurs.

The flying dinosaurs are still with us, but all other flying reptiles went extinct, and left no descendants. For millions of years, a large group of winged reptiles, known collectively as pterosaurs ("winged lizards"), dominated the skies. The first vertebrates to fly, they had light hollow bones, and leathery wings that stretched between the long fourth finger-bones and the tail. They were not dinosaurs and they were not birds. They ranged in size from the pigeon-sized *Pterodactylus kochi* to the incredible *Quetzalcoatlus,* which had a longer wingspan than a World War II fighter plane; they were probably warm-blooded, as birds and bats are today. We do not know if they soared or flapped their wings, and so far, we cannot figure out how such ungainly animals could have walked on land—if they walked at all. Some species survived for 10 million years; many appear to have died out 144 million years ago, a date that does not coincide with any known extraterrestrial impact or mass extinction. *Quetzalcoatlus,* with its 50-foot wingspan, was one of the last of the pterosaurs, disappearing roughly at the time of the K-T extinction, 65 million years ago. Once the pterosaurs were gone, nothing like them was ever seen again.

Terrestrial extinctions are certainly as puzzling as their marine or avian counterparts. Although we are unable to comprehend the big picture, that is, why so many species have become extinct, we might be better able to understand those cases where we have more clues, or where the extinctions occurred more recently. (Of course, when we ourselves killed all the individuals of a species, as we did with the passenger pigeon or the Tasmanian tiger, we only have to look in a mirror to identify the perpetrator.)

Around 12,000 years ago, human beings crossed the Bering Land

Bridge from Asia to North America. Some think they walked across, others that they came by boat, but they managed to get there one way or another. For millions of years before their arrival, the plains of North America had been populated by large mammals such as mammoths, wooly rhinos, sabertooth cats, cave bears, ground sloths, giant deer, huge camels, and wild horses. Coincidental with the arrival of humans, the animals began to disappear; within a couple of thousand years they were all gone. Some are preserved as fossils; others are known because they were trapped in tar pits such as La Brea in California. It is easy to draw the connection between the extinction of these Pleistocene mammals and the arrival of hungry humans. Indeed, many paleontologists are of the opinion that man-the-hunter killed them all off—or at least reduced their populations so that they were unable to propagate their species. Other scientists believe it was Ice-Age climate changes that did in the megafauna, while still others argue that a combination of "over-kill" and "over-chill" was responsible for the extinction of the large mammalian fauna.

Recently, a new possibility has been suggested: some sort of hyper-disease pathogens, perhaps carried by humans, perhaps by the dogs that accompanied them, may have been introduced from Asia into North America, and infected animals that had no immunity to them. Before you dismiss this idea—known facetiously as the "over-ill" theory—consider the AIDS virus, which is believed to have jumped from chimpanzees to humans; as of this writing, it has killed more than 23 million people worldwide. If modern medicine cannot find a cure for AIDS, imagine what would have happened if an epizootic was introduced 10,000 years ago. The North American megafauna is extinct, and has been replaced by other mammals, not necessarily the descendants of the Pleistocene species. *Something* killed off all the mammoths, cave bears, and sabertooth cats, but we don't know what it was.

In recent years, diseases of various kinds have affected disparate groups of animals; one amphibian species, the golden toad of Costa Rica, is believed to have been wiped out by a fungus. No other recent species has been made extinct by disease, but enough species have been devastated to raise the possibility that an epidemic could indeed eliminate an entire species. In the 1890s, the morbillivirus that causes the herbivore disease known as rinderpest killed two million domestic cattle every year in sub-Saharan Africa, and also infected wild buffalos and various antelope species. Black-footed ferrets in the American West, which have been the subject of intense conservation efforts, are fatally susceptible to canine distemper virus (CDV)—also a morbillivirus—which can be transmitted by dogs and coyotes. Lions and wild dogs in Tanzania have also died of CDV, after coming into contact with domestic dogs.

But nowhere has mutated CDV been more destructive than in the oceans. Believed to have been introduced into Lake Baikal by the discarded carcasses of sled dogs that had died of the virus, CDV killed 70 percent of all the Baikal seals, and then somehow leapt from this enclosed body of water into the sea. Grey seals and harbor seals died by the tens of thousands in western European waters in the late 1980s; after a lull of some twenty years, CDV returned. Infecting seals, the virus mutates into phocine distemper virus (PDV), which has affected various species from the North Sea to the Mediterranean. The last Caribbean monk seal was seen alive in 1952, and the species is now believed to be extinct. Similar in everything except geography, the Mediterranean monk seal is considered one of the most endangered of all large mammals. If any of the 1,000 remaining Mediterranean seals should contract PDV, it could spell extinction for the entire species. Viruses seem to be particularly effective in the water, and PDV insidiously mutated again, this time to infect dolphins. Common and bottlenose dolphins began washing ashore on European beaches in 1988 and in September 1990, hundreds of striped dolphins appeared as well on the shores of France and Spain. Specialists concluded that the dolphins were the victims of a viral epizootic that was spreading to both the North African and the French Mediterranean coasts. West Nile Virus begins in birds, but has now been known to kill people as well.

I own many books about "the history of life," or about some subset of creatures that roamed the earth before the present. I will choose one at random: *The Marshall Illustrated Encyclopedia of Dinosaurs & Prehistoric Animals*, published in 1988, and reissued in 1999. The drawings are most impressive. The book contains color pictures of over 500 different vertebrates, beginning with the jawless fishes, and working through amphibians, marine reptiles, flying reptiles, "ruling reptiles," mammal-like reptiles, birds, and finally mammals. I have another inclusive book, *Prehistoric Life*, by David Norman, that includes the Ediacaran fauna and other ancient invertebrates such as brachiopods, trilobites, crabs, worms, jellyfishes, cephalopods, etc. The point is the same: Every single animal in these books is extinct, except the last one, *Homo sapiens*.

No other creature in the three-billion year history of life has had such an effect on his fellow passengers on Spaceship Earth as the upright, bipedal, hairless, big-brained hominid that calls himself *Homo sapiens*. We began modestly—and recently—killing off many large animals for food or self-defense. But in recent years we have ratcheted up the rate; we are now mowing down entire species with terrible and reckless abandon. That various whale species survived what can only be described as a war is a testimony to man's inefficiency; during the worldwide whaling frenzy of the nineteenth and twentieth centuries, it certainly looked as if we were trying to eliminate all large cetaceans from the face of the earth. Mankind has already destroyed countless species on land and sea, and is hard at work to raise those numbers. In many instances, we do not directly target the particular species, but if we poison or destroy the habitat in which it lives, the result is the same.

For the most part, we have eliminated animal species because we ate them out of existence, but we also killed millions of creatures for their fur coats or their oil. There were some, such as the Tasmanian tiger (thylacine), that we eliminated because we perceived it as a threat—not to us, but to our sheep. Fortunately there have been some instances in which our anomalously large brain has enabled (some of) us to reflect on what we were doing, and had us sound the alarm for creatures not quite extinct. It is too late to save the dodo, the passenger pigeon, the Carolina parakeet, the great auk, the ivory-billed woodpecker, Steller's sea cow, the Caribbean monk seal, and other less charismatic species, but when it appeared that we were on the brink of losing forever the sea turtles, the California condor, the whooping crane, the black-footed ferret, the American bison, the giant panda, Przewalski's horse, and Père David's deer, a few people who had finally realized that the big brain could be used for something other than destructive purposes stepped in and tried to turn the tide. Curiosity, they say, killed the cat, but it has also been the impetus for a sort of antidote to extinction, the discovery of many heretofore unexpected animals, such as the coelacanth, the megamouth shark, the Indopacific beaked whale, and even a few creatures with weirdly unfamiliar names—the bonobo, the okapi, and the saola.

Recent discovery, however, is far from a guarantee of protection. Gorillas were first reported by Europeans around 1859, but the subspecies that lives in the mountains of Rwanda is now seriously endangered; in fact, all gorillas are now imperiled by the combined threats of disease (the Ebola virus) and humans who kill them for food, which then becomes "bushmeat." The bonobo (previously known as the pygmy chimpanzee) was first described by science in 1933, and it is already endangered by forest reclamation and bushmeat hunters. The coelacanth, a primitive fish once thought to be extinct since the days of the land dinosaurs, was discovered alive off South Africa in 1938; because of its rarity and its place in the evolutionary history of vertebrates, it is being collected by museums to the point that it is now considered endangered.

Documenting the 3 billion-year history of extinction is an exercise in triage. I have had to choose which creatures to include in this account and which to leave out. An attempt to list—never mind to discuss—the vast number of creatures that have disappeared from the earth and the oceans would have been futile. I am guilty of speciesism, emphasizing some species over others, so you will read very little about the destruction of trees and other plants, even though their loss is significantly more important in the great scheme of life than the loss of a couple of birds or monkeys that lived in the treetops. There have been any number of "parallel" extinctions, where more than one similar species was eliminated, and I have had to select from the better-known or more dramatic accounts. Thus, only a couple of the hundreds of species of land dinosaurs are mentioned; the flying reptiles are represented by a few spectacular examples; and the great marine reptiles receive only the most perfunctory of discussions.

This, of course, raises the question of how many extinct species are known at all. The fossil record is scanty at best, and there are millions of species whose existence we do not even suspect. New paleontological discoveries are a source of never-ending surprise (think of the feathered dinosaurs), and a book like this can only serve as an introduction to a subject that is as vast as life itself. Extinction has rolled through the history of life on earth like a juggernaut. There was nobody to protect the dinosaurs or the pterosaurs, and our guilty, feeble efforts to protect species that we ourselves have endangered may be too little and too late. It has been more than adequately demonstrated that there is no effective

way to fend off the inevitability of extinction, though some of us inno-
cently believe we can do it.

In the process of writing what amounts to a single, continuous book
about life in the sea, I found that certain subjects had to be incorporated
into more than one volume. For example, I first wrote the unfortunate
history of Steller's sea cow (*Hydrodamalis gigas*) for *Monsters of the Sea*
(1994), because this animal was related to the living sirenians (manatees
and dugongs), which some people believe gave rise to the myth of mermaids. Somewhat modified, the story appeared in *Aquagenesis: The Origin
and Evolution of Life in the Sea* (2001), and again in *The Empty Ocean* (2003).
The destruction of the great sea cows, only 28 years after they were discovered wallowing in the shallows off a remote island in the western
North Pacific, is the paradigm of anthropogenic extinction, and even
though I recognize that I am repeating myself by including it in almost
every book since 1994, I feel that I have no choice but to put it in, and it appears in this book too. The sea turtles did not appear in *Monsters,* but they
showed up in *Aquagenesis* and *The Empty Ocean,* and can be found again
here, swimming toward their rendezvous with extinction. As for the marine reptiles, they made a cameo appearance in *Monsters* (the Loch Ness
Monster is believed by some to be a 100 million-year-old plesiosaur), a
brief debut in *Aquagenesis,* and finally, they were featured in their very
own book, *Sea Dragons,* published just before this one. So if some of the
material in this book about extinction looks familiar, that is because I had
to use it in almost all those other contexts.

The most frequently recycled story I have ever written is that of *Carcharodon megalodon,* the gigantic extinct relative of the great white shark. I
researched this subject first in 1973, and an early version appeared in my
Book of Sharks, published in 1975. When John McCosker and I wrote *Great
White Shark* (1991), we expanded the megalodon material, and added
many more illustrations. Not surprisingly, there is a discussion of megalodon in *Monsters of the Sea,* and another in *Aquagenesis.* When Steve Alten's novel *Meg* appeared in 1997, I reviewed it for the *Los Angeles Times,*
and used much of my own material for reference. Because *The Empty
Ocean* is about the recent depletion of marine life, megalodon didn't make
that cut, but its extinction earns it a prominent place in this book.

Because of this "recycling," the advice and assistance originally provided for those other projects has been incorporated into this one. People

who may have helped, say, with discussions of whales, turtles, or sharks, will find their assistance acknowledged again here, even though they may have forgotten our correspondence. Among those who have contributed to this perpetually unfinished product are James Atz, Peter Benchley, Mike Caldwell, Eugenie Clark, Arthur C. Clarke, Jonathan Cobb, Vincent Courtillot, Joel Cracraft, Darryl Domning, Niles Eldredge, Bill Evans, Ash Green, Clare Flemming, Richard Fortey, Rodney Fox, Errol Fuller, John Hare, David Hill, Tony Juniper, Ole Lindquist, Ross MacPhee, John McCosker, Chris McGowan, Colin McHenry, Ryosuke Motani, Darren Naish, Betsy Nicholls, Leslie Noè, Mark Norell, Mike Novacek, Seiji Ohsumi, Charles Repenning, Carl Safina, Leighton Taylor, Ron and Valerie Taylor, Arthur and Ellen Wagner, Don Walsh, Scott Weidensaul, and Wendy Wolf.

Hugh van Dusen edited this book because my clever agent Carl Brandt managed to talk him into publishing it. Stephanie Guest and Walter the parrot stood by (not always patiently), offering advice when needed, and sometimes silence when required. As I write this, a brand-new person has arrived on earth, and while I understand that extinction may not be the ideal subject with which to welcome a newcomer, I am pleased to dedicate this book to my granddaughter, Stella Juniper Adams.

I

INTRODUCTION

Extinction (Sort of) Explained

Standard textbooks on evolutionary biology and paleontology hardly mention extinction. Much is said about the origin of species and the evolution of species once they are formed, but discussions of extinction are usually limited to casual references and the enigma of the great mass extinctions. On causes of extinction we are apt to read, "Species become extinct when population sizes drop to zero," or "Species die out if they are unable to adapt to changing conditions." These statements are true, of course, but are virtually devoid of content.

—David Raup, 1991

Everybody knows what extinction is. The dictionary defines it as "the act of extinguishing, or, the fact of becoming extinguished or extinct." (*Extinguish* is in turn defined as "to put out a fire, a light, or to bring an end to.") More to our point, Thain and Hickman's *Penguin Dictionary of Biology* (1996) defines *extinction* as "Termination of a genealogical lineage. Used most frequently in the context of a species, but applicable also to populations and to taxa higher than species." Thus the fundamental precept of extinction is self-evident: a species (or population, genus, or family) is extinct when its last member has died.

Yet extinction theory is greatly complicated by a number of factors, among them the inability of biologists and paleontologists to agree on exactly what a species is. "Alarmingly," noted Purvis, Jones, and Mace (2000), "there are over 20 species concepts now in common use." For living animals, we recognize as separate species those that are morphologically similar but cannot interbreed. The same criterion obviously cannot be applied to fossils, so paleontologists have to make use of anatomical differences and similarities—when there is enough fossil material to make a determination.

In recent times, however, with the introduction of DNA analysis, what was long believed to be a single species can now be fragmented into two or more. Killer whales, the most widely distributed of all cetaceans, found from southern polar waters to the Arctic and many places in between, were once thought to be a single worldwide species—*Orcinus orca.* New observations and analyses have shown that there may be an Antarctic species that is quite different from its northern counterparts in coloration and behavior; it has been provisionally named *O. glacialis.* Where there were once believed to be six species of balaenopterid whales (blue, fin, sei, Bryde's, and two species of minke), two more were added as of 2003 (see pp. 254–55). The two distinct species of elephant—African and Asian—have been subdivided into three with the recent addition of the genetically distinct Bornean elephant; there may actually be as many as six different subspecies. The gorilla may indeed be not one (*Gorilla gorilla*) but two closely related species. In a *New Scientist* article dated November 22, 2003, Bob Holmes and Jeff Hecht wrote, "A new trend is to delineate species as evolutionarily separate lineages, including separated populations that are evolving in divergent ways. This has already happened for albatrosses. There are 13 recognized species, but the IUCN* lists 21 threatened lineages." On the other hand, it has recently been shown that the animal known as the red wolf, which was once awarded the species name of *Canis rufus,* is not a separate species at all, but a hybrid of the gray wolf and the coyote.

Even if we are able to plug in an acceptable definition of a species, however, identifying the moment that it became extinct is much more problematic. For example, rumored sightings of such animals as the ivory-billed woodpecker and the Tasmanian tiger continue to circulate, and while these animals are generally considered extinct, it is impossible to state unequivocally that a few stragglers may not be found in their often inaccessible habitats. The 1938 discovery of the coelacanth, thought to have been extinct for 75 million years, is the paradigmatic case of a rediscovered "lost" animal, and the unexpected appearance of the previously unknown megamouth shark in Hawaiian waters in 1975 indicates how difficult it is to make categorical statements about existence or nonexistence. If it is so difficult with modern animals, imagine how hard it is to decide from fossil evidence alone that a particular species became extinct at a particular

---

* World Conservation Union, formerly the International Union for the Conservation of Nature.

moment. There are no more trilobites, pterosaurs, or ichthyosaurs, but when did the last one die?

Extinction is one of the most powerful forces on earth, and one of the most enigmatic. It affects every species that has ever lived, and has eliminated most of them. In time, it will eliminate us too. Despite its tremendous importance, however, nobody is quite sure what it is or how it works. We know that there have been countless numbers of living things that have walked, run, crawled, flown, swam, or just remained stationary over the past 3 billion years, and that the great majority of them are gone; but beyond that, we know very little. The fossil record not only supports the all-encompassing theory of evolution, demonstrating conclusively that life changes over time, but it is also our primary evidence for extinction. Because so many creatures are no longer viable, extinction can be clearly read in the fossil record, although the actual evidence of evolution—"change over time"—is only infrequently revealed.

Until the nineteenth century, almost everybody—scientists included—accepted the traditional Christian view that the Bible was to be taken literally, and that God had made the sun, the moon, the earth, and the oceans. He also made all the mammals, birds, alligators, snakes, fishes, and insects, but his crowning achievement was "to make man in our own image, after our likeness; and let them have dominion over the fish of the sea, and over the fowl of the air, and over cattle, and over every creeping thing that creepeth upon the earth." Even Aristotle believed that the animals had been divinely arranged in a ladder, with humans confidently perched on the top rung, the epitome of life.

In the sixteenth century, there were only about 150 kinds of mammals known to Europeans, approximately the same number of birds, and perhaps thirty kinds of snakes. For the first edition of his *Systema Naturae*, published in 1735, Carolus Linnaeus, a Swedish botanist, categorized all the known creatures and plants into an arrangement in which every living thing was given a binomial name, corresponding to its genus and species. (He tried the same system with minerals, but it didn't work as well.) Linnaeus was born in 1707, so by the time he had developed his classification, he was able to include some newly discovered beasts; but he still firmly believed that all known species were unchanging creations of God, who had chosen to arrange things so that Man resided at the top of the pyramid. He is reputed to have said, "God created, but Linnaeus classified."

As human horizons widened, animals that Linnaeus never knew of began to appear. In North America, there were raccoons, pronghorns, grizzly bears, and mountain lions; South America had weird and wonderful monkeys, sloths, anteaters, llamas, and armadillos; and there was an entire continent in the southern sea populated by the strangest fauna of all: kangaroos, wallabies, wombats, koalas, and platypuses—mammals that raised their young in a pouch. The "dark continent" of Africa had giraffes, lions, zebras, baboons, chimpanzees, and gorillas, not one of which was mentioned in the Biblical story of the Great Flood. Whales and dolphins had been known to humans ever since they demonstrated their unfortunate inclination to beach themselves. Two thousand years ago, Aristotle wrote that "It is not known for what reason they run themselves aground on dry land; at all events, it is said that they do so at times, and for no obvious reason." Until whalers and other seafarers took to the sea, the only cetaceans that could be known were those that washed ashore.

True, Europeans had seen strange fossils since ancient Greece, which suggested that there were some life-forms that were no longer with us, but for European and American minds, this meant only that the Flood described in the Bible must have wiped out those species that Noah didn't load onto the Ark. (How the fishes, whales, and dolphins managed to board the Ark was never explained; maybe they swam along in its wake.)

Fossils—the word comes from the Latin *fossilis,* meaning "dug up"—originally referred to any natural object that was pulled out of the ground. The discovery of the first fossils has not been recorded, but Adrienne Mayor has written a book, *The First Fossil Hunters,* in which she suggests that when ancient Greeks, Romans, Persians, and Chinese found fossilized material they simply incorporated it into their mythology; one example is the ancestral giraffe (*Samotherium*) skull that appears almost intact on a black-figured vase that is believed to show Heracles fighting a legendary monster. To the ancients, extinction was not an issue; Mayor says they seamlessly placed the beasts represented by skulls of long-gone mammals and even dinosaurs into their legends and myths.

When fossilized shark teeth were first discovered on land, long before any were brought up from the oceans' depths, their origin was a complete enigma. Pliny, the great Roman student of nature, believed that they had fallen from the sky during eclipses of the moon. They were later thought to be the tongues of serpents that St. Paul had turned to stone while he was visiting the island of Malta; in consequence they acquired the name *glossopetrae* ("tongue stones"). They were believed to have magical properties, especially as counteragents to the bites of poisonous snakes; to that end they were often worn as talismans.

In 1565, the Swiss zoologist Konrad Gesner wrote *On Fossil Objects,* a little book that included everything from gemstones and crystals to ammonites, belemnites, and sharks' teeth. However, Gesner hardly differentiated the fossils from one another, and did not concern himself with what had happened to the animals who originally occupied his "organic" fossils. Robert Hooke (1635–1703), the British architect who helped rebuild London after the Great Fire of 1666, who described the shadow made on saturn by the planet's rings, and who made a map of the moon's craters, has been described as "London's Leonardo" (Bennett *et al.* 2003), because like the more famous Italian polymath, he understood and described certain phenomena long before anyone else. Like Leonardo da Vinci, he was an instrument maker; he specialized in clockwork mechanics, devising the spiral spring in watches. However, Hooke is not as well known as Leonardo or his own contemporary Isaac Newton, because, as Richard Stone (2003) wrote, "set against the brilliance of Isaac Newton, Hooke has tended to shine like a 60-watt lightbulb." It did not help Hooke's reputation that Newton "denied many of Hooke's contributions and tried to obliterate them from history." Recently, with new books being published on Hooke's substantial accomplishments, and symposia convened to discuss his contributions, his reputation is on the ascendant. Hooke believed, wrote Stone, "that the biblical flood could not be taken literally." He also wrote, "There have been many other Species of Creatures in former Ages of which we can find none at present"; in other words, he recognized the concept of extinction long before anyone else.

In 1695, John Woodward (1665–1728), a teacher of physics at Gresham College in London, published his *Essay Toward the Natural History of the Earth: and Terrestrial Bodies, especially Minerals: as also of the Seas, Rivers and Springs. With an Account of the Universal Deluge: and of the Effects that it had upon the Earth.* Woodward espoused the idea that Noah's Flood had not only changed the face of the earth, it had created fossils as well. For some of Woodward's contemporaries, including the Danish naturalist Nicolaus Steno, fossils showed that momentous changes had taken place over time, with mountains thrown up and land conveyed from one place to another.

For Woodward, the fossils demonstrated that the planet had been completely reconfigured by the Flood. "The *Terraqueous Globe*," he wrote, "is to *this Day* nearly in the same *Condition* that the *Universal Deluge* left it; being also likely to *continue* so till the Time of its final *Ruin* and *Dissolution*, preserved to the *same End* for which 'twas first *formed*."

Woodward believed that the floodwaters had receded with such violence that great mountains and valleys were raised up or gouged out; what we now know as fossils were actually organisms forced into the stone by the violence of the waters. Among his supporters was Baron Gottfried von Leibniz (1646–1716), better known as a mathematician and philosopher. Leibniz believed in the Flood, but recognized that so much water could not have come from rain alone. He proposed that the water had been contained in hollows within the earth, and squeezed out to flood the planet. But it was a Swiss physician named Johann Jacob Scheuchzer (1672–1733) who would become the most prominent advocate of the deluge theory; in 1708, he wrote *Piscium querelae et vindiciae* ("Complaints and Justifications of the Fishes"), in which he argued—through the voice of a Latin-speaking fish—that fossils are the remains of various sea creatures that had been carried to the mountains by the Flood.

Except perhaps to Robert Hooke, the origin of fossils was a complete mystery to everyone during the seventeenth and eighteenth centuries, but there was no shortage of explanations for the mysterious appearance of various animal forms in places where their presence could not be easily explained, such as fishes on mountaintops. The Victorian naturalist Philip Gosse maintained that fossils were strewn around by God as part of the spontaneous creation of a complete world; evidence of a history that had never really occurred. A God who could make something as intricate as a man would have no trouble with a few stony bones that resembled a fish. Others believed that it was just coincidence that the fossils resembled living animal and plant forms; they had never actually been alive. Almost everybody, however, subscribed to the idea that when Noah's Flood submerged the earth, all the animals were brought aboard the ark, and it was their descendants—and *only* their descendants—that populated the earth. To have believed otherwise would have been sacrilegious.

However, this dogmatic view of the Gospel truth was beginning to change. In *The Map That Changed the World* (2001), Simon Winchester wrote:

In addition it had not escaped the notice of some collectors that many of the figured stones they found represented animals and plants that did not seem currently to exist. This suggested, in other words, that if indeed the stones were relics, they were relics of living creatures that were no longer around and had since become extinct. Since extinction was an impossible, unthinkable event in any divinely created cosmos, then this notion too was invalid, inappropriate, and wholly wrong.

In other words, it was difficult enough to understand how unknown animals could somehow be turned to stone, or why seashells might appear on mountaintops, but it was nearly impossible for early philosophers to comprehend that God's mercy might include the whimsical elimination of some of his creations from the face of the earth. The very concept of extinction presented an almost undecipherable conundrum.

Jean-Baptiste Pierre Antoine de Monet, Chevalier de Lamarck, was born in Picardy, in northern France, on August 1, 1744. He served for several years in the infantry, but when he developed an interest in plants, he resigned from the army to study botany. In 1778, he published *Flore françoise*, a manual for the identification of French plants, and when the Musée National d'Histoire Naturelle was founded in 1793, he was placed in charge of the invertebrate section. In 1800, he announced a revision of Linnaeus's classification of the lower animals, which is largely in use today. From his work with invertebrates, he envisioned a vast sequence of life-forms, moving as if on a staircase, from the simplest upward to the most complex forms, which he detailed in his *Philosophie zoologique* in 1809. In this seminal study, he explained that organs were strengthened with use and weakened with disuse, and that such environmentally determined changes could be passed along to subsequent generations—the process known as the inheritance of acquired characteristics. (It was here that the lengthening of the giraffe's neck was first discussed.) Lamarck's ideas were accepted by many (including Charles Darwin), not only because they explained the differences between animals, but because Lamarck believed that every animal "must change insensibly in its organization and its form," and therefore his evolutionary theory excluded the possibility of extinction. Even though it explained many of the biological mysteries of the day, Lamarck's theory had to be rejected, when it was found that modifications made to living creatures—such as cutting off the tails of generations of rats—cannot effect the genes, and therefore cannot be passed along to future generations.*

In 1819, the Rev. George Young found a fossil "fish-lizard" at Whitby, in Yorkshire. Young described it as having the features of a crocodile, a fish, and a dolphin, and said, "To what class of animals this skeleton and others found at Whitby should be assigned, it is difficult to determine." From the drawing in his 1820 description, the animal has been identified as *Leptopterygius acutirostris,* the ichthyosaur now known as *Temnodontosaurus.* (The actual fossil cannot be found, so the identification is based on the drawing in Young's paper.) Young opined that ichthyosaurs might still be found swimming in uncharted waters, and wrote,

> It is not unlikely, however, that as the science of Natural History enlarges its bounds, some animal of the same genus may be discovered in some parts of the world . . . and when the seas and large rivers of our globe shall have been more fully explored, many animals may be brought to the knowledge of the naturalist, which at present are known only in the state of fossils.

Since they have been extinct for 85 million years, no living ichthyosaurs were ever found in the "seas and large rivers of our globe," but, in the fossilized skeleton of the ichthyosaur, Young saw a glimmer of the idea that Darwin would publish forty years later: that animal species were "designed" for their particular function. Not surprisingly, he assigned the design function to God:

> Some have alleged, in proof of the pre-Adamite theory, that in tracing the beds upwards, we discern among the inclosed bodies

---

* Neither Lamarck nor Darwin had any idea that there were such things as genes, and therefore could not have known why acquired characteristics could not be inherited. It was not until Gregor Mendel's experiments with peas that the mechanisms for heredity were discovered in 1865, but even then, Mendel's work remained in obscurity until it was "rediscovered" in 1900. And although DNA was actually discovered in 1869 by Friedrich Miescher, that biologist never suspected its actual function. Only in 1962 did James Watson and Francis Crick work out the mechanisms whereby the role and structure of deoxyribonucleic acid (DNA) were explained, allowing for an understanding of genetic storage, transmission, and evolution.

a gradual progress from the more rude and simple creatures, to the more perfect and completely organized; as if the Creator's skill had improved by practice. But for this strange idea there is no foundation: creatures of the most perfect organization occur in the lower beds as well as the higher.

As Simon Winchester wrote, "The Reverend Young could not, however, go any further than this. The forces ranged against him—of custom, history, doctrine and common acceptance—were just too formidable." The concept of extinction was completely alien to people of Young's time and disposition.

Georges Léopold Chrétien Frédéric Dagobert, Baron Cuvier (1769–1832), was the founder of comparative anatomy and probably the first paleontologist. As the "Magician of the Charnel House," he demonstrated that he could reconstruct an entire animal from a single fossilized bone. He believed that fossil sequences were the result of periodic catastrophes, where groups of animals were replaced by new forms in successive creations. He thought that each new form was a step in the progressive sequence that would eventually lead to man, the most sublime of God's creations. In addition to the Biblical Flood, he believed, the earth has been subjected to a succession of natural catastrophes throughout its history. Cuvier died thirty years before Darwin published his theory of evolution; throughout his life, he believed that the extinct types were the ones that had been swept away by successive disasters, and that fossils were the irrefutable evidence of these disasters. In reference to a series of imagined floods, he wrote,

These repeated [advances] and retreats of the sea have neither been slow nor gradual; most of the catastrophes which have occasioned them have been sudden; and this is easily proved, especially with regard to the last of them, the traces of which are most conspicuous. . . . Life in those times was often disturbed by these frightful events. Numberless living things were victims of such catastrophes; some, inhabitants of dry land, were engulfed in deluges; others, living in the heart of the seas, were left stranded when the ocean floor was suddenly raised up again; and whole races were destroyed forever, leaving only a few relics which the naturalist can scarcely recognize.

No matter how it happened, though, Cuvier was among the first to recognize that certain species were no longer among the living. Of course, there was always the possibility that they might turn up in some remote corner of Africa or South America, but exploration in those regions was making it more and more unlikely. After comparing the skulls of the Indian and African elephants, Cuvier announced ("with remarkable self-assurance—some might term it arrogance," wrote Martin Rudwick in 1997) that they were different species, and furthermore, that the "mammoth," recently unearthed in Ohio, was distinct from either of the two living elephants. He believed that the Ohio mastodon was extinct, as were the previous owners of many fossil animal parts, but he was unable to determine why so many living species survived the catastrophes while the extinct species (*espèces perdues*) did not. Cuvier wrote (translated in Rudwick, 1972), "The most important question being to discover if the species that then existed have been entirely destroyed, or if they have merely been modified in their form, or simply transported from one climate into another."

Considered by many to have been one of Britain's notable eccentrics, William Buckland, born in Axminster in 1784, was in fact of sound mind, and certainly not all that eccentric, for he was appointed Dean of Westminster Abbey in 1845. His reputation for strange behavior is in part attributed to his keeping wild animals in his house, such as bears, jackals, birds, snakes, and Jacko the monkey. He also decided to eat his way through the animal kingdom, serving his dinner guests hedgehog, snake, crocodile, or mice, but, says Winchester, "reserving the viler things for himself—he declared that he found mole perfectly horrible, and the only thing worse was that fat English housefly known as the bluebottle." Once while lost, he dismounted from the carriage, picked up a handful of dirt and ate it, able to tell from the taste that they were in Uxbridge. He was the first professor of geology (which he privately referred to as "undergroundology") at Oxford, and an enormously popular lecturer. He was also one of the first of a long list of eminent clergymen who studied natural history and geology in a scientific fashion.

In 1821, he crawled into the Kirkdale Cave in Yorkshire, where he found the bones of animals then found in England, such as foxes, weasels, deer, rabbits, mice, and various birds; but he also found the bones of exotic species, certainly not found in his England: hyenas, elephants, rhinoceroses, hippopotamuses, lions, and bears. Because he believed implicitly in Noah's Flood, he expected that these assorted bones would have been tumbled about by the raging torrents that followed the forty days and nights of rain, but they were not much disturbed. And to confuse things even more, some were of elephants and rhinoceroses that not only didn't live in England, they no longer lived anywhere—Baron Cuvier had shown them to be extinct. Buckland realized that some of the bones had gouges on them that corresponded to the teeth of hyenas. "To confirm his interpretations, Buckland, the consummate empiricist, kept a spotted hyena, named Billy, as a house guest for a short while," and fed him fresh ox bones (McGowan, 2001). The bite marks on the bones that Billy had chewed closely matched those on the Kirkland Cave bones. Forced to explain how these bones could have survived the Flood, Buckland decided that the hyenas were driven from the cave by the rushing waters about 6,000 years earlier, and could not return because the cave was flooded. That neither the bones nor the cave showed signs of a flood did not bother Buckland, for the evidence had to fit his preconceived notions, not the other way around. The cave paleontology described in his *Reliquiae Diluvianae* became part of the concept of "diluvialism," which attributed a variety of geological surface phenomena to the waters of the Flood, just as Cuvier had suggested.

In 1824, Buckland became the first to name and scientifically describe a dinosaur, although he did not call it that; the term would not appear until Richard Owen coined it some eighteen years later. Buckland had a fossil lower jaw (with teeth in place), several vertebrae, and some fragments of the pelvis and shoulder bones of certain large, unknown animals that he had been collecting for about a decade from the Stonesfield quarries near Oxford. In his paper ("*Notice on the* Megalosaurus *or great fossil lizard of Stonesfield*"), he identified it as a species of giant, extinct lizard (*megalosaurus* means "great lizard"), that exceeded forty feet in length and had a bulk equal to that of a large elephant. In his later years, Buckland became profoundly upset at the way his guiding philosophies had all been rejected or overthrown, and he seems to have gone mad, either from depression over his failed career or from an unspecified disease. He spent his final years in a lunatic asylum in Clapham near London, and died in 1856. As Deborah Cadbury wrote in 2000,

He had wrestled with opponents of the new science who supported the biblical interpretations, as each new wave of evidence highlighted fresh anomalies. During his career, geologists had shown that the earth was not six thousand years old, but of much greater antiquity. Life was not made in a single week; the six days of Creation had become "geological ages" covering vast periods of time. There had been no worldwide Deluge; Noah's Flood was increasingly seen as an unimportant regional event and many of the phenomena that Buckland had used to explain it were now thought to be due to glaciation. There was no evidence that creatures had populated the earth from a single site on their release from Noah's Ark. Rather there appeared to be centers of creation on different continents. Even the superb design of creatures in which Buckland had seen the hand of God did not fit easily with the progression of life forms in the fossil record. The relentless onslaught of new evidence, the endless attempts to bridge an unbridgeable gap, seemed to have finally taken their toll.

Cuvier's idea that catastrophism was the operative force in the history of the earth's creatures contrasts with the uniformitarianism of geologist Charles Lyell (1797–1875), whose first volume of *Principles of Geology* Darwin took with him aboard the *Beagle* in 1831 and whose second volume reached Darwin by shipboard mail in Montevideo in September 1832). Darwin was intellectually indebted to Lyell—"My ideas come half out of Lyell's brain," he wrote. He accepted Lyell's concept of slow and gradual change, though he eventually disagreed that species themselves were permanent. According to Lyell, the present is the key to the past: the same forces that shaped the earth in ancient times are still at work and can be observed, as in rivers carrying sediments, or ocean waves wearing down rocks. Lyell's theory of uniformitarianism was applied to sedimentary rocks, river deltas, and eroded landscapes, but Darwin realized that slow, steady changes over long periods of time could also explain major modifications in animal forms.

Darwin signed aboard the *Beagle* as naturalist and companion to Captain Robert FitzRoy, and spent the years from 1831 to 1836 circling the globe, visiting such places as Patagonia, the Galápagos Islands, Australia, and Tahiti, collecting specimens and thinking about how animal species

might have evolved. Upon the *Beagle's* return, Darwin retired to a com-
fortable house in the English countryside, and spent the next twenty years
working on *The Origin of Species,* which was published in 1859. While Darwin was working, Alfred Russel Wallace (1823–1913), studying the wildlife
of Malaysia, came up with an almost identical theory of evolution; Wallace, unaware that Darwin was working on the same theory, wrote to
Darwin to ask his opinion of his ideas. Wallace published his theory of
evolution in a British journal in 1855, but it went virtually unnoticed.
Afraid of being eclipsed by Wallace, Darwin rushed his *Origin of Species*
into print. It is Darwin who is remembered as the author of the ground-
breaking treatment of the evolutionary process; until recently, Wallace
was relegated to a sometimes lengthy footnote.

   In recent years, however, Wallace's contributions have begun to receive the recognition they deserve. David Quammen devoted a significant
proportion of *The Song of the Dodo* (1996) to Wallace's life and accomplishments, writing that "Wallace and Darwin announced the theory [of
natural selection as the driving force of evolution] simultaneously. For a
variety of reasons, some good, some shabby, Darwin received most of the
recognition, and Wallace, in consequence, is famous for being obscured."
Wallace's role is discussed at length in Penny van Oosterzie's 1997 book,
Where Worlds Collide: The Wallace Line. In 2001, Peter Raby published a
lengthy biography, *Alfred Russel Wallace: A Life,* which should go a long
way toward resuscitating the reputation of the man that Quammen called
"the patriarch of biogeography." Like everyone else in the nineteenth century, Wallace could not understand why creatures became extinct, but he
foresaw their destruction at the hands of man. In his 1869 *Malay Archipelago,* he wrote (of the birds of the Aru Islands):

> It seems sad that on the one hand such exquisite creatures should
> live out their lives and exhibit their charms only in these wild, in-
> hospitable regions, doomed for ages yet to come to hopeless bar-
> barism, while on the other hand, should civilized men ever reach
> these distant lands, and bring moral, intellectual, and physical light
> into the recesses of these virgin forests, we may well be sure that
> he will so disturb the nicely balanced relations of organic and inor-
> ganic nature as to cause the disappearance, and finally the extinc-
> tion of these very beings whose wonderful structure and beauty he

alone is fitted to appreciate and enjoy. This consideration must surely tell us that all living things were *not* made for man.

Although Thomas Malthus (1776–1834) was concerned with *human* overcrowding, Darwin applied Malthus's approach to his own study of all animals. He concluded that the reproductive powers of animals are much greater than is required to maintain their numbers; therefore, a large number of offspring must somehow be eliminated if the numbers are to remain constant. There must, he said, be a "struggle for existence" between members of a species, and also between species, in competition for available resources.

In his book, whose full title is *The Origin of Species by Natural Selection or the Preservation of Favoured Races in the Struggle for Life,* Darwin recognized that the best—if not the only—way to account for the vast variety in living things was by what he called "descent with modification." (The word *evolution* does not appear in the first edition of *The Origin of Species,* and *evolved* appears as the last word in later editions.) The similarities as well as the differences between animal species was explainable if animals had common ancestors, he wrote; they could therefore not be independent creations. Lamarck had already postulated this, suggesting that the changes were brought about by the inheritance of acquired characteristics, a mechanism that Darwin himself eventually came to accept, because he was unable to explain the changes in any other manner. Darwin's book described the gradual process where species that are marginally better adapted to their environment reproduce more successfully than those who do not possess such a marginal advantage. By this process—which he called "natural selection"—the less well adapted species gradually disappear, and are replaced by the more fit. This is the definition of "survival of the fittest."

In the section of *Origin of Species* titled "On Extinction," Darwin wrote that "the extinction of old forms and the production of new and improved forms are intimately connected together. The old notion of the inhabitants of the earth having been swept away by catastrophes at successive periods is very generally given up, even by those geologists . . . whose general views would naturally lead them to that conclusion." If Darwin rejected Cuvier's catastrophist views (and those of the other geologists), how did he explain the disappearance of species? They morphed into other forms by the process of natural selection. He wrote, "as species or groups of species gradually disappear, one after another, first from one spot, then from another, and finally from the world," so the earlier forms were replaced by the better-adapted ones. He wrote: "The theory of natural selection is grounded on the belief that each new variety and ultimately each new species, is produced and maintained by having some advantage over those with which it comes into competition; and the consequent extinction of the less-favoured forms almost inevitably follows."

Darwin made no claim to understand fully the mystery of extinction. He said:

> We need not marvel at extinction; if we must marvel, let it be at our own presumption in imagining for a moment that we understand the many complex contingencies on which the existence of each species depends. If we forget for an instant that each species tends to increase inordinately, and that some check is always in action, yet seldom perceived by us, the whole economy of nature will be utterly obscured. Whenever we can precisely say why this species is more abundant in individuals than that; why this species and not another can be naturalised in a given country; then, and not until then, may we justly feel surprise why we cannot account for the extinction of any species or group of species.

In the seminal chapter on "Natural Selection," he wrote, "Though Nature grants long periods of time for the work of natural selection, she does not grant an indefinite period; for as all organic beings are striving to seize on each place in the economy of nature, if any one species does not become modified and improved in a corresponding degree with its competitors, it will be exterminated."

Since Darwin presented them in 1859, there have been numerous revisions of his theories of evolution. Few theorists can now be counted as strict Darwinists, attributing all evolution to the gradual modification of life forms as they become better adapted to the environmental conditions around them, leading to the survival of the fittest and the disappearance of the unfit. It is now recognized—as it was by Darwin—that such a random, gradual process could not possibly have done the job in the allotted time, even if we concede that multicellular life has been on earth for more than half a billion years. In response to the impossibility of the gradual time frame, Stephen J. Gould and Niles Eldredge developed their theory of punctuated equilibrium (1977), where species remain stable for a long time, and then suddenly (in the geological and paleontological sense) disappear as the old form is replaced by a new one. Their theory—abbreviated as "punk eek"—was developed to explain anomalies in the fossil record, but it has also been applied to living species.

One of the major problems in evolutionary theory—and the *sine qua non* of extinction theory—is trying to understand what happened to all those species that didn't make it. David Raup, perhaps the foremost extinction theorist of our day, has estimated that there have been on earth "between 5 and 50 billion species. And only an estimated 5 to 50 million are alive today." More than 99.9 percent of all the species that have ever lived on earth are extinct, including every single one of our hominid ancestors. And 100 percent of evolutionary biologists do not know what causes a species to become extinct—except when humans kill all the members of that species.

Extinction is part of the evolutionary process (or perhaps evolution is part of the extinction process), but how the two intersect is not clearly understood. Indeed, evolution and extinction are equally mysterious; there is no question that both have been occurring for eons, but how they work, either together or separately, is not fully evident. In a 1996 book, David Archibald wrote, "The only way a steady state of species numbers could be maintained, even while evolution is occurring, is for other species to become extinct. Thus extinction, rather than being a rare and negative event in human time frames and sensibilities, is actually a very common and positive counterpoint to evolution." Most extinctions occurred in the distant past, but there are some areas where the process is painfully visible. In his 1991 book, *Extinction: Bad Genes or Bad Luck?* Raup wrote,

> If we accept that turnover in species is merely nature's way, just as nature has given humans a limited life span, then there is nothing in species extinction worthy of wonder. But there is absolutely no basis for equating the life spans of species with those of individual humans. There is no evidence of aging in species or any known reason why a species could not live forever. In fact,

virtual immortality has been claimed for the so-called living fossils (cockroaches and sharks, for example).

Raup then said:

The disturbing reality is that for none of the thousands of well-documented extinctions in the geological past do we have a solid explanation of why the extinction occurred. We have many proposals in specific cases, of course: trilobites died out because of competition from newly evolved fish; dinosaurs were too big or too stupid; the antlers of Irish elk became too cumbersome. These are all plausible scenarios, but no matter how plausible, they cannot be shown to be true beyond a reasonable doubt. Equally plausible scenarios can be invented with ease, and none has predictive power in the sense that it can show a priori that a given species or anatomical type was destined to go extinct.

In a later essay (1996a), Raup wrote, "In the years since Alvarez *et al.* (1980) published their bold claim that the end-Cretaceous (K-T) mass extinction was triggered by a comet or asteroid impact, research on the causes or consequences of species extinction has flourished. However, despite quantities of new field data and the availability of large data-bases, a general model of extinction has not emerged."

Like other scientists, Raup and John Sepkoski (1984, 1986) suspected that there might be some sort of pattern. They noticed that there was a certain periodicity to the mass extinctions; in the past 250 million years, they seemed to be occurring every 26 million years. These periodic extinctions included three of the so-called big five, the end-Permian, end-Triassic, and end-Cretaceous (K-T),* as well as several minor extinction events. "This startling conclusion," which Anthony Hallam and Paul Wignall (1997) described as "potentially the most significant discovery in the Earth sciences since the plate tectonics revolution," cried out for a causative agent. Raup was ready to provide it: he postulated a companion star to the Sun, which he called "Nemesis," the Death Star.

---

* *K-T* is an abbreviation for *Cretaceous-Tertiary*; Why the *K? Cretaceous* derives from *creta*, Latin for "chalk," which is *kreide* in German.

Raup suggested that Nemesis had a fixed orbit of 26 million years; every time it passes through the Oort cloud, an envelope of comets that also orbits the sun, "its own gravitational force will deflect some of the comet orbits in random ways. Most of them will be thrown out of the Solar System, but some will be sent in towards us. As a result, one or more of the comets will collide with the Earth. And we know from the geological records of Earth history that such collisions can be devastating. One incident killed the dinosaurs and another got the last of the crab-like creatures called trilobites." The fact that nobody has ever seen Nemesis, and that there is no evidence that our sun has a companion star, is not the point. The existence of the death star is a scientific construct, used to provide possible answers to some rather difficult questions. In *The Nemesis Affair*, Raup wrote:

> Jack [John Sepkoski] presented the results of our number crunching and concluded that the 26-million-year periodicity for the extinctions of the past 250 million years was real. He also developed the notion that the clocklike spacing of extinctions could be explained more easily by calling on extraterrestrial processes than earthbound ones. . . . In the environment of space, lots of bodies are circling other bodies at pretty regular rates.

Of all the species that have lived on the earth since life first appeared here 3 billion years ago, only about one in a thousand is still living today. The vast majority became extinct, typically within 10 million years or so of their first appearance, though some hung around for much longer. There is no reason, however, to believe that the species alive at present are special in any way. Presumably they too will become extinct within the next ten million years, and make way for successors themselves.

It is reasonable to differentiate background (ordinary) from mass extinctions, because even though they produce similar results—the permanent disappearance of one or more species—the mechanism might be completely different. People write about mass extinctions because they can understand catastrophic events like extraterrestrial impacts, climate change, destruction of habitat, or even accidental or intentional elimination by humans. But they do not write much about background, everyday extinctions, because nobody knows why they might occur. Stephen J. Gould wrote (1994):

> We use the same word—extinction—for such events of death across all scales, but does such an amalgamation help or obfuscate our search for causes? I finally concluded, with some regret, that the unity was false, and the compendium both forced and harmful. We have an explanatory habit in science, born of reductionism, that leads us to render events at grand scale in terms of causes operating at smaller, observable levels. Thus nearly all the participants [in a symposium on extinction] assumed that the smallest scale events dubbed "extinction" in the vernacular must be models or prototypes for the great geological episodes of mass-dying (also called "extinctions"). But the causes may be quite different (ordinary natural selection in the biotic and competitive mode at small scale, with fortuity of survival through environmental catastrophe in at least some events at grandest scale)— even though death be a common result. By using the same word across all scales, and by assuming that small must extrapolate to large, our quest for unity was derailed by a false unity.

With Gould's prescient observations in mind, then, I would like to introduce a new terminology to differentiate mass extinctions, which I hereby designate "macroextinctions," from background or commonplace extinctions, which I shall call "microextinctions." These terms, along the lines of the accepted words "macroevolution" and "microevolution," will probably not receive much in the way of employment, but in the future, we will be able to use one term when we talk about the extinction of a single species, and another term when we refer to events that include the simultaneous disappearance of entire phyla or families, or of thousands of species.

Paleobiologist Steven Stanley's book *Extinction* (1987) is devoted almost entirely to mass extinctions, not the everyday kind that paleontologists refer to as background extinctions (because they are taking place all the time—in the background, as it were). Everyday extinctions are not characterized by violent geological or extraterrestrial events, and are far less dramatic than the mass extinctions, which we do not understand all that well either. In a passing reference to noncatastrophic extinctions, Stanley wrote, "It must also be admitted that, although millions of species have died out in the geologic past, for only a handful do we know with a high degree of certainty the actual cause of the extinction. For the vast majority of lost species, we simply cannot reconstruct either demographic history or relevant environmental changes with sufficient detail to understand what happened."

Because Niles Eldredge's 1991 book *The Miner's Canary* is subtitled *Unraveling the Mysteries of Extinction*, we expect that he will actually unravel these mysteries. In instead, he tells us what extinction *isn't*:

> Evolution does not bring with it a built-in extinction mechanism, eliminating the old at the advent of the new. When something new comes along in evolutionary history, as when single-celled eukaryotes first appear, they do not by any means automatically supplant those organisms that retain the ancestral features. Even if evolution often represents a new, and, in some instances, even an arguably better (i.e., more efficient) way of doing things, that still is by no means a guarantee that the older, more primitive forms will inevitably drop out, becoming extinct in the face of overwhelming odds in general competition with the new, improved model.

Eldredge then veers off in the direction of mass extinctions:

> There is most definitely a ticking of the extinction clock that indeed may be fairly regular over long spans of time within groups such as, say, trilobites, dinosaurs, and hominids (our own lineage). I stress this here because we are about to review the last 670 million years of the history of life with heavy emphasis on extinction. And the extinctions I will stress, quite naturally, will be the prodigious ones—the ones that attracted everyone's attention even in the pre-Darwinian days of the late 1700s and early 1800s. . . . The danger, of course, in such an approach is creating the impression that all extinction occurs only as mass extinction.

In another book also published in 1991, *Fossils: The Evolution and Extinction of Species*, the prolific Eldredge wrote that species change in response to changing environmental conditions, and those that cannot change, become extinct:

The old Darwinian projection more or less assumed that, as environments change, organisms will, in effect, sit there, grin and bear it; if there is a suitable variation for selection to work on, over the generations species will become adapted to the newly instituted conditions. What really happens most of the time, though, is that organisms seek out conditions for which they are best suited, abandoning habitats no longer to their (evolutionarily imbued) liking. So this process of habitat tracking is actually a form of natural selection! But in this case, the organisms doing the surviving and reproducing tend to be the ones that have found the recognizable living conditions to which they are already adapted. Natural selection is for the *status quo*. Even in the face of prolonged environmental change, species will not change much as long as their members can find suitable living conditions. *The alternative is extinction, which is far more likely to occur than progressive modification in response to new environmental challenges.* (My italics.)

So extinction, the mysterious and powerful converse (or complement) of evolution, can sometimes be explained in Darwinian terms. If there is no adaptation to compensate for an environmental change, various species will be unable to function effectively in their modified environment, and in time will die out. These cases do not require global warming or cooling, overspecialization, deadly infectious disease, or extraterrestrial impact to explain extinction (although these can certainly speed up the process), but only the inability of the creatures to adapt to their changing world. So far, there is no equation that can predict with any certainty when a particular species will become extinct. However, the near universality of extinction suggests strongly that everything will eventually die out, whatever the time sequence or the cause.

At a 1988 symposium jointly sponsored by the Linnean and Royal Societies of London, Antoni Hoffman of the Polish Academy of Sciences presented a paper he called "What, if anything, are mass extinctions?" He argued that "many phenomena that have been traditionally been called 'mass extinctions' are in fact clusters of extinction episodes roughly associated in geological time." A large variety of physical environmental events could lead to extinctions: bolide impacts, volcanic activity, climate change, sea-level fluctuations, oceanic anoxic events; because each of these events occurs randomly in time, once every 50 million years on average, "there is a greater than 50% chance that, within a period of 100 million years, two or more events will roughly coincide in time (that is, will occur during a 2-million-year time interval), and there is almost certainty that coincidence during a 4-million-year time interval will take place within a period of 250 million years."

Thus, concludes Hoffman, mass extinctions are actually "sets of lesser episodes of extinction" that, on the geological time scale, appear to occur approximately at the same time. Of the recognized mass extinctions, Hoffman believes that only the late Permian extinction, where 95 percent of living species died out, was caused by a single factor.* According to Hoffman, the introduction of oxygen to the sea floor (by the sinking of nutrient-rich sediments to the bottom), and the resulting nutrient deficiency at the surface, caused mass extinction in the sea.

It is possible to discuss mass extinctions without even mentioning asteroids, craters, or even dinosaurs. Before the development of a backbone, when all life was restricted to the sea, there occurred the first of the great dyings, probably the most severe in the 600 million years from the Cambrian to the Cretaceous. Some 438 million years ago, creatures that lived in the world's cold oceans were decimated by a drop in temperature in the water in which they lived. After the Cambrian, all trilobites had the ability to roll up like pill bugs, perhaps to protect themselves from nautiloids, the new predators of this age. Many-tentacled, and equipped with powerful beaks, these cephalopods were evidently the scourge of the trilobites, and may have hastened their extinction. One species of nautiloid can still be

---

* In 1999, David Raup wrote that Otto Schindewolf had proposed that the Permian mass extinction was caused by an exploding supernova. Raup considered that idea a long shot; the star would have to have been at most ten light-years away to have done the damage that Schindewolf predicted. "Far more important," wrote Raup, "Schindewolf had no independent evidence at all for an exploding star in the Permian. He made the proposal because he found the suddenness and intensity of the mass extinction inexplicable in any other way."

seen today; the chambered nautilus (*Nautilus pompilius*) is the delicate deepwater descendant of Ordovician giants, some of which had straight shells that were ten feet long.

According to Steven Stanley, global cooling, brought about by the movement of continental masses, was the dominant agent in this killing. "Late in Ordovician time," wrote Stanley, "the supercontinent Gondwanaland moved over the South Pole. In a polar position, the land mass cooled to the degree that it accumulated glaciers. In fact, it is not unlikely that every time a continent has moved over a pole during the last billion years of earth history it has become glaciated." As glaciers increase, sea levels decrease, because the water is removed from the sea and locked up on land. The late Ordovician extinction cleared out 85 percent of known species (all of which were marine), including rugose corals, conodonts, trilobites, brachiopods, and nautiloids. Gondwanaland would move over the South Pole again in another 71 million years, precipitating the great Devonian extinction, the second of the "big five," but before that happened, the ammonites evolved, and also the first of the jawed fishes. The record of jawless fishes extends all the way back to the Cambrian, but the appearance of jaws made it possible for these later fishes to eat larger prey than their jawless predecessors; they blossomed into the armored placoderms and arthrodires.

Stanley believes that "since the environmental effects of an impact would have been relatively short-lived, it cannot completely account for a crisis that probably lasted for at least two million years. Furthermore, even if the Alvarez hypothesis should prove to be correct, there is little evidence that asteroid impacts might account for the many other crises in the history of the oceans." He also does not think that decreases in shallow sea floor acreage—brought about by the exposure of continental shelves by falling sea levels—is much of a factor either. What then has been responsible for mass extinctions in the ocean? The cooling of the sea:

> To summarize, the Cretaceous crisis was not a single brief event. Different groups of organisms declined and became extinct at different times, over a period of at least two million years. The sequence of the disappearances is also significant, because it contradicts the old idea, adopted by the Alvarezes, that extinctions might begin at the bottom of the food web and propagate upward

in a kind of domino effect. The lowly plankton suffered at the very end of the Cretaceous crisis, after the decline of many plankton-eating mollusk groups and after the total disappearance of the carnivorous ammonoids.

But why the oceans might have cooled, with such disastrous and final results for the trilobites, graptolites, inoceramids, ammonites, and mosasaurs, Stanley can't say. It might have been a result of the periodic advance of glaciers from the poles, or a reduction in the sun's radiation, but other factors could be involved as well: "Although continued examination of the fossil record should reveal further evidence of the link between cooling and marine extinctions, it may never be possible to understand fully the reasons for climatic changes millions or hundreds of millions of years ago."

In the section "Patterns of Extinction: Retail and Wholesale" in *Darwin's Spectre: Evolutionary Biology in the Modern World* (1998), Michael Rose suggests that extinction can be defined by its inevitability: "Almost all species go extinct within a few million years. This is the natural result of a living world in which species are generated according to their own particular population genetics, as opposed to some cosmic ordering principle, be it aesthetic or moral. Beautiful birds of paradise are annihilated, while cockroaches thrive. Not all species will have the attributes required to reproduce at a rate sufficient to allow their preservation in the face of the numerous sources of mortality that face every species."

Before addressing the subject of mass extinctions in *Beyond Natural Selection* (1997), Robert Wesson wrote, "There are many guesses why great numbers of families and orders die out, including changes of sea level, climate, volcanism, disease, and the wounding of the earth by a comet or asteroid. But none makes clear why all of some classes—all dinosaurs and all pterosaurs, and so forth—should succumb, while other classes survived, some of them apparently undiminished." Virtually the only extinctions that anyone talks about are mass extinctions, or those that have been directly engineered by humans.

In a recent paper, the title of which is simply "Extinction," Purvis, Jones, and Mace (2000) said, "Species are lost from a system either through perturbations imposed on it from the outside (extrinsic causes, which may be biotic or abiotic) or because of evolutionary changes in the numbers (intrinsic causes)." They believe that "explaining why some taxa are affected and some escape is a major goal of conservation biology." They designed their study because there was "progress towards explanatory models for the current crisis . . . but the development of a truly predictive model is hampered by the formidable difficulties of integrating present and past extinctions." In other words, it would be helpful to be able to use information about past extinctions to explain (and maybe even resolve) the current crisis, but there isn't enough information in the fossil record to provide a baseline from which we might draw helpful observations or conclusions.

Darwin recognized that in any given species, some of the members naturally acquire modifications that make them a little better at making a living in their particular environment. If there is no change in that environment, there will be no noticeable change in the species. If, however, the environment changes even a little, as a result, say, of a gradual change in temperature that affects the plant life, the animals must adjust to this change; those that already have a modification that gives them a little edge will survive to produce offspring that have the advantageous modification. Those that—by chance—are not so advantaged will be unable to compete with the better-adapted, and will eventually die off. In a given population, this will not be visible, for the species will remain constant over time, but if one element of the population is somehow separated from another and left to develop on its own, a new species might appear.

This concept, long understood but rarely observed, was brought into sharp focus by the evolutionary biologists Rosemary and Peter Grant, in their study of various closely related species of finches on the little Galápagos island of Daphne Major. Their observations have been extensively documented in their own published papers, but in 1994, Jonathan Weiner won the Pulitzer Prize for his brilliant discussion of the Grants' findings and their implications. In *The Beak of the Finch*, Weiner tells of the Grants' experiences with some thirteen species of "Darwin's finches," including one that is flightless; one that cohabits with marine iguanas; one (the vampire finch) that lives on blood; and another (the cactus finch) that makes tools.

The Grants caught and banded thousands of finches and traced their elaborate lineage, enabling them to document the adjustments that the individual species made, especially in their beaks, in response to environmental changes. During prolonged drought, for instance, those with even slightly longer beaks are better able to reach the tiniest of seeds. Even more fascinating, the Grants have documented changes in DNA among their birds, leading Weiner to declare that "Darwin did not know the strength of his own theory. He vastly underestimated the power of natural selection. Its action is neither rare nor slow. It leads to evolution daily and hourly, all around us, and we can watch."

In recent years speciation at an astonishingly rapid rate has been observed in cichlid fishes, especially those that live in Africa's Lakes Victoria, Malawi, and Tanganyika. Molecular phylogenetic techniques have enabled scientists—particularly George Barlow of the University of California at Berkeley—to identify around 500 species that derived from a single ancestral lineage in less than 14,000 years, the apparent age of the lakes. There are species that have developed to feed only on the scales of other cichlids, others that tend and feed on their own gardens of algae, some that have evolved to suck the young out of the mouths of brooding mothers of other species, and one species that lies on the bottom, mimicking a dead fish, and pounces on other fish that come to feed on the "corpse." "Cichlid fishes," wrote Barlow (2000), "are amazing creatures. In terms of sheer number of species they are the most successful of all vertebrate animals, and the extent and speed with which they have evolved in some African lakes has made them the darlings of evolutionary biologists."

We don't know very much about the aging process, except that every living multicelled creature eventually ages and dies, and they usually do so in a proscribed time frame. Mayflies may live for a day, some other insects for a week, some birds for a year or two, and parrots for seventy-five. Despite great advances in health care, human beings rarely make it to a century. Some tortoises have been known to live for two centuries, and the tube worm *Lamellibranchia* is known to live even longer. There are trees that can be dated at three thousand years of age, but ultimately they too weaken and die. Something within the genes causes the cells to die, the connections to weaken, the bones (or branches, or exoskeleton) to become desiccated and fragile. No multicelled organism can escape the ravages and results of aging. There may be some program deep within the DNA of living things that signals when the animal's time is up, but we will probably see 500-year-old men before we decipher the extinction code.

If one rejects the concept that there has been evolutionary "progress" from one form to another, with minute improvements accumulating until "better-adapted" species emerge, how then can we explain the disappearance of so many species and the survival of so many others? The old idea of a hierarchical ladder (sometimes called the *Scala Natura* or "Great Chain of Being"), which led ever upward toward improvement, has been replaced by a concept of a multibranching bush. Some forms die out for a variety of reasons (competition, climate change, maladaptation, overspecialization, catastrophe, and so on); others survive, perhaps because they were lucky and happened to be in the right place at the right time, or happened to have just the right equipment to survive.

This might explain the abundance of fossils of creatures that look as if they were destined for failure—they often were. We see such wondrous animals as ammonites with shells six feet across; a shark with a circular saw in its mouth; a walruslike dolphin with a single tusk pointing toward its tail; and a whole class of semiaquatic animals called desmostylians that had teeth like hay-rakes and feet that looked as if they were put on backwards. They are all long gone, but in today's oceans, there are creatures that are equally as wonderful, such as the sixty-foot-long giant squid with eyes as big as dinner plates; fishes that light up; the sperm whale, with its gigantic, oil-filled nose; the gigantic whale shark that feeds only on plankton; birds that have lost the power of flight except underwater; and the narwhal, with its single, spiraled tooth projecting eight feet out of its upper jaw.

We are accustomed to think of strange-looking extinct animals as "failed experiments," but this is not at all the case. Every species that we know of—and more that we will never know of—has had the opportunity to strut and fret its hour upon the stage and then be heard no more, but the fossil evidence tells us that some of them strutted for a very long time. The nonavian dinosaurs arose some 220 million years ago, and while many dinosaur species went extinct during their 150-million-year "reign," the group lasted for approximately twice as long as the mammals have so far, and therefore represent one of the great success stories of life on earth. The trilobites, which lasted 150 million years *longer* than the dinosaurs, were even more successful, but became just as extinct. Some of the long-necked plesiosaurs, pterodactyls, ichthyosaurs, tyrannosaurs, horned dinosaurs, and eighty-ton sauropods were larger than most of the creatures that came later, but the blue whale, the largest animal that has ever lived on earth, is alive and swimming today. If they weren't so familiar, the rhinos, hippos, and elephants would surely be regarded as bizarre holdovers from ancient times. Anomalies like aardvarks, pangolins, anteaters, walruses, manatees, porcupines, platypuses, echidnas, and naked mole rats are so weird-looking, it is hard to believe that they didn't die out millions of years ago. The most anomalous creature by far is the hairless, bipedal primate with the big brain and the power of speech. Of all the species that have ever lived on earth, *Homo sapiens* is the only one with the power to eliminate other species at will.

During our brief "dominion over the fish of the sea, and over the fowl of the air, and over every living thing that moveth upon the earth" [*Genesis* 1:28], we have witnessed the exacerbated extinction of all sorts of animals. We—that is, *Homo sapiens* the not-so-wise—killed the last of the sea cows and the great auks; we watched helplessly as Martha, the last passenger pigeon, died in a cage, the sole survivor of a species once so numerous that its flocks darkened the skies. We know why they—and hundreds of other species—are listed among the missing, but we don't know what happened to all those animals that vanished long before we arrived to wreak our special kind of havoc.

It might be that species that were *almost* as well-adapted died off, more as a result of happenstance (the wrong modification accidentally appearing at a disadvantageous moment) than any general factor like climate change or overspecialization. What if man-induced extinctions are the equivalent of the nonanthropogenic extinctions of the past? What if "survival of the fittest" means only that some species are (or were) fortunate enough to survive, when others, almost exactly like them, didn't have the one characteristic that was necessary for survival? Luck would then play a much greater role than we recognize. "Extrinsic driving forces," wrote Purvis *et al.* (2000), "geological change, climate change, faunal mixing and now anthropogenic disturbances can swamp the effects of the smaller scale evolutionary processes. If the entire habitat of a species is eliminated, the adaptive potential of an organism may have little bearing on its ability to persist. To that extent, survival may well be a matter of luck."

It may not be that one form changed into another, but rather that the form with a miniscule advantage—accidentally acquired through a random genetic modification—survived, while those without the accidental advantage fell by the wayside. This, of course, is more or less what Darwin said. In a 2001 article discussing the possibility that prey animals might be unfamiliar with reintroduced predators, J. L. Gittleman and M. E. Gompper identify the "Evil Quartet" of extinction causes—habitat destruction, overexploitation, introduced species, and secondary extinctions." They conclude their discussion with these words: "Like most ecological and evolutionary problems, extinctions rarely have a single cause. Extinction results from a series of factors, interactions among factors, and the multiplicative effects of these factors."

We know precisely why there are no more giant sea cows, passenger pigeons, Tasmanian tigers, and Carolina parakeets, and why there may soon be no more California condors, giant pandas, black rhinos, Galápagos tortoises, or Chinese river dolphins. We destroyed their habitat so they had no place to live, or we killed them, every last one. With this obvious exception, there is no inclusive theory as to why extinction happens. The tyrannosaurs and ichthyosaurs are long gone, as are the trilobites, ammonites, mammoths, sabertooth tigers, and millions of other, lesser-known creatures. What happened to them? We don't know. Most scientists agree, however, that throughout the earth's long history, certain events characterized as "mass extinctions" have occurred. On some available evidence, we are able to speculate as to the causes of these mass extinctions. Indeed, this is one of the few areas of extinction theory about which there is any way of speculating at all.

Mass Extinctions

In his 1991 book *Extinction: Bad Genes or Bad Luck,* David Raup wrote, "Mass extinction is box office, a darling of the popular press, the subject of cover stories and television documentaries, many books, and even a rock song." (The song is "Nemesis" by the [extinct] group known as Shriek-back). The most spectacular of these mass extinctions—although not the deadliest—was the so-called K-T extinction, which occurred some 65 million years ago, and completely wiped out the nonavian dinosaurs. (Most paleontologists today believe that living birds are directly descended from dinosaurs, and that "avian" dinosaurs are thus not extinct at all.) During the Cretaceous period, which lasted roughly from 140 to 65 million years ago, many new terrestrial forms appeared, including the dinosaurs, mammals, flying reptiles, frogs, and turtles. Also at this time, major radiations occurred in several established groups, including the marine reptiles, rudist and other bivalves, ammonites, belemnites, scleractinian (stony) corals, and brachiopods.

During the K-T extinction, 85 percent of all species disappeared, making it second only to the Permian mass extinction as the largest mass extinction event in geological history. The K-T extinction event has generated considerable public interest, primarily because of the strong possibility that a massive extraterrestrial body slammed into the earth creating such environmental havoc—earthquakes, volcanoes, tsunamis, acid rain, worldwide wildfires, vaporization of the atmosphere, and so on—that the dinosaurs and many other life forms were killed off. Although nonavian dinosaurs were among the unfortunate victims to perish in the K-T extinction, we also lost the mosasaurs, pterosaurs, belemnites, ammonites, rudist bivalves, and many species of plants (though many ferns and seed-producing plants were spared). Remarkably, most of the mammals, birds, turtles, crocodiles, lizards, snakes, and amphibians that were alive at that moment were unaffected by the event, which does raise some questions about its effectiveness as a global catastrophe.

It is therefore possible that all recent mammals, birds, etc., are the descendants—with major modifications—of K-T survivors that were somehow spared, because they happened to be out of the range of the firestorm, or underground, or in some other circumstance that enabled them to live through the catastrophe. While all the nonavian dinosaurs, mosasaurs, pterosaurs, belemnoids, ammonoids, and rudist bivalves died, it only required a few—or at worst, a pair—of survivors to repropagate each of the surviving species, which then had 65 million years to develop into the vast variety of forms that now exist, or have subsequently faded into extinction.

The Precambrian era was the period in earth history before the evolution of hard-bodied and complex organisms. Throughout this period, the dominant organisms were soft-bodied, simple, and entirely marine. (Hard-bodied organisms, that is, those with skeletons or shells, did not appear until the beginning of the Cambrian, 510 to 540 million years ago.) About 650 million years ago, 70 percent of the dominant Precambrian flora and fauna perished in the first great extinction. This extinction seems to have been the determining factor that encouraged the diversification of succeeding life forms. However, this distinct Precambrian fauna, which consisted of animals that resembled modern-day soft-bodied organisms such as sea pens, jellyfish, and segmented worms, also experienced a second extinction event, at the close of the Vendian, the period that followed the Precambrian. The first extinction of the Precambrian has been correlated with a large glaciation event that occurred about 600 million years ago; it was so severe that almost all micro-organisms were wiped out. Some paleontologists believe that the Vendian fauna were the progenitors of the Cambrian fauna, while others believe that the Vendian animals have no living representatives. Under this latter hypothesis, the

Vendian fauna is believed to have undergone a mass extinction—for reasons unknown—after which the Cambrian fauna evolved.

During the Cambrian period, the world was largely covered by water, and almost all living things were marine organisms. At the beginning of the period, hard-bodied organisms included only small skeletonized sponges and mollusks; but by about the middle of the Cambrian, diversification of the shelly fauna occurred. The most important phyla present in Cambrian communities included trilobites, archaeocyathids, brachiopods, mollusks, and echinoderms. At least four major extinctions occurred during the Cambrian. The first extinction took out the oldest group of trilobites (the olenellids), as well as the primary reef-building organisms, the archaeocyathids. The remaining three extinctions were irregularly distributed around the late Cambrian epoch boundary; they severely affected trilobites, brachiopods, and conodonts.

Although organisms already present in the Cambrian were numerous in the Ordovician (about 440 to 450 million years ago), a variety of new types flourished, including cephalopods, corals (including rugose and tabulate forms), bryozoans, crinoids, graptolites, gastropods, and bivalves. Ordovician communities typically were more complex than their Cambrian counterparts, largely because of the greater diversity of organisms; but life in the Ordovician continued to be restricted to the seas. The Ordovician extinction is generally regarded as the second most devastating extinction of marine communities in earth history. It saw the disappearance of one-third of all brachiopod and bryozoan families, as well as numerous groups of conodonts, trilobites, and graptolites. Much of the reef-building fauna was also decimated. In total, more than one hundred families of marine invertebrates perished.

What caused the Ordovician mass extinction? As with the others, we really don't know, but paleontologists have theorized that it might have been the result of a single event: the glaciation of the continent Gondwana at the end of the period. Evidence for this glaciation was discovered by geologists in the Sahara Desert, where rock magnetism evidence and glacial deposit data have provided a possible scenario. When Gondwana passed over the South Pole in the Ordovician, intense global cooling caused large-scale continental climate change, resulting in widespread glaciation. Glaciation caused a lowering of the sea level worldwide as large amounts of water became tied up in ice sheets. The lowering of the sea level, which reduced available ecospace on continental shelves, and the cooling caused by the glaciation itself, are the likely driving agents for the Ordovician mass extinction.

The Permian mass extinction occurred about 250 million years ago and was the greatest mass extinction ever recorded in earth history. It was significantly larger than the Ordovician and Devonian crises, and much more devastating than the better-known K-T extinction that felled the dinosaurs. According to Karen Wright (2001), one of the reasons the Permian extinction is not better known is that "no one knows how the story ends." Until recently, paleontologists believed that the late Permian event lasted for five to eight million years; it was associated primarily with massive global changes, such as the tectonic activity that brought about the new continent of Pangea, producing enormous effects on climate and sea level. As the plates moved and ground against each other, magma exploding through the earth's crust may have released enough gases to blot out the sun for years, depriving plants of the ability to photosynthesize. If little oxygen was pumped into the atmosphere, CO_2 and other greenhouse gases would have accumulated and raised global temperatures enough to affect every living thing. In his 1993 book about the Permian extinction, Smithsonian paleobiologist Douglas Erwin wrote:

> Undoubtedly the most important geologic event in the Permian was the formation of the great supercontinent Pangea. This landmass included virtually all of the large continental fragments, although some microplates never collided with Pangea. The southern continents (what are now South America, Africa, India, Antarctica, Australia, and parts of the Middle East and Southeast Asia) were united into a large continent known as Gondwana during the Paleozoic. North of Gondwana lay Laurasia, which included North America, Europe, the Russian and Siberian platforms, and Kazakhstan.

With the formation of the supercontinent Pangea, land area exceeded oceanic area for the first time in geological history. The result of this new global configuration was the extensive development and diversification of terrestrial vertebrate fauna, and the concurrent reduction of marine communities. On land, sail-backed creatures known as pelycosaurs appeared, along with the other synapsid forms that were originally classified as "mammal-like reptiles," but are no longer believed to have been true reptiles; they were not mammals either. Life in the seas was similar to that found in middle Devonian communities, and included the brachiopods, ammonoids, gastropods, crinoids, bony fishes, and sharks. Corals and trilobites were also present, but in greatly diminished numbers. During this time, we encounter the first signs of land animals that returned to the water. The three-foot-long mesosaurs were shaped rather like crocodiles, though their long, thin jaws were equipped with needlelike teeth that formed a kind of sieve allowing them to strain small organisms out of the water, rather like baleen whales.

Ninety to ninety-five percent of marine species were eliminated as a result of the Permian event. The primary marine and terrestrial victims included the fusulinid foraminifera, trilobites, rugose and tabulate corals, blastoids, acanthodians, placoderms, and pelycosaurs, none of which survived beyond the Permian boundary. Other groups that were substantially reduced included the bryozoans, brachiopods, ammonoids, sharks, bony fishes, crinoids, eurypterids, ostracods, and echinoderms. In *The Great Paleozoic Crisis*, his detailed study of the Permian extinction, Douglas Erwin compared the mystery to Agatha Christie's *Death on the Nile*, where *all* the suspects had a hand in the murder. He wrote,

> I believe that the extinction cannot be traced to a single cause, but rather a multitude of events occurring together, in particular the increased climatic and ecologic instability associated with the regression and a combination of greenhouse warming and possible oceanic anoxia from increased atmospheric CO_2. One of the most striking aspects of the latest Permian is the diversity of inputs of light carbon, including juvenile volcanic CO_2, oxidation of organic carbon on shallow marine shelves, and release of gas hydrates. The last two sources will increase as the regression progresses. I would suggest that we need to look at the extinction in three parts: the terrestrial events, the marine events along the margin of Pangea, and the events in South China. The terrestrial extinction was not insignificant, but does not appear to be as extensive as the marine extinctions. Here global warming, increased seasonality, and a possible reduction in habitat diversity

associated with harsh environmental conditions in the interior of
Pangea seem sufficient to cause the extinction. The major phase
of the marine extinction appears to have begun earlier on the
mainland of Pangea with the onset of the regression and the corresponding loss in habitat diversity and reduction of shelf area.
The major extinctions were triggered by the dramatic increase in
CO_2 and methane from oxidation of organic material, gas hy-
drates, and possibly other sources. This triggered global warming
and perhaps an increase in oceanic anoxia. By this point the effects were sufficiently far-reaching to affect various refugia, including South China, eliminating many of them.

There are now those who suggest that it happened rapidly, in less
than half a million years. In a study published in 2000, several scientists
from the Nanjing Institute of Geology and Paleontology (along with
Erwin) suggested that the event took place almost overnight (geologically
speaking), "coincident with the eruption of the Siberian flood basalts." In
500,000 years, volcanoes spread over Siberia from the Ural Mountains to
Lake Baikal, pouring out lava that covered the land in a blanket two miles
thick. The lava would have killed everything it covered, but much more
damaging would have been the debris and noxious gases (such as CO_2)
generated by the ongoing eruptions. But to some, the Permian extinction
happened even faster, meaning that not even volcanic eruptions could explain the widespread annihilation of life, and they continue to search for
evidence of an extraterrestrial impact.

All previous theories about the causative agent of the great Permian
extinction had to be discarded—or at least severely modified—in February
2001, when Luann Becker of the University of Washington, Robert Poreda
of the University of Rochester, and their colleagues published an article in
Science. Their analysis of rocks at the Permian-Triassic (P-T) boundary revealed evidence of a massive impact that, like the K-T comet or asteroid
impact that triggered the extinction of the dinosaurs 65 million years ago,
initiated the "mother of all extinctions" 251 million years ago. In response
to their publication, the New York Times of February 23, 2001, announced,
"Meteor crashes led to extinctions in era before dinosaurs."

The evidence consists of the noble gases helium and argon apparently trapped in the soccer-ball-shaped molecular "cages" of fullerenes

(or "buckyballs," both named for Buckminster Fuller). These gases did not resemble any found on earth, but did resemble those found in meteorites. (This is parallel to the so-called "iridium spike" from the K-T impact at Chicxulub that wrought such worldwide havoc 65 million years ago.) Fullerenes had already been associated with the formation of the 1.85-billion-year-old Sudbury Crater in Ontario, and even with the K-T boundary layer in various locations. Becker, Poreda *et al.* found the compounds at Meishan in southern China, Sasayama in southwest Japan, and the Bükk Mountains of Hungary; they noted that the noble gases in the fullerenes were fifty times more prevalent at the P-T boundary than anywhere else, which strongly argues for an extraterrestrial event. They wrote, "Only stars or collapsing gas clouds have sufficient helium pressures and provide an environment conducive to fullerene synthesis. . . . Thus it would appear that ET [extra-terrestrial] fullerenes were delivered to Earth at the PTB [P-T boundary] possibly related to a cometary or asteroidal impact event." The impact, which probably had an almost instantaneous effect, killed off 90 percent of the marine species and perhaps 70 percent of the terrestrial species alive at that moment. It marked the seminal moment when life on earth shifted from the mostly passive life forms to the more active predatory types that searched for food instead of waiting for it to drift by.

In November 2003 Luann Becker, with her colleagues Basu, Petaev, Poreda, and Jacobsen, published further evidence of a meteor strike at the P-T boundary, when they found fullerenes in meteor fragments in Antarctica. In the paper (also published in *Science*) they wrote, "In one sample, the meteoric fragments are accompanied by more abundant discrete metal grains, which are also found in an end-Permian bed at Meishan, southern China. We discuss the implications of this finding for a suggested global impact event at the Permian-Triassic boundary."

Becker's theory was supported by Kunio Kaiho of Tohoku University and his colleagues, who claimed (2001) to have found evidence in southern China that a massive impact converted huge amounts of solid sulfur into sulfur-rich gases. They wrote, "These data suggest that an asteroid or comet hit the ocean at the end of the Permian time and caused a rapid and massive release of sulfur from the mantle to the ocean atmosphere system, leading to significant oxygen consumption, acid rain and the most severe biotic crisis in the history of life on Earth." The released sulfur could have consumed 20-40 percent of the atmosphere's oxygen, and generated enough acid rain to raise the acidity of the ocean's surface waters temporarily to that of lemon juice. Kaiho's team found sulfate in end-Permian limestone, marl, and shale rocks formed from shallow seafloor sediments. The rocks also have a nickel-rich layer; the nickel could have been carried by an impacting meteorite. Moreover, within the nickel-rich layer the researchers detected a sudden change in the relative amounts of different sulfur isotopes (whose atoms have slightly different masses). If a giant meteorite impact vaporized a large area of sulfur-containing rock where it struck the seabed, it would probably have ejected the lighter of sulfur's two common natural isotopes into the air, changing the isotope ratio of the remaining rocks. From the size of the isotope ratio shift, Kaiho's group estimated that the meteorite could have been up to sixty kilometers (thirty-seven miles) across. The K-T (Chicxulub) meteorite that is believed to have taken out the last of the nonavian dinosaurs was probably less than ten kilometers (six miles) across. If Kaiho's team is correct, we will have found evidence that at least two of the five mass extinctions were caused by extraterrestrial impacts. (An impact in the ocean is harder to find, but the Chicxulub asteroid hit in the ocean in the vicinity of what is now the Yucatán Peninsula, and was at least partially responsible for the K-T extinction, which marked the disappearance of the dinosaurs—and numerous other life forms—65 million years ago.)

But as with all mass-extinction theories, Becker's has come under fire. In a 2003 article in *New Scientist* called "Wipeout," Michael Benton wrote, "Later in 2001, several learned journals published critiques in which other geochemists tried and failed to replicate Becker's results. Other samples from the same boundary bed in China apparently contained no buckyballs, argon or helium, and a Japanese geologist, Yukio Isozaki of the University of Tokyo, pointed out that the Japanese sample came from a rock formation that didn't even include the Permian-Triassic boundary." There is strong evidence that the end of the Permian was characterized by massive volcanic eruptions, particularly in the area known as the Siberian Traps, causing major disruptions in the atmosphere, and covering two and a half million square miles of what is now eastern Russia in mile-thick lava. In 1997, British geologists Anthony Hallam and Paul Wignall published *Mass Extinctions and Their Aftermath*, in which they showed how the eruption of the Siberian Traps would have caused acid rain, a high emission of carbon dioxide and sulfur dioxide, global warming, global darkness, and eventually, mass extinction. Benton concludes: "It took 20 or 30 million years for coral reefs to reestablish themselves, and for the forests to regrow. In some settings, it took 50 million years for full ecosystem complexity to recover. Geologists and paleontologists are only beginning to get to grips with this most profound of crises." Michael Benton himself wrote a book about the Permian extinction, which he called *When Life Nearly Died: The Greatest Mass Extinction of All Time* (2003). He argued that the event was not initiated by an extraterrestrial agent at all, but rather by massive volcanism, which released methane gases into the sea and poisonous gases into the atmosphere, snuffing out life forms in both biomes.

On August 12, 1986, a cloudy mixture of carbon dioxide (CO_2) and water droplets rose violently from Lake Nyos in Cameroon, in western Africa. Normally, CO_2 remains in solution—as with the bubbles in an unopened bottle of soda—but something forced the gas to the surface (the lake sits atop an old volcano) and it rose and burst from the surface releasing a deadly cloud of heavier-than-air poisonous gas. Thousands of cattle, innumerable other animals, and 1,800 people were asphyxiated. These deaths hardly count as a mass extinction, but the event was one of the factors that prompted Gregory Ryskin of Northwestern University to suggest that a massive eruption of methane gas (CH_4) from the ocean might have exploded and incinerated the land dwellers of the Permian. In *Geology* (2003), he wrote,

> The consequences of a methane-driven oceanic eruption for marine and terrestrial life are likely to be catastrophic. Figuratively speaking, the erupting region "boils over," ejecting a large amount of methane and other gases (e.g., CO_2, H_2S [hydrogen sulfide]) into the atmosphere, and flooding large areas of land. Whereas pure methane is lighter than air, methane loaded with water droplets is much heavier, and thus spreads over the land, mixing with air in the process (and losing water as rain.) The air-methane mixture is explosive at methane concentrations between 5% and 15%; as such mixtures form in different locations near the ground and are ignited by lightning, explosions and conflagrations destroy most of the terrestrial life, and also produce great amounts of smoke and carbon dioxide.

Such a methane explosion might have brought on the Permian extinction by killing the land dwellers and also killing the land plants. "The paleogeography of the Permian," wrote Ryskin, "may have led to the development of a large number of stagnant anoxic regions, and thus to accumulation of very large amounts of dissolved methane. The unusual severity of the Permian-Triassic extinction may have been the result of chance as several different oceanic locations erupted in succession," or, what Raup would categorize as "bad luck."

Wiped out during the great Permian extinction were the mammal-like reptiles known as synapsids (or theraspids), represented by such unusual creatures as the ten-foot-long carnivorous pelycosaur *Dimetrodon*, which had such a great elongation of the neural spines that they are believed to have carried a sail. (The function of this dorsal fin—which could not be raised or lowered—has been hotly debated; it probably had some sort of a thermoregulatory function, but whether it served to cool the animal or to provide a greater surface for the absorption of heat has not been determined.) It was the synapsid *Lystrosaurus,* found by Edwin Colbert in the Antarctic in 1969, that led him to conclude that Antarctica and Africa had once been connected as Gondwanaland. *Dicynodon* ("two dog teeth") was a strange-looking creature with the body of a hippopotamus, the beak of a turtle, and canine tusks in its upper jaw. It was originally found in 220-million-year-old late Permian deposits in South Africa. Recently (2002), paleontologists Tony Thulborn and Susan Turner of Monash University in Australia described a fossil dicynodont that appears to have come from strata only 105 million years old. If their analysis is correct, some of the pelycosaurs hung around much longer than anyone suspected—at least in Australia. Most of the synapsids disappeared long before the end-Permian event—whatever it was; in their place arose the earliest mammals—the shrew-sized moganuconodonts and triconodonts. It would require another 215 million years and a series of catastrophic mass extinctions before the mammals could begin the long journey that would take them to dominance over the planet.

Gregory Retallack, Roger Smith, and Peter Ward (2003) also think that the release of oceanic gases might have contributed to the Permian extinction, but not by igniting and burning up the land and all the animal life. Methane, they suggested, would have reduced the relative amount of oxygen in the air, triggering pulmonary and cerebral edema, the same condition that affects mountaineers at high altitudes. Land animals would have suffocated, and marine life would have been similarly affected by oxygen-poor water. Those animals that survived the end-Permian event were built to cope with low oxygen levels. The dog-sized *Lystrosaurus* and the larger *Dicynodon,* both of which survived the Permian extinction, were built low to the ground, and may have been burrowers, which might have contributed to their survival in a global atmospheric crisis. Of course, it may have been luck, too.

At the conclusion of the Triassic period, 208 million years ago, something happened that drove a large proportion of all living things extinct, including some species of cephalopods, brachiopods, sponges, and conodonts, primitive marine creatures that many believe were the first vertebrates. In a recent study, Ward, Haggart *et al.* (2001) suggest that the end-Triassic event, which "led to the demise of as many as 80% of all living species," happened in a geological instant, and was probably caused by an extraterrestrial impact, a sudden burst of volcanism, or maybe both. The synapsids were eliminated, but the dinosaurs' ancestors managed to survive the cataclysm, paving the way for their terrestrial dominance after the Triassic-Jurassic (T-J) event. In a *New York Times* article (Chang, 2001), Ward is quoted as saying, "For whatever reason the dinosaurs squeak through, and the mammal-like reptiles don't. It really opens the world to the age of dinosaurs as we know it." It has been suggested that an astral body (or bodies) of some sort struck the earth about 214 million years ago, Manicouagan Crater in Quebec, a deeply eroded circular crater about fifty-three miles across, may be evidence of the end-Triassic mass extinction event. Manicouagan and four other impact structures (Rochechouart in France, Saint Martin in Canada, Obolon in the Ukraine, and Red Wing in North Dakota) show evidence for virtually simultaneous impacts. These structures, which occur along two parallel paths, may represent the remains of a crater chain at least 2,776 miles long. In their 1998 study, "Evidence for a late Triassic multiple impact event on Earth," Spray, Kelley, and Rowley suggested that "the five impact structures were formed at the same time (within hours) during a multiple impact event caused by a fragmented comet or asteroid colliding with Earth."

II

WHERE DID
EVERYBODY GO?

Evolution, as we understand it, modifies various animal forms over time until they metamorphose into other forms or vanish altogether, becoming extinct. The recognizable ancestors of most animals today did not look very much like their living descendants, except in those particulars that enable us to identify them as early versions.

But there are some creatures, known by the oxymoronic term "living fossils" (fossils, by definition, cannot be alive) that have somehow managed to avoid the evolutionary process altogether, and have come through the ages in almost the same form they acquired 100 or 200 million years ago. The concept of a "living fossil" was first introduced in 1859 by Darwin in *The Origin of Species*. He wrote, "As we here and there see a thin straggling branch springing from a fork low down on the tree, and which by some chance has been favoured and is still alive at its summit, so we occasionally see an animal like *Ornithorhynchus* [the platypus] or *Lepidosiren* [the lungfish], which in some small degree connects by its affinities two large branches of life, and which has apparently been saved from fatal competition by having inhabited a protected station."

How could some creatures make it through the mass extinctions, climatic anomalies, swooping temperature variations, wobbling of the earth on its axis, extraterrestrial bombardments, and other perturbations that wiped out nearly everything else? If there is no adaptation to compensate for environmental change, then a species will be unable to function effectively in a modified environment, and in time will die out. In contrast, animals whose environments remained stable, or who could engineer minor changes that enabled them to survive for hundreds of millions of years—think of the horseshoe crabs, crocodiles, lungfishes, coelacanths, and chambered nautiluses—have been awarded the honorary appellation of living fossils.

Sharks, probably the animals most frequently referred to as living fossils, do not really qualify for the epithet. Cartilaginous fishes have indeed been around since the Devonian, 300 million years ago, but the sharks we know today, including the big dangerous ones and also the host of smaller species that are utterly harmless to people, are dramatically different from the ones that swam in ancient seas. Then, there were "sharks" with teeth on top of their heads, others with a circular-saw blade in their lower jaws, and still others that had an apparatus growing out of their heads that looked for all the world like antlers. Some modern sharks are as weird as their ancient counterparts, like the lanternsharks that glow in the dark or the goblin shark with its strange, blade-like protuberance where its nose ought to be; but these creatures are not holdovers from the distant past.

The idea of living fossils—the term *bradytely,* meaning "evolving at a slow rate" is often applied—is more than a little difficult for evolutionary theorists to accept. Environments change; animals are *supposed* to evolve. In his essay in the 1984 collection *Living Fossils,* Niles Eldredge wrote, "In the context of Darwin's own founding conceptions, and certainly from the perspective of the modern synthesis, living fossils are something of an enigma, if not an embarrassment."

In *Tempo and Mode in Evolution* (1944), George Gaylord Simpson wrote,

> Although, again, the factual data are wholly inadequate, population size must almost necessarily be a primary factor in bradytely. Bradytely involves not only exceptionally low rates of evolution but also survival for extraordinarily long periods of time. Such survival with little change implies the almost complete elimination of random modifications and, of course, excludes chance extinction, requirements most unlikely to be met continuously for such long periods unless the population is large. For many groups

of particularly slow evolution the existence of large breeding populations has been demonstrated or is a probable inference. There are apparent and possible exceptions, notably *Latimeria,* of which only one individual has ever been found, and *Sphenodon,* the population of which is now near the vanishing point. But we have no reason to believe that *Latimeria* may not have been or may not even now be abundant in its own habitat, which is little known. As regards *Sphenodon,* its reduction in numbers is recent, and it was abundant so few generations ago that time has not yet sufficed for any pronounced small-population effects.

Simpson wrote *Tempo and Mode* in 1944, when only one coelacanth (*Latimeria*) had been found. They are not, in fact abundant, but since the first discovery in 1938, hundreds of sightings have been made in the Comoro Islands off Madagascar, and a completely unexpected population has been found in Indonesian waters. *Sphenodon* (the tuatara) is the sole survivor of the order of reptiles known as Rhynchocephalia ("beak-heads"), whose many species flourished during the Triassic period, from 245 to 208 million years ago. All but the tuatara were extinct by the late Cretaceous, about the time the last of the dinosaurs disappeared.

Today, the tuatara lives only in New Zealand. Its name comes from a Maori word that means "spiny back," a description of the erectile crest that runs down the back, more prominent in males than females. There are two species of these lizard-shaped reptiles, *Sphenodon punctatus,* found on the northern islands, and *S. guntheri,* found only on Brothers Island. Tuataras may grow to a length of about twenty-eight inches (seventy centimeters), and can take twenty years to reach maturity. Tuataras have a very low metabolic rate, and can live for a hundred years. They are solitary, burrowing animals, feeding mostly at night on insects, lizards, snails, and bird chicks and eggs. Unlike all other living toothed reptiles, the tuatara's teeth are "acrodont," meaning that they are fused to the jawbones. The animal lacks external ear holes, has a diapsid skull (two openings on either side), and possesses a "parietal eye" on the top of the head which contains a retina and is similar to a normal eye, although its function is not clearly understood and scales grow over it in adults.

The tuatara has been falsely called a living fossil, but while it is very similar to its extinct ancestors, it has features unique to its own modern species. Indeed, because it is the only surviving representative of its kind, there are those who would place it high on the "triage" list of animals that have to be saved. "Far more valuable than a panda or a rhino," says Robert May (quoted in Gibbs, 2001), "are relic life forms like the tuatara. Just two species of tuatara remain from the group that branched off from the main stem of the reptilian evolutionary tree so long ago that this couple make up a genus, an order, and almost a subclass all by themselves."

The fact that whole classes of animals disappeared without leaving a descendant in today's world does not mean that they were not successful. It only means—tautologically—that they became extinct. Consider the trilobites. They endured for 300 million years, and died out before the dinosaurs even made their debut. "Who are we johnny-come-latelies," asks Richard Fortey (2000), "to label them as either 'primitive' or 'unsuccessful'? Men have so far survived half a per cent as long." Trilobites were extremely plentiful, and represent more than half of the known fossils from the Cambrian period, 500 million years ago.

OK, What *Really* Happened to the Dinosaurs?

The ancient marine reptiles were not dinosaurs; nor were the ptero-dactyls and pterosaurs. Birds may have been (and may still be), but in the popular parlance, dinosaurs can be defined as "all animals descended from the first dinosaur, the common ancestor" (Norell, Gaffney, and Dingus, 1995). Kevin Padian's 1997 definition in the *Encyclopedia of Dinosaurs* reads as follows:

> To be a dinosaur, then, according to current definitions within the phylogenetic system, a given animal must be a member of the groups descended from the most recent common ancestor of birds and *Triceratops*. The diagnosis of this group, and its membership, will change as we learn more about the included taxa and modify the distributions and synapomorphies accordingly.

To date, scientists have identified some 450 dinosaur genera based on fossil body parts, some of which are well founded, while others are dubious, known only from fragments. Some, in fact, are relegated to the category of *nomina dubia*, which means "dubious names," and includes those animals described from inadequate bone fragments, or whose names have been replaced, for one reason or another, by a different name, as is the case where *Apatosaurus* was substituted for the hoary old *Brontosaurus*.

A great many more dinosaur genera populated the Mesozoic from the middle Triassic through the end of the Cretaceous; but it has been estimated that as many as 90 percent of these forms lived in "nondepositional" regions, where their fossils would not be preserved, and we are highly unlikely ever to find them. Also, many of these dinosaurs were small, birdlike forms that are seldom found because of their size and fragility, even in regions favorable to fossilization. Altogether, several thousand dinosaur genera were alive at any one time, which is entirely consistent with estimates for extant mammals, birds, and reptiles. Some dinosaurs were huge—the largest terrestrial animals in history—and some were no larger than chickens. The largest dinosaur so far recorded was *Seismosaurus,* whose estimated length was 150 feet, and the smallest—so far—was *Compsognathus,* which was about the size of a raven. (Of course, if we consider living birds as dinosaurs, the smallest would be the bee hummingbird, which weighs less than one-tenth of an ounce.) Some dinosaurs had smooth skin and some had scales; some had bumps, horns, frills, spines, or spikes; some had feathers. Some had beaks, and some had a mouthful of huge, serrated teeth. Some were cold-blooded, like today's reptiles, but some were warm-blooded, like today's birds and mammals.

When Sir Richard Owen coined the term "dinosaur" in 1842, he intended for it to apply to fossils of giant reptiles that had been unearthed in England and were different from any known reptiles. His definition reads as follows:

> The combination of such characters, some, as the sacral bones altogether peculiar among Reptiles, others borrowed, as it were, from groups now distinct from each other, and all manifested by creatures far surpassing in size the largest of existing reptiles, will, it is presumed, be deemed sufficient ground for establishing a distinct tribe or suborder of the Saurian Reptiles, for which I would propose the name *Dinosauria.*

Owen's Dinosauria ("fearfully great lizards") had the five sacral vertebrae fused, and the hips structured so the animals had to have walked upright; they could not have sprawled like living lizards and crocodiles. Later discoveries, particularly in Europe and North America, revealed that the dinosaurs not only walked upright, they usually did it on their hind legs.

A feature that is essential to the Dinosauria is the "hole in the hip socket." As described by Dingus *et al.* (1995), "Dinosaurs arose from a common ancestor that had a hole in the middle of the hip socket. This feature was apparently related to the evolution of an erect posture, in which the hind legs are oriented vertically beneath the hips instead of sprawling out to the side, as in turtles and lizards." Around 1887, Harry Govier Seeley, a former student of Owen's, reclassified the Dinosauria into two groups, the Ornithischia ("bird-hipped") and the Saurischia ("lizard-hipped"), and by the 1970s, arguments were being made for the erection of a new class of vertebrates, to include all dinosaurs and birds, based largely on their common upright stance and what some considered their warm-bloodedness (Bakker and Galton, 1974).

With the introduction of cladistic analysis, where relationships are established on the basis of shared derived characteristics (as opposed to observed morphological similarities), and organisms are grouped according to their common evolutionary history, the division of dinosaurs into ornithischian and saurischian has been abandoned by some. "Indeed," wrote Michael Benton, "the very definition of the Dinosauria has become subject to revision. . . . It becomes clear that, as for all major taxa, the term *dinosaur* has a significant arbitrary component. There is a monophyletic assemblage that includes all animals that we choose to call dinosaurs, but where we choose to draw the boundary between dinosaurs and non-dinosaurs is a subjective matter, not a given fact in nature." In his *Historical Atlas of the Dinosaurs*, published in 1996, however, Benton reverted to the earlier interpretation, writing, "The Order Dinosauria ('terrible reptiles') falls into a number of subgroupings, each of them corresponding to a major split that happened during the evolution of the group. Soon after the origin of the dinosaurs, more than 230 million years ago, the group split into two, the suborders Saurischia ("reptile hip") and Ornithischia ("bird hip"). Saurischians are distinguished by features of their skulls, backbones, and arms, while ornithischians have a specialized set of hip bones, where the pubis, the front bone, runs back parallel to the ischium."

However the classification is determined, there are two recognized types of saurischian dinosaurs: theropods and sauropods. The theropods ("beast feet") include all the carnivorous dinosaurs, most of which stood up on their hind legs and used their forelegs for grasping, which placed some of them in a separate category, the Maniraptora. The best known of the theropods—especially after *Jurassic Park*—are *Tyrannosaurus* and *Velociraptor*, but there are many other less formidable members of the group. Also classified as theropods are the lightly built ornithomimids ("bird mimics") such as *Struthiomimus* ("ostrich mimic"), named for its long legs, evidently designed for running; and *Deinocheirus*, which was similar to *Struthiomimus*, except that its forelegs were enormous. *Archaeopteryx* and *Confuciusornis* were theropods—and so are all living birds. As Henry Gee wrote (in *Nature* in 2000),

> Theropod dinosaurs are among the most successful of land animals. The key to their 230-million-year success story could lie in the fact that they were, from the very beginning, bipeds. Even way back in the distant Triassic period, theropods such as *Eoraptor* walked on their hind legs. From tiny *Compsognathus* to the gigantic *Gigantosaurus* and terrifying *Tyrannosaurus*, theropods strode two-leggedly across the world. To be sure, the mass-extinction at the end of the Cretaceous Period, 65 million years ago, set them back a bit. But theropods soon bounced back, and the world today resounds with the wingbeats of thousands of species of feathered, flying theropods—the birds.

The sauropods ("lizard feet") were for the most part very large herbivorous dinosaurs with a long neck surmounted by a small head, a long tail, and thick columnar legs. They had twelve or more cervical vertebrae and large nostrils high on the skull. Among the more popular large sauropods are *Apatosaurus* (formerly known as *Brontosaurus*), *Diplodocus*, *Brachiosaurus*, and the truly gigantic *Seismosaurus*, nicknamed "the earth shaker." When the massive bones of the sauropods were first discovered, it seemed obvious that such gigantic animals could not possibly have walked on land, and it was therefore postulated that they spent most of their lives in water, which could afford them some support. This turns out to be dead wrong, of course; even the largest sauropods lived, walked, and fed on land. Moreover, the image of *Diplodocus* or *Apatosaurus* (née *Brontosaurus*) dragging its tail on the ground as it lumbered ponderously along is also incorrect. Numerous sauropod trackways show no evidence of tail-dragging, and from a careful analysis of the tail vertebrae, paleontologists have concluded that these animals held their tails straight out behind them, and may even have used them as weapons.

The ornithischian dinosaurs were so named because the arrangement of their hip bones was originally believed to resemble that of birds, but if birds are lizard-hipped, it certainly confuses things, and means that the ornithischian dinosaurs are not really bird-hipped at all. In any case, the ornithischian category includes the armored dinosaurs, such as *Protoceratops, Triceratops, Pachycephalosaurus,* and *Stegosaurus;* the duck-billed dinosaurs known as hadrosaurs; and the iguanadonts, the first dinosaurs ever discovered.

George Gaylord Simpson, writing in 1953, said, "No one knows exactly why the dinosaurs and a host of other ancient forms became extinct. This is not because there is anything mysterious or metaphysical about extinction or because the possible causes are unknown. It is just because there are many reasonable possibilities, and the record does not enable us to say in the particular case which of them were actually involved. All we can say is that something changed and the dinosaurs did not."

In his 1968 book *Before the Deluge,* Herbert Wendt quoted "the skeptical and witty naturalist" Fritz Kahn:

> There has been much discussion about the causes of the rise and fall of the saurians. But why? Are not two hundred million years enough for the dominance of a family, and a hundred million years for the despotism of giants? Is not Olympus vacant now and the Parthenon a ruin? Why do we expect any stock to be immortal? Everyone knows that everything mortal is mortal. . . . Families die out; so do nations; races; so too have the saurians disappeared. . . . Grandeur comes and grandeur goes; and so with the saurians, a grand creative era in the history of life came to its end—the baroque era of zoology.

One of the first suggestions that something from outer space might have contributed to the dinosaurs' disappearance came from Max de Laubenfels, a specialist in invertebrates from Oregon State University, who presented "one more hypothesis" in the *Journal of Paleontology* in 1956. He wrote:

Attention is called to the great destruction that resulted from a meteorite impact in Siberia in 1908. A larger impact would cause more widespread destruction. Several larger impacts may have occurred in geologic time. The survivals and extinctions at the close of the Cretaceous are such as might have been expected to result from intensely hot winds such as would be generated by extra large meteoric or planetismal impacts. It is suggested that, when the various hypotheses to dinosaur extinction are being considered, this one be added to the others.

In a 1964 article in *American Scientist* entitled "Riddles of the Terrible Lizards," Princeton University paleontologist Glenn Jepsen summed up the solutions that had been proposed for the extinction of the dinosaurs:

Authors with varying competence have suggested that dinosaurs disappeared because the climate deteriorated (became suddenly or slowly too hot or too cold or too dry or too wet), or that the diet did (with too much or not enough of such substances as fern oil; from poisons in water or plants or ingested minerals; by bankruptcy of calcium or other necessary elements). Other writers have put the blame on disease, parasites, wars, anatomical or metabolic disorders (slipped vertebral disks, malfunction or imbalance of hormone or endocrine systems, dwindling brain and consequent stupidity, heat, sterilization, effects of being warm-blooded in the Mesozoic world), racial old age, evolutionary drift into senescent overspecialization, changes in pressure or composition of the atmosphere, poisonous gases, volcanic dust, excessive oxygen from plants, meteorites, comets, gene pool drainage by little mammalian egg-eaters, overkill capacity by predators, fluctuation of gravitational constants, development of psychotic suicide factors, entropy, cosmic radiation, shift of the Earth's poles, floods, continental drift, extraction of the moon from the Pacific Basin, drainage of swamp and lake environments, sunspots, God's will, mountain building, raids by little green hunters in flying saucers, lack of even standing room in Noah's Ark, and paleo-weltschmerz.

According to geological and paleontological conventions, the Cretaceous period ended and the Tertiary period began 65 million years ago. Strata below the boundary between the two—usually referred to as the "K/T boundary"—reach into the time that elapsed from 245 to 65 million years ago, inclusively known as the Mesozoic period. It is usually subdivided into the Triassic (245 to 208 million years ago), the Jurassic (208 to 144 million years ago), and the Cretaceous, which lasted from 144 to 65 million years ago. During the Mesozoic, as the world heated up, reptiles took advantage of the changing climate and came to dominate the land (as dinosaurs), the air (as pterosaurs), and the seas (as ichthyosaurs, mosasaurs, and plesiosaurs). In fact, the Mesozoic is often referred to as the "Age of Reptiles." No dinosaurs, pterosaurs, ichthyosaurs, mosasaurs, or plesiosaurs are known to have lived before or after this 180-million-year period.

The dinosaurs and pterosaurs laid their eggs on land, but many—perhaps all—of the marine reptiles gave birth to live young in the water. (Turtles and crocodiles, which are reptiles but not dinosaurs, lay their eggs on land.) The boundary between the Cretaceous and the Tertiary is definitive and geologically abrupt. In the wink of an eye, the reign of the reptiles ended, and although some endured to the present—there are still lizards, turtles, crocodiles, snakes, and the peculiar creature known as the tuatara—the terrestrial dinosaurs, the flying reptiles, and the great marine reptiles disappeared. (At least the living ones did; we document their existence by the fossil evidence.) Sixty-five million years ago an enormous object from outer space crashed into the earth wreaking such havoc that it changed the nature of life on earth forever.

In 1980, in the journal *Science,* physicist Luis Alvarez, his geologist son Walter, and chemists Frank Asaro and Helene Michel published "Extraterrestrial Cause for the Cretaceous-Tertiary Extinction." The fourteen-page article was summarized (by the authors) as follows:

> Platinum metals are depleted in the earth's crust relative to their cosmic abundance; concentrations of these elements in deep-sea sediments may thus indicate influxes of extraterrestrial material. Deep-sea limestones exposed in Italy, Denmark, and New Zealand show iridium increases of about 30, 160, and 20 times, respectively, above the background level at precisely the time of the

Cretaceous-Tertiary extinctions, 65 million years ago. Reasons are given to indicate that this iridium is of extraterrestrial origin, but did not come from a nearby supernova. A hypothesis is suggested which accounts for the extinctions and the iridium observations. Impact of a large earth-crossing asteroid would inject about 60 times the object's mass into the atmosphere as pulverized rock; a fraction of this dust would stay in the stratosphere for several years and be distributed worldwide. The resulting darkness would suppress photosynthesis, and the expected biological consequences match quite closely the extinctions observed in the paleontological record. One prediction of this hypothesis has been verified: the chemical composition of the boundary clay, which is thought to come from the stratospheric dust, is markedly different from that of clay mixed with the Cretaceous and Tertiary limestones, which are chemically similar to each other. Four different independent estimates of the diameter of the asteroid give values that lie in the range 10 ± 4 kilometers.

They were saying that they had found evidence of an extraterrestrial body crashing into the earth some 65 million years ago, creating such environmental havoc that various life forms were extinguished. Then Luis Alvarez, a Nobel laureate in physics, gave a paper at the National Academy of Sciences in Washington, D.C., on April 28, 1982, that was even more compelling. Written as a talk to fellow scientists rather than a scientific paper (though it was published in 1983 as "Experimental evidence that an asteroid impact led to the extinction of many species 65 million years ago," in the *Proceedings of the National Academy of Sciences*), the arguments were powerful and provocative. Thanks to the Alvarezes' 1980 report, our way of thinking about the sudden disappearance of the dinosaurs was changed forever. A discussion of the impact theory can be found in Walter Alvarez's popular 1997 book with the wonderful title, *T. rex and the Crater of Doom*.

On cliff faces outside the medieval Umbrian town of Gubbio, a centimeter-thick layer of clay marks the transition between the lower layer (the Cretaceous, which abounds in planktonic foraminiferans, and the layers above (the Tertiary), which contain only sparse and poorly formed fossils. If there were any clues to the mysteries of the K-T extinction, they might be found in this thin layer of clay. Alvarez, Alvarez, Asaro, and Michel found high concentrations of iridium in the Gubbio clay, and then decided to find out if this "iridium anomaly" was a local Italian phenomenon, or if similar results could be obtained from other sites of similar age. They traveled to Stevns Klint in Denmark, another location known for cliffs that clearly showed the K-T boundary, and once again, high concentrations of iridium showed up. Because iridium is very rare in the earth's crust, but quite common in meteorites, they concluded that "the pattern of elemental abundances in the Gubbio sections is compatible with an extraterrestrial source for the anomalous iridium and incompatible with a crustal source." In other words, the thin layer that marked the transition between the Cretaceous and the Tertiary contained a substance that could only have come from outer space.

In the early papers, Alvarez & Co. wrote that they "would like to find the crater produced by the impacting object," but even without it, they believed they had "developed a hypothesis that appears to offer a satisfactory explanation for nearly all the paleontological and physical evidence." Within a decade, they had found the crater, the exact "smoking gun" predicted by the collision hypothesis.

As early as the 1950s, geologists working for the Mexican national oil company Pemex had discovered rocks that had been shocked and melted, hidden away beneath the sediments of the Yucatán peninsula. The Chicxulub Crater (named for an ancient Mayan village near the site, and pronounced cheek-zhu-loob), is buried under about 600 feet of sedimentary rock that has accumulated since the impact. The irregular crater is 125 to 185 miles across; early on, at least one Mexican petroleum engineer suspected it might be an impact crater. To make a hole this size, an object with the approximate mass of Mount Everest would have had to hit at a speed of 60,000 miles an hour, instantly varporizing more than 125,000 cubic miles of the earth's crust.

By 1990, Mexican and American scientists had examined samples from the Chicxulub drill sites that showed evidence of shock metamorphism, a condition that could only have come from a massive impact. Melt rocks and spherules of impact glass were entirely consistent with the 65-million-year-old date. It has been determined that the Chicxulub crater is the largest known impact site on earth.

Supporters of the Alvarez *et al.* theory of dinosaur destruction—and there are many—believe that the impact induced a global environment collapse of such magnitude that it culminated in biological devastation. They believe that the uppermost layers of rock where the asteroid hit were heated to such temperatures that carbon dioxide and sulfate aerosols were released into the atmosphere, creating a worldwide climate of acid rain and smog—not to mention the darkening effect of dust clouds circling the earth, which lowered temperatures drastically in a so-called impact winter. Fiery debris ignited continent-sized fires that burned for years. A cataclysm of this magnitude brings out the most hyperbolic in writers, even if they are scientists. In *Night Comes to the Cretaceous* (1998), James L. Powell, geologist and director of the Los Angeles County Museum of Natural History, wrote:

> A few minutes later, the mixture of vaporized meteorite and rock, still traveling at ballistic velocities of 5 km/sec to 10 km/sec, began to reenter the atmosphere. The individual globules were traveling so fast that they ignited, producing a literal rain of fire. Over the entire globe, successively later the greater the distance from the target, the lower atmosphere burst into a wall of flame, igniting everything below. . . . Everything that could burn did.

For the nongeologist and nonastronomer who might have trouble visualizing the events that brought about the demise of the terrestrial dinosaurs, *T. rex and the Crater of Doom* includes artists' renditions of the dinosaurs, the impact, the tsunami, the crater after the impact, and the impact winter. Here is the sizzling jacket copy, written (by the publishers) to entice a potential customer to buy the book:

> Picture a day sixty-five million years ago. On the morning of that day, *Tyrannosaurus rex* stood at the apex of creation. But by nightfall a gigantic comet or asteroid, as big as Mount Everest, had slammed into the Yucatán Peninsula. The explosion on impact, which took less time than it takes to read this paragraph, was equivalent to the detonation of a hundred million hydrogen bombs. It produced a cloud of roiling debris that blackened the sky for months as well as other geological disasters—and triggered the demise of *T. rex.*

It is hard to resist such persuasive prose, but there are those who have managed to do so. They say that meteorites approach the earth all the time, and while some have actually landed with a significant impact, they have had little effect on the earth's atmosphere.* And while the eruption of the Philippine volcano Tambora in 1815 may have darkened the skies and killed off the year's corn crop, it otherwise did no permanent damage. Besides, they say, the fossil record does not show that the land dinosaurs died within a couple of years; it took a much longer period, perhaps tens or even hundreds of thousands of years. There is little doubt that a meteorite struck the earth about 65 million years ago, because shocked quartz, which is an indicator of high-temperature impact, and iridium-laden layers have now been found around the world at the corresponding stratigraphic levels. There was certainly an impact at Chicxulub, and because it coincided with the mass extinction of the dinosaurs and many other kinds of animals, it is more than a little difficult to deny a relationship between the two events, but the nature of that connection remains unresolved.

Australian biologist Tim Flannery wrote (2001) that "the death dealing lump of rock was at least ten kilometres in diameter. That's a piece of celestial real estate two or three times the length of Australia's Uluru (formerly Ayers Rock), the largest monolith on our planet." Because the available terms are so hyperbolic and the events themselves so conducive to what would otherwise be wild exaggeration, it seems hard to resist the temptation to describe the impact of a large extraterrestrial body on the earth. In his description of the creation of the 600-million-year-old crater at Ancraman, South Australia, Paul Davies 1999 has a go:

> The incoming body, typically several kilometers across, might weigh a hundred billion tons. Traveling at a speed of perhaps

---

* In addition to the smaller meteorites that fall to earth quite often, there have been meteorite impacts that have made a substantial impression on the planet's surface. One that landed in the Arizona desert about 22,000 years ago excavated what is now Meteor Crater, 4,000 feet across and 600 feet deep. The impact ejected nearly 200 million tons of rock, including fragments of nickel-iron alloy that could only have come from a meteorite. Near Lake Manicouagan in the Canadian province of Quebec is the largest meteorite impact crater where actual meteoric debris has been found. It is about 11,000 feet in diameter, and now contains a lake whose surface is 500 feet below the crater rim.

twenty or thirty kilometers per second, it delivers a blow equivalent to at least a hundred million megatons of TNT, far more than all the world's nuclear weapons put together. When it enters the atmosphere, the object displaces a vast column of air, creating a powerful shock wave that circles the globe. On hitting the ground, the meteor, along with much of the material at the impact site, is instantly vaporized. Huge quantities of rock are excavated from the surrounding terrain and hurled high into the air, even into space, leaving a gigantic crater. Large chunks of ejected rock rain back down again, hundreds or even thousands of kilometers away, glowing fiercely and igniting vegetation. The ground shock produced by the primary impact creates the most violent earthquakes, wreaking still more damage. If the meteor falls into the sea, tsunamis many kilometers high devastate the ocean rim, inundating immense tracts of land. The dust thrown up by the impact blankets the planet, blotting out the Sun for months, poisoning land and sea with acid rain. The deadly aftermath proves too much for many living species, and they are quickly driven to extinction.

Chicxulub was located off the Yucatán Peninsula when the crater was found; 65 million years ago, when the meteor hit, the entire peninsula was underwater. At that time, the continents were already arranged more or less the way they are now, but the level of the sea was considerably higher—perhaps 500 feet above present-day levels—and large portions of what is now land was underwater. A good portion of the southeastern United States, including what is now Arkansas, Louisiana, Mississippi, Alabama, Georgia, Florida, South Carolina, North Carolina, Virginia, and Delaware, was submerged (Hallam and Wignall 1997). North America was bisected by the Western Interior Basin (also known as the Cretaceous Seaway), which, at its greatest extent, covered with water everything in a broad swath from the Arctic Ocean to the Gulf of Mexico, from the eastern foothills of the Rockies to the Mississippi. The South Atlantic Ocean was much narrower, and Brazil and West Africa were joined by a tenuous landbridge. Large portions of North Africa were submerged, and most of Europe and western Asia were beneath the waters of the Tethys Sea. India was an island east of Africa heading for southern Asia, where the collision would buckle both plates and raise up the Himalayas (Scotese, Gahagan, and Larson 1988).

The impact of a giant object in the ocean, even a shallow sea like the Caribbean of 65 million years ago, some say, produced tsunamis of such unimaginable magnitude—estimates run as high as 1,000 feet—that millions of living things were killed instantly. The identification of tsunami run-up debris far inland in Alabama and Texas (Bourgeois *et al.* 1988) lent support to the idea that the object hit at Chicxulub, directly across the Gulf of Mexico. For those dinosaurs that survived the floods and the fires, perpetual night killed the plants on which the herbivores fed; with the prey animals gone, the predators had nothing to feed on, so they died too.

In his 1998 book, *Night Comes to the Cretaceous,* James L. Powell is less reserved about the global significance of the impact theory than Walter Alvarez; he calls Alvarez's story "a brilliant example of how scientists challenge and overthrow orthodoxy." Powell points out that diamonds, which were previously known to occur only deep within the earth where heat and pressure are very high, have been found at virtually every impact site studied, and only *in* the K-T boundary, not above or below it. "The finding of billions of diamonds," he wrote, "must rank as the most surprising and important of the unexpected discoveries triggered by the Alvarez theory." Because he is trying to give a balanced account, Powell devotes considerably more space than Alvarez to arguments against the impact theory, particularly those of Charles Officer. Officer wrote article after article disputing Alvarez's conclusions, and then summarized his objections in a book written with science writer Jake Page, *The Great Dinosaur Extinction Controversy* (1996).

If geology or paleontology were conducted by consensus, the Alvarez impact theory of dinosaur destruction would be taught as fact. But, because paleontology is a science of interpretations, two people looking at the same data or the same fossils can come up with completely different explanations. And while one hopes that personal agendas would not affect the results, scientists are only human, and not immune to emotional attachments to their own theories, often at the expense of objectivity. The Alvarez impact theory of dinosaur extinction is a case in point. Charles Officer, Charles Drake, Gerta Keller, and numerous others disagree with the Alvarezes, and have published articles and books detailing the areas where they believe they are right and the Alvarezes are wrong.

Prior to the discovery of Chicxulub, one of the strongest objections to the Alvarez theory was that there was no crater where the iridium-rich meteorite might have landed. In a 1985 paper, "Terminal Cretaceous Environmental Events," Officer and Drake wrote, "Whatever the origin of the major K/T environmental events that led to the observed floral and faunal changes, it would be encouraging to find some direct legacy of the event itself. For the asteroid impact hypothesis, this would be the crater itself or some evidence of it."

Officer, Drake, and many other people believe that the volcanic eruption in India that produced the formation now known as the Deccan Traps is at least one of the agents responsible for the K-T extinction. In an article in *Scientific American*,* geophysicist Vincent Courtillot wrote,

> The sheer size of the Deccan Traps suggests that their formation must have been an important event in the earth's history. Individual lava flows extend well over 10,000 cubic kilometers. The thickness of the flows averages from 10 to 50 meters and sometimes reaches 150 meters. In western India the accumulation of lava flows is 2,400 meters thick (a quarter the height of Mt. Everest). The flows may have originally covered more than two million square kilometers, and the total volume may have exceeded two million cubic kilometers.

In 1999, Cambridge University Press published Courtillot's *Evolutionary Catastrophes: The Science of Mass Extinctions*. The book was originally supposed to be about the collision between India and Asia 50 million years ago when the Deccan Traps erupted, but it evolved into a discussion of mass extinctions in general. As might be expected, Courtillot's prejudices as a volcanologist shine through; he says that hundreds of huge, if brief, lava flows so upset the atmospheric balance, and therefore the food chains, that the dinosaurs were gradually eliminated. The Chicxulub impact and the Deccan Traps eruptions obviously occurred at the same time, but Courtillot believes that volcanism was the moving force behind

---

* In the April 1990 issue, the editors of *Scientific American* asked the question, "What caused the great extinctions?" Walter Alvarez and Frank Asaro answered with "An Extraterrestrial Impact," while Courtillot provided the opposing point of view with "A Volcanic Eruption."

all other mass extinctions. He notes that Officer and Drake "attacked the impact hypothesis and argued that the events of the K-T boundary might have lasted at least 10,000 to 100,000 years. . . . They even noted that the Deccan Traps had undoubtedly been produced quite rapidly (though they could not say just how rapidly) and at the right period." Courtillot adds,

> The age and duration of the laying down of the Deccan Traps, and the considerable climatic disturbances that must have resulted, suggest a plausible scenario of an ecological catastrophe likely to result in the massive extinctions of the K-T boundary. This scenario is furthermore quite close to the one proposed by the asteroid advocates—except its duration is counted in tens or hundreds of thousands of years, not days or months—and just as terrifying.

Courtillot recognizes that the selectivity of the extinctions cannot easily be explained by the impact scenario: "Those species that pulled through—placental mammals, birds, amphibians—undoubtedly did so for a variety of reasons, among which we may presume were simply their small size and the concomitant very large numbers of individuals, as well as their nocturnal habits, their tolerance of temperature changes, their underground habitat, and what they ate (as root eaters, carrion scavengers, eaters of organic matter in various forms, etc.). But it remains a mystery why some reptiles with a certain lifestyle survived, yet all species of dinosaurs, with quite a similar lifestyle, did not." He concludes:

> This picture of our planet's climate 65 million years ago, this poisoning of the atmosphere with volcanic gases for several hundred to thousand centuries, these mass extinctions of species that had dominated the Earth and the waters for so long, all seem to me to be compatible with many of the records that geologists, geochemists, and geophysicists have extracted since the 1960s from the archives preserved in rock.

Another mystery appears when you look at reconstructions of the earth at the moment of the impact, 65 million years ago. As mentioned earlier, the continents were arranged more or less the way they are now, but because the level of the sea was higher, large portions of the landmasses were underwater. But if India was an island en route to an appointment with southern Asia, were its volcanoes erupting as it chugged northward? Courtillot believes that the volcanic eruptions of the Deccan produced poisonous gases, particularly sulfur dioxide (SO_2), that contributed mightily to the destruction of species, but where was India when it was erupting? I asked Courtillot about this; he answered:

> In my view . . . the Deccan took place at 65 Ma [Millennia ago] as India was on its voyage northward and unrelated to the collision. . . . India was probably much larger to the North than it is today and this extension (called greater India) implies that India probably touched Asia earlier than generally assumed, and possibly as early as KT time! But contact seen by paleontologists does not mean collision as seen by the geophysicist or tectonicist, implying major frontal contact, large stresses and deformation, which probably did not occur until 55 or 50 Ma.

The two geologic events may have been related. There is always the possibility that an asteroid impacting the earth created such seismic havoc that it somehow initiated the movement of the crustal plates and even set off the volcanoes. Because there is evidence that both events occurred, more or less simultaneously, some would attribute the extinction of the dinosaurs to this unfortunate congruence of disasters. In a summary of earth history published as the twentieth century was coming to a close, the editors of *New Scientist* wrote,

> The twilight of the Cretaceous period brought the demise of the largest creatures that ever walked our planet—the dinosaurs. A giant fireball came screaming out of the sky as an asteroid or comet about 15 kilometres wide hit the Earth near Mexico. The dust thrown up by the impact darkened the sky for months. Many plants died out, and the giant reptiles starved to death. The same fate befell land animals weighing over 25 kilograms, along with their marine cousins, toothed birds, ammonites and corals. Many more species were finished off soon afterwards by intense volcanic activity, as the impact triggered lava flows over 1 million square kilometres of the Earth's surface.

Many people are confused by the apparent contradiction between the propositions that the dinosaurs were all eradicated by the K-T event, but birds, the actual lineal descendants of dinosaurs, somehow survived the catastrophe; all the dinosaurs were *not* killed off by the impact. Jason Lillegraven and Jaelyn Eberle (1999) examined the "faunal changes through one of the thickest and one of the most complete records of terrestrial vertebrates spanning Lancian (~latest Cretaceous) and Puercan (~earliest Paleocene) ages, the Ferris Formation in the Hannah Basin, southern Wyoming"; they found an abundance of dinosaur fossils (*Ornitholestes, Dromaeosaurus, Struthiomimus, Tyrannosaurus, Ankylosaurus,* etc.) below the Lancian-Puercan boundary, and only mammals above it. They wrote, "At least in this part of the North American western interior, the first evolutionary radiation of condylarths was subsequent to the last appearance of dinosaurs, not synchronous or prior to it. . . . We recognize no evidence that mammals were 'recovering' from events that led to the demise of the dinosaurs."

Vertebrate paleontologists Lowell Dingus and Timothy Rowe wrote an entire book in 1997, called *The Mistaken Extinction: Dinosaur Evolution and the Origin of Birds,* addressing the subject of the nonextinction of birds during the K-T event. In the chapter "Crossing the Boundary," they ask, "Did the birds rebound from a catastrophic near miss with extinction? Did the history of birds parallel that of mammals in an explosion of Tertiary diversification?" Basing their answers on the work of American Museum of Natural History paleontologist Mark Norell, they suggest that a phylogenetic map of bird lineages shows that while birds are indeed descended from a common late Jurassic ancestor that lived around the time of *Archaeopteryx,* about 150 million years ago, the fossil record does not so far support this suggestion of a rebound or explosion. They fill in the missing pieces with "ghost lineages," which "predict what the fossil record should eventually yield." But a few bird fossils from the Cretaceous do exist, including the Hesperornithiformes (large aquatic birds that had greatly reduced wings and paddle-like feet); Ichthyornithiformes (seabirds that were accomplished flyers); and Enatiornithines (finch-sized birds that lived alongside the terrestrial dinosaurs).

To the best of our limited knowledge, all these birds were extinct 5 million years before the end of the Cretaceous. So if these birds didn't make it through, what did? Alan Feduccia of the University of North

Carolina argues that only a single lineage of birds made it through, but, like the Phoenix, the proliferation of modern birds rose from the ashes. The birds for which we have pre-K-T fossils are extinct, but since there are plenty of birds around today, the ghost lineages—that is, species for which there is no fossil evidence whatsoever—are the ancestors of modern birds. Dingus and Rowe believe that "dinosaurs not only crossed the K-T boundary, they survive today in great abundance." Those who claim that the dinosaurs became extinct are making the mistake of their title. Similarly, at the conclusion of their 1998 article, "The Origin of Birds and Their Flight," Kevin Padian and Luis Chiappe wrote,

> Most lineages of birds that evolved during the Cretaceous died out during that period, although there is no evidence that they perished suddenly. Researchers may never know whether the birds that disappeared were outcompeted by newer forms, were killed by an environmental catastrophe, or were just unable to adapt to changes in their world. There is no reasonable doubt, however, that all groups of birds, living and extinct, are descended from small, meat-eating dinosaurs. . . . In fact, living birds are nothing less than small, feathered, short-tailed theropod dinosaurs.

In a comprehensive article on extinctions in the June 1989 issue of *National Geographic,* an elaborate chart shows the abrupt end of the dinosaur line and the explosion of mammals after the K-T event, but there is no mention whatsoever of birds. Surely the most unlikely scenario is that birds, delicate creatures that they are (and presumably were for the most part, although some early birds were gigantic, hulking things), survived the massive conflagrations, while their oversized relatives, the dinosaurs, got starved, fried, drowned, or asphyxiated. The tiny mammals may have hidden in burrows as acid rain fell and flames roared overhead, but where was a bird to go if all the trees (not to mention the food and the air) was reduced to ashes? The survival of some bird species may be strong evidence *against* the worldwide firestorm that the *Geographic* article describes as "an awesome terrifying orgy of flame . . . [when] the entire world caught fire."

Some of the small mammals, which lived alongside the dinosaurs and would eventually inherit the earth from the ruling reptiles, seem to have escaped more or less unscathed, as did the land plants, insects, frogs, lizards, snakes, and crocodiles that were the forerunners of today's representative species. It is possible to envision at least some proportion of the chipmunk-sized Cretaceous mammals taking refuge in caves or burrows, and thus avoiding the conflagrations going on around them, but most birds, whether or not they are descended from dinosaurs, are not usually given to hiding in burrows or caves,* and would have had an even harder time surviving firestorms, floods, and the destruction of plant life. Given the magnitude of the explosion, the tsunamis, the radiation, the fires, and the darkening cloud, how could a group of small animals miraculously survive the equivalent of an explosion millions of times greater than the one that destroyed Hiroshima?

If, as most paleontologists now believe, birds are directly descended from dinosaurs, then all the dinosaurs did *not* die in the K-T extinction; many continued to evolve as birds. Even if we define extinction as Darwin would, that is, a species disappears as it evolves into another, we still cannot easily explain the ability of birds to pass unscathed—or even much reduced in numbers—through the K-T boundary. In any case perhaps because there is so little evidence, birds usually get short shrift in discussions of the Cretaceous extinction. The maxim "absence of evidence is not evidence of absence" is surely applicable here. In *T. rex and the Crater of Doom* (1997), Walter Alvarez suggests that the mammals, crocodiles, and turtles, "being smaller and thus more numerous, would increase their chances of survival, and this may help explain the survival of birds as well." Then in his 1991 book, *Extinction: Bad Genes or Bad Luck?*, David Raup diagrams "the evolution of birds from Jurassic dinosaurs," showing the line of birds breaking off from the line of dinosaurs sometime in the Jurassic. While the dinosaur line ends abruptly at the K-T boundary, the bird line goes on uninterrupted to the present day. "This pattern," writes Raup,

---

* Some birds live in burrows, sometimes of their own making. Penguins of the genus *Spheniscus* (the African, Humboldt, Magellanic, and Galápagos) lay their eggs in burrows, and raise their chicks therein. Certain shearwaters, particularly the short-tailed (*Puffinus tenuirostris*), lay their eggs in burrows; the chicks mature underground before setting off on their transhemispheric migrations. Cave-dwelling birds are much less common, but the oilbird (*Steatornis caripensis*) of northern South America spends its days in caves, coming out at night to feed on the oily fruits of palm trees.

"has led some to argue that dinosaurs did not die out at the end of the Cretaceous, that they merely evolved wings in the Jurassic and flew away." In the midst of a worldwide firestorm, where could they have flown *to*?

Whether there was a single, extraterrestrial impact, multiple impacts, volcanic action, or some synergistic combination, many kinds of animals did *not* die out at the time of the K-T extinction. Ancestral frogs, salamanders, fishes, sharks, turtles, lizards, mammals, crocodiles, and birds survived to propagate their kind. This does not necessarily show that the extinction event was selective, only that it was not universal. From the geological evidence, it is clear that a massive object hit at Chicxulub. The Deccan Traps are also incontrovertible evidence that there was enormous—and enormously destructive—volcanic activity at this time. It is possible that the Chicxulub impact rained death and destruction on portions of North America, eliminating the nonavian dinosaurs there, among other things, while the volcanoes of the Deccan produced poisonous gases that killed off vast numbers of other animals, and even entire species. The last of the dinosaurs seem to have had the misfortune to be living in North America when the asteroid hit, but what if the gases, the firestorm, the ejecta, or the tsunamis did not reach Gondwana? Today's incipient threats to life and wildlife, such as acid rain or the reduction of greenhouse gases, are still geographically limited; while the massive destruction of the Amazon rain forest is certainly a tragedy of worldwide proportions that indirectly affects all biodiversity, it does not immediately affect the lives of cheetahs, muskoxen, or kangaroos in other regions. In the end, all modern animals—including us—will become extinct, but the immediate causes might not be so easy to identify.

Whether the impact killed them, or they died of unknown causes, the great terrestrial dinosaurs were all gone after the dust cleared from whatever it was that happened 65 million years ago. Contrary to popular conceptions, all the dinosaurs were not marching around as the Cretaceous came to its abrupt halt. Most of them were extinct anyway, and many had been extinct for tens of millions of years. For example, the carnivore *Allosaurus* was gone 135 million years ago, as was *Apatosaurus,* the great sauropod that used to be known as *Brontosaurus.* The feathered reptile *Archaeopteryx*, originally found in shales from 150 million years ago, was gone 30 million years later, and *Iguanadon,* the first discovered dinosaur, vanished 110 million years ago. *Deinonychus*, the dinosaur described as warm-blooded by John Ostrom in 1969, disappeared from the fossil record 100 million years ago, and *Proto-ceratops*, the predominant dinosaur of the Gobi Desert, was last recorded from strata dated at 75 million years old. In fact, as far as we can tell from the scanty fossil record, almost all the dinosaurs had died out except for those in western North America. *Saltosaurus, Ankylosaurus, Pachycephalosaurus, Gallomimus, Triceratops,* and the great *Tyrannosaurus* were the last remnants of the line of giant reptiles that had dominated the land for 150 million years. When the asteroid hit, it was these dinosaurs that were somehow eliminated. After the impact, no terrestrial dinosaurs have been recorded. For the most part, their fossils have been found in the Hell Creek region of Montana, and while this does not mean that this was the last refuge of the last of the dinosaurs, it does mean that we have not been able to find very many fossils elsewhere. It was also at this time that the last of the mosasaurs died out. By the time the asteroid struck, the ichthyosaur "fish-lizards" and the plesiosaurs had been extinct for 20 million years.

Because massive volcanism was occurring at the time of the K-T extinctions, it is not difficult to associate the two events. "Radiometric and paleomagnetic analyses have shown that the flood basalts were

Tyrannosaurus rex

deposited over a relatively short period of time around the K-T boundary," wrote Officer and Page in 1996. Erupting volcanoes like Thera, Tambora, and Krakatau can produce darkened skies, atmospheric cooling, and sulfurous acid rains, not unlike the effects suggested for a meteorite impact. (It has also been suggested that an impact might have triggered volcanic eruptions, although how this might have happened is unclear.) An iridium layer, so important to the impact theorists, has also been found in the basalts of the Deccan Traps, deposited before the putative impact; anyway, Officer and Drake (1985) do not believe that the iridium layer was laid down instantaneously, "but during a time interval of some 10,000 to 100,000 years." (Unaccounted for in the volcanism theory is the shocked quartz granules that could only have come from an impact.)

The evidence is spotty, however, and what there is does not provide universally acceptable answers one way or the other, so the causes of the K-T extinction will probably never be known to any degree of certainty. "I am not sanguine," wrote Archibald (1996), "that we will ever confidently be able to answer whether the dinosaurs departed with a bang or a whimper." David Raup asks and answers the question: "Did the K-T extinction last several millions of years or was it over in a few minutes? This is a vexing and important question, to which the fairest answer is that we don't know." He explains the difficulties:

> Geologic dating is often so uncertain that one cannot be sure that the rocks at different sites are the same age. Even if the K-T boundary can be identified at each site—not always possible—the boundary itself may not be the same age at all sites. Suppose that at one site the rocks deposited during the last two or three million years of the Cretaceous were eroded away before the start of the Tertiary. The K-T boundary, defined as the upper surface of the younger Cretaceous rocks, would be several million years older than the actual age of the K-T transition. . . . Because of erosional and other gaps in the rock record, it is extremely difficult for us to amass a sufficient quantity of high-quality data to say anything more than that the extinctions occurred sometime near the end of the Cretaceous.

To exemplify the title of his provocative book, *Extinction: Bad Genes or Bad Luck?* Raup introduces the trilobites, who lived for more than 300 million years and died about 245 million years ago. He asks: "Why? Did the trilobites do something wrong? Were they fundamentally inferior organisms? Were they stupid? Or did they just have the bad luck to be in the wrong place at the wrong time?" The book jacket calls Raup a "statistical paleontologist," which suggests more of an interest in number crunching than fossil digging, and his book includes more graphs and charts than most books about dinosaurs. In the introduction, Stephen J. Gould says that if his friend Dave Raup "has any motto, it can only be: Think the unthinkable (and then make a mathematical model to show how it might work); take an outrageous idea with a limited sphere of validity and see if it might be expanded to explain everything." In this case, Raup's "outrageous idea" (developed with Jack Sepkoski) was that meteors and comets have been crashing into earth since the planet was formed, and that they are somehow related to the principal extinction events, which have occurred approximately every 26 million years. The connection between mass extinctions and extraplanetary collisions is not supported by sufficient evidence, but Raup and Sepkoski believe that there might be one. "My own view," wrote Raup, "is that periodicity is alive and well as a description of extinction history during the past 250 million years—despite the lack of a viable mechanism."

Raup's ideas about all extinctions—and that includes "background" as well as mass extinctions—are equally outrageous. He argues that normal environmental stress (temperature changes, raising and lowering of sea level, wandering continents, etc.) usually produces adaptations or migrations, not extinctions. He writes, "Most plants and animals have evolved defenses against the normal vicissitudes of their environment." But a stress that has never been experienced previously, and for which the species has no defense, such as epidemic disease as suggested by Robert Bakker, can certainly devastate a species, and perhaps even lead to its extinction. (Fred Hoyle and Chandra Wickramasinghe combined the asteroid and disease theories in 1978, saying that the deadly viruses were brought to earth as passengers on extraterrestrial impactors.) Because the earth has been bombarded by a steady stream of icy comets and stony asteroids and meteorites, it is conceivable that all extinctions are connected to visits of space debris. In conclusion, Raup answers the question posed by the title of his book:

> Extinction is evidently a combination of bad genes and bad luck. Some species die out because they cannot cope in their normal habitat or because superior competitors or predators push them out. But as is surely clear from this book, I feel that most species die out because they are unlucky. They die because they are subjected to biological stresses not anticipated in their prior evolution and because time is not available for Darwinian natural selection to help them adapt.

We do know that 65 million years ago, at the end of the Mesozoic, as the grand finale of the Age of Dinosaurs played out, a major episode of volcanic activity spewed out lava across a huge region of India—wherever that region was at the time. This is apparent from the thick flows of basalt that cover much of the southwest of that subcontinent. Around the same time, a large asteroid or comet struck the earth at Chicxulub, near what would rise from the sea as the Yucatán Peninsula. Either event could have lowered temperatures, reduced sunlight, destroyed the ozone layer, generated acid rain, decimated plant life, and eventually killed off the large dinosaurs as well as many other organisms, both on land and in the seas.

The lethal effects of the extraterrestrial impact are thought to have acted over a period of a few months, whereas the effects of the volcanic activity could have lasted for millions of years. To show that the dinosaurs were killed off rapidly by the impact, we would need to be able to pinpoint events on a year-by-year basis. But for the period at the end of the Mesozoic, we can date events only with a margin of error of plus or minus 30,000 to 100,000 years. Thus, we have no way of knowing for sure whether it was the impact, the volcanic event, or both that caused the large nonavian dinosaurs to become extinct. We cannot even tell if all the other types of animals that became extinct did so at the exactly the same time. Whatever caused it, the Mesozoic Age of Dinosaurs came to an end, making way for the Cenozoic Age of Mammals. (Mammals actually arose at roughly the same time as the dinosaurs, but they were overshadowed—literally and figuratively—by the dinosaurs.) Most paleontologists believe that the avian dinosaurs are not extinct. They believe that birds evolved from small carnivorous dinosaurs closely related to *Deinonychus* or *Velociraptor;* living birds are therefore dinosaurs.

Either the pendulum has swung completely in favor of the extraterrestrial impact explanation of the disappearance of the dinosaurs, or David Kring and Daniel Durda have managed to persuade the editors of *Scientific American* that it has. The assertive opening sentence of their December 2003 article reads as follows: "By now it is common knowledge that the impact of an asteroid or comet brought the age of the dinosaurs to an abrupt end." No equivocation there; indeed, they say that the asteroid or comet was "so large that when its leading edge made contact with the ground, its trailing edge was at least as high as the cruising altitude of a commercial airliner. It produced an explosion equivalent to 100 trillion tons of TNT, a greater release of energy than any event on our planet in the 65 million years since then." The impact, they say, superheated the atmosphere to such an extent that most of the plants on the earth's surface spontaneously burst into flame, causing massive wildfires that "wrecked the base of the continental food chain and contributed to a global shutdown of photosynthesis." Some creatures survived because not all areas were equally affected. The fires in the vicinity of the impact spread, and as the earth rotated under the ash cloud, fires ignited on a vast portion of the Southern Hemisphere, affecting central South America, Central Africa, the Indian subcontinent, and Southeast Asia. (Soot from the wildfires has been found in most of these areas.) "The firestorm created by the Chicxulub impact and the subsequent pollution were devastating," conclude Kring and Durda. "But it was probably the combination of so many environmental affects that proved to be so deadly. . . . Life's diversity was its salvation. Although multitudes of species and countless individual organisms were lost, some forms of life survived and proliferated."

Should we mourn the passing of the terrestrial dinosaurs? "Well," said Paul and Anne Ehrlich in their 1981 book *Extinction,* "the answer is yes and no." Presaging the publication of Michael Crichton's *Jurassic Park* by more than a decade, they wrote, "What a thrill it would be if in national parks people could see great lumbering brontosauruses weighing forty of fifty tons grazing across the landscape, or herds of ceratopsian dinosaurs roaming like rhinoceri with three gigantic horns! With luck one might even get to watch an attack on a grazer by that mighty predator *Tyrannosaurus.*" Their answer is "yes" because we could not (until the advent of computer-generated images) see the mighty dinosaurs in action, but "no" because their extinction meant the development of mammals to replace them:

> If, on the other hand, the dinosaurs had become extinct and the mammals had not evolved to take over the roles that dinosaurs had played, it would be a very different world indeed. Since we too are mammals, there would of course be no people; that would be the most important difference from our point of view! The principal reason, then, that people don't miss the dinosaurs or other groups of long extinct organisms is because replacements for them have evolved.

Lucky for us, then, that the dinosaurs vanished. The Ehrlichs also point out that extinction is the natural order of things; species have gone extinct since life began. Replacement species do not always arise, and available niches are not always filled. We are not sure what part the diverse ammonites, some of which had shells six feet across, played in prehistoric ocean ecology; and though they introduced eyes to the world, the trilobites, which lasted for hundreds of millions of years and absolutely dominated the seafloor community, left no descendants. There were no replacements to play their parts. The roles played by the large marine reptiles, and for that matter the flying reptiles, remain unoccupied. Killer whales and great white sharks are of course today's oceanic apex predators, but they are considerably smaller than the largest pliosaurs and mosasaurs. There are large predatory sharks and fishes, of course, but their origins predate those of the ichthyosaurs, plesiosaurs, and mosasaurs, so they cannot be considered "replacements," and while there are large flying birds, like condors and albatrosses, nothing comes close to the gigantic *Queztalcoatlus*, a pterosaur with a forty-foot wingspan.

The Dinosaurs Are *Not* Extinct After All

Prehistoric skies were filled with swooping, gliding, diving, flying reptiles, ranging in size from the blackbird-sized *Anurognathus* to *Queztalcoatlus*, with its forty-foot wingspan, the largest flying creature in history. Every single species of flying reptile (except for birds) is extinct. The most famous of the flying reptiles was of course *Archaeopteryx*, a long-extinct contemporary of the dinosaurs and plesiosaurs, but the paleontological jury is still arguing about what its descendants might be, or if its kind became extinct without passing along its genes. (Its full name, *Archaeopteryx lithographica*, comes from the stone of the quarries at Solnhofen, in Bavaria, which stone was so fine that it was—and still is—used for making lithographic plates.) Even though *Archaeopteryx* was clearly a reptile, Dingus and Rowe (1998) unequivocally refer to it as "the oldest known bird." Most paleontologists are prepared to classify *Archaeopteryx* as a bird—Peter Wellnhofer, considered the world's foremost authority on *Archaeopteryx*, also calls it "the oldest bird known to man"—and its position in the evolution of modern birds is stable indeed.

In the 1870s, on the basis of his studies of the dinosaur *Megalosaurus* and the living ostrich, Thomas Henry Huxley proposed that birds and dinosaurs were closely related; indeed, the bulk of subsequent evidence has only strengthened that assessment. Demonstrating just how close birds and small dinosaurs are in their gross overall morphology, a couple of

Archaeopteryx

Archaeopteryx skeletons were originally misidentified as the small dinosaur *Compsognathus,* and the first *Archaeopteryx* specimen found (in 1855) was classified as a pterodactyl until finally recognized for what it was by John Ostrom of Yale in 1970. Ostrom said that birds were the direct descendants of theropod dinosaurs. Chatterjee (1997) wrote, "Because of its evolutionary importance, *Archaeopteryx* was the subject of an international symposium at Eichstätt, Germany, in 1984. An attempt was made to resolve some of the controversies, but there was little consensus about its status, ancestry, relationships, mode of life, and flight capabilities." Nonetheless, it is unquestionably paleontology's crown jewel; the fossil that tells us more—albeit somewhat ambiguously—than any other. In 1990, Wellnhofer wrote, "With its reptilian body and tail yet undeniably birdlike wings and feathers, *Archaeopteryx* provides paleontologists with their most conclusive evidence for the evolution of birds from reptiles. This pigeon-sized prehistoric bird is known from only six fossil skeletons and the imprint of one lone feather, but paleontologists have deduced a wealth of information from these few specimens."

It has now been established that some theropod dinosaurs bore feathers during at least some stages of their lives. The dinosaurs collectively known as theropods include all carnivorous dinosaurs, such as tyrannosaurs (e.g., *T. rex*), allosaurs (*Allosaurus*), coelurosaurs (*Struthiomimus*), maniraptors (*Velociraptor, Ornitholestes*), and of course, *Archaeopteryx*. And all living birds. As Dingus, Gaffney, Norell, and Sampson wrote in 1995,

> Just as people belong to their families because they are descended from their parents and grandparents, birds belong to the group called dinosaurs because they descended from a dinosaur. Consider some of the features that birds inherited from their dinosaurian ancestors. Their upright stance came from the very first dinosaur. They inherited their three-toed hind feet and hollow bones from the common ancestor of all meat-eating dinosaurs. The elongated arm bones that support the wind evolved from the first coelurosaur, which probably looked like *Ornitholestes* . . . Birds are feathered dinosaurs, just as humans are large-brained primates that walk upright.

In the above explanation, the authors included this sentence: "As far as we know, feathers are unique to birds, but this doesn't mean that birds are not dinosaurs." In an astonishing demonstration of how quickly paleontological speculation can become paleontological fact, no sooner had Dingus *et al.* published their description of theropods in 1995, feathered dinosaur fossils were discovered in China. As paleontologist Philip Currie describes it:

> In September 1996, I went to Beijing to spend a few working days with my longtime colleague Dong Zhiming on some dinosaurs he had collected in northwestern China. Shortly after arriving, he showed me a Chinese newspaper report of a beautiful little dinosaur skeleton found in Liaoning, a province northeast of the capital city. . . . The specimen was beautiful, exposed from the tip of the nose to the tip of the tail on a small slab of rock. But that is not what caught my attention: it was the rim of structures that surrounded almost the entire body. They were real and they belonged! I knew instantly that the first "feathered" dinosaur had indeed been discovered.

Younger by 20 to 30 million years than *Archaeopteryx,* the chicken-sized fossils, again similar in size and shape to *Compsognathus,* have what appear to be rudimentary feathers, although the nature of these structures has been hotly debated. (Some say they are indeed feathers; others argue that they are hairlike quills like those of the kiwi; and some say they are merely subcutaneous structures.) Early bird or transitional dinosaur, *Sinosauropteryx* is clearly related to *Compsognathus;* it is considered one of the most important finds in recent paleontological history. The "feathered" fossil was featured on the cover of the November 14, 1997, issue of *Nature* and a painting of the reconstructed animal appeared as the cover illustration for the 1997 *Encyclopedia of Dinosaurs.* The entry for "Feathered Dinosaurs," written by Philip Currie, reads as follows:

> In 1996, several skeletons of a 1-meter-long animal were found in Liaoning (People's Republic of China) that show featherlike structures covering the head, trunk, tail, arms, and legs. Named *Sinosauropteryx prima,* this animal is closely related to *Compsognathus* from the Upper Jurassic of Europe. The integumentary structures were simpler than true feathers, and each seems to be composed of a central rachis and branching barbs but lacks the aerodynamic quality of avian feathers. The longest "feathers" were about 3 cm in length, and it has been suggested that they were more suitable for insulation than they were as display structures. The discovery of these specimens has given additional support to the hypotheses that theropod dinosaurs were the direct ancestors of birds, and that some theropods were endothermic [warm-blooded].

This long-tailed little creature, dead for 130 million years, has become the center of a fierce paleontological firestorm. Some believe that the interesting "integumentary structures" are feathers, while others are not convinced. Even before the publication of the initial description (Chen *et al.* 1998), the Academy of Natural Sciences of Philadelphia sent a group of scientists to China to study the fossil. The group included John Ostrom of Yale, the discoverer of the dinosaur *Deinonychus* and the man who introduced the theory of warm-blooded dinosaurs; Larry Martin of the University of Kansas, who does not agree with Ostrom (and the majority of modern paleontologists) that birds are descended from dinosaurs; Peter Wellnhofer, the *Archaeopteryx* expert from the Bavarian State Museum in Munich; and Alan Brush of the University of Connecticut, an expert on feathers. They were given unlimited access to the three known specimens of *Sinosauropteryx* by the Beijing Geological Museum and the Nanjing Institute of Geology and Paleontology. After three days of examination (not nearly enough, they stressed), they agreed only that the dark-colored ridge running down the back didn't really look like modern feathers.

Close-up photographs of two specimens showed that the structures are "spiny feathers" or at least hairlike protofeathers, and not fibrous muscle tissue. (They are also not accidents of preparation as has been argued by some researchers.) *Sinosauropteryx* retains several primitive features—such as very short arms—indicating that it is the least birdlike of the theropod dinosaurs preserved in this ancient Chinese lakebed. Although individual feathers in *Sinosauropteryx* are hard to pick out, they all appear about the same size and shape, and appear to be composed of fine filaments branching from hollow quills, rather like down feathers in birds today. This contrasts with feathers on the hand (remiges), tail (retrices), and body (contour) that evolved later in birds. These feathers can vary considerably in size and shape on different parts of the body, and unlike *Sinosauropteryx* feathers, the filaments in "aerodynamic" feathers of living birds are tightly bound together by tiny hooks, thus forming clean, sharp-edged outlines and broad, fixed, aerodynamic surfaces. (To confuse matters, however, barbed feathers have also been found on dinosaurs that were obvious non-flyers. It may be that the barbs help to lock—and thus straighten out—ruffled feathers.)

Among the more recent discoveries are the Chinese forms *Sinornithosaurus millenii* and *Protarchaeopteryx robusta,* both of which have been classified as dromaeosaurs; and *Beipiaosaurus inexpectus* and *Caudipteryx zoui,* all dinosaurs with feathers—or were they actually birds? Hundreds of fossils of the pigeon-sized *Confuciusornis sanctus* ("holy Confucius bird") from the early Cretaceous, 120 million years ago, as well as many other birds of the same size have been found in Liaoning Province. *Confuciusornis* is the oldest-known beaked bird, and one of the oldest known birds—only *Archaeopteryx* is older. In one striking respect, this primitive bird had taken an evolutionary step that seems to anticipate every bird of today: its beak was completely toothless. (*Archaeopteryx* had teeth, but

Confuciusornis was the earliest bird known to have lost the toothy jaws of its reptilian ancestors.) Because of the large number of specimens, the fossil record for *Confuciusornis* is more extensive than that for most other fossil birds. The condition of preservation shows that the specimens died suddenly and were embedded in the sediment. The birds probably lived socially near freshwater lakes and ponds. They might have been able to swim, or they might have nested in trees. Their anatomy reveals a surprising mixture of primitive and derived features. According to Peters and Qiang (1999),

> The large wings leave no room for doubt that *Confuciusornis* could fly; soaring was probably its preferred mode of flight. The three fingers could be moved independently of each other. The mobility of the second (big) finger was apparently restricted. This finger bore the distal primaries, its claw is small. The functions of the first and third fingers are not clear. The third finger has a very short basal phalange and was highly flexible. It probably served as a sort of landing flap when it was stretched out caudally, thus improving the flight properties of the wing, all the more as the alula of the first finger was lacking. The function of the enormous claws on the first and third finger is unknown. The assumption that these claws were climbing devices is implausible. . . . The configuration of the toes is far from that of typical perching birds.

Even though asymmetrical feathers have been found in nonflyers, asymmetry is a necessary component of wing design in flying birds, and the symmetrical feathers on the arms and tail of *Protarchaeopteryx* meant that it probably wasn't much of a flyer. Most of its body was covered with downlike filaments, and it had a large fan of feathers at the end of the tail (Qiang, Currie, Norell, and Shu-An 1998).

Of all living creatures, only birds have feathers, so it was natural to assume that feathers were exclusive to birds. Now that dinosaurs like *Sinosauropteryx, Beipiasaurus, Caudipteryx,* and—of course—*Archaeopteryx*— have been shown to have sported a feathery integument, some of the earlier assumptions have to be revised. *Archaeopteryx* was originally identified as a bird *because* it had feathers, but when two feathered dinosaurs (*Caudipteryx* and *Protarchaeopteryx*) from Liaoning Province were shown to predate *Archaeopteryx*, it was obvious that this identification was too simplistic, and that birds evolved from "feathered, ground-living, bipedal dinosaurs" (Ji *et al.* 1998). The feathers of *Caudipteryx* and *Protarchaeopteryx* are symmetrical, and may have even had interlocking barbs, but their arms were not nearly long enough to have functioned as wings. According to Padian (1998). "The available evidence suggests that structurally airworthy feathers may have evolved before they were long enough, or their possessors able to use them for flight. The work of Ji *et al.* should lay to rest any remaining doubts that birds evolved from small coelurosaurian dinosaurs."

All birds are dinosaurs, but not all dinosaurs are birds. Among the feathered dinosaurs that were not birds is the recently unearthed fossil from the early Cretaceous Yixian Formation of China, described as a fully feathered dinosaur, which therefore indicates that "feathers pre-dated the origin and flight of birds" (Sues 2001). The fossil, dubbed *Sinornithosaurus* ("Chinese bird-lizard"), found in sedimentary rock in Liaoning Province, was first described by Xu, Wang, and Wu in 1999. Subsequently, another fossil feather-covered dinosaur was found in Liaoning Province; and Norell, in an interview (Anon. 2001) said it may also be *Sinornithosaurus*. The new fossil, identified as NGMC 91, has the sickle claw present only in dromaeosaurs, and elongated connections between the tail segments, which served as stiffening rods. This fossil demonstrates that birds are the living descendants of theropod dinosaurs, and that birds are not the only creatures with feathers. But these little dinosaurs obviously could not fly, so what purpose could feathers have served in nonflying animals? Some paleontological theorists believe that they were somehow connected with warm-bloodedness, and were likely to have served as insulation.

When *Archaeopteryx* was first discovered, it was mistaken for a small *Compsognathus,* which was found in the same depositional environment. Of course, the absence of evidence is not the same thing as the evidence of absence; just because feathers were not found on the *Compsognathus* fossils, that does not prove that the little dinosaurs were featherless in life. Specimens of *Compsognathus* have been buried for 150 million years, and in that time it is possible—even likely—that all elements of the integument—

scaled, furred, or feathered—would be lost, even when the fossilized bones remain. As Greg Paul (2002) wrote, "The absence of plumage or skin associated with the *Compsognathus* skeletons preserved in fine-grained sediments does not rule out the presence of body feathers any more than the absence of same in complete urvogel [*Archaeopteryx*] specimens from the same lagoonal sediments." In other words, not all the *Archaeopteryx* specimens are feathered, but no one doubts that the actual animal was. Adult tyrannosaurs were probably not covered in feathers like gigantic peacocks or cassowaries, but it is likely that some juvenile theropods—including the formidable *T. rex*—might have had a coat of fuzz or feathers, and that some of the adults might have retained some feathers as well.*

All the *Archaeopteryx* fossils have come from the Solnhofen limestones of Germany, while almost all of the recently discovered "feathered dinosaur" specimens have been found in the Yixian and Jiufotang Formations of Liaoning Province. Does this mean that feathered dinosaurs lived only in what is now Germany and China? Not really. It means that the fossils that have been found in both locations were so well-preserved that details like downy feathers, tiny bones of the feet, and hair and teeth, could be identified with precision. The Solnhofen Formations (150 million years old), and Liaonang Formations (125 to

---

* The purported presence of feathers in theropod dinosaurs has ignited an explosion of imaginative, colorful, and often bizarre interpretations of the appearance of these animals at the hands of paleo-artists and other interpreters of Mesozoic life-forms. The 1999 *National Geographic* article in which the faked *Archaeoraptor* fossil was presented was entitled "Feathers for T. rex?" It included a painting of an unfeathered adult and a fuzzy juvenile tyrannosaur. In the 2002 *Dinosaurs of the Air*, Greg Paul, a respected author and artist, drew an adult *Tarbosaurus* (a tyrannosaur) with little feather gauntlets on its tiny forelimbs, but by 2003, a *Scientific American* cover story on "Dinosaur Feathers" shows all the adult, carnivorous, bipedal dinosaurs covered with feathers. In their 2003 *A Field Guide to Dinosaurs*, Henry Gee and paleo-artist Luis Rey depict all the predatory dinosaurs as feathered, garishly colored creatures, that resemble parrots more than the dull theropods of earlier illustrators like Charles R. Knight. In defense of their color schemes, Gee and Rey wrote, "We'd guess that our dinosaurs are livelier and more colorful than some you might have seen illustrated in other works. Yet we maintain that their bold colors are well within the bounds of scientific possibility. Many birds and reptiles—unlike most mammals—have color vision, so it seems reasonable to presume that dinosaurs would have responded to the colorful shades of their fellows."

128 million years old) have yielded more than feathered dinosaurs, of course; there have been birds (such as *Archaeopteryx*), small mammals, lizards, fish, and various plants that enable paleontologists to identify the environment in which these animals lived and died. The fossil beds consist of fine-grained sandstones that were formed in a lake or riverine environment; the animals died and were buried in quiet conditions. The saline water was probably low in oxygen, and perhaps even toxic; since not even bacteria could survive in these conditions, the carcasses remained undisturbed. Over time—and 150 million years is a very long time indeed—these structures were replaced by minerals, but because there were no currents to disturb the incipient fossils, they were preserved in extraordinary detail. The entombing yellowish rocks were deposited in almost perfectly horizontal layers, with almost no irregularities. When found by German and Chinese fossil hunters, the layers could often be split apart, revealing positive and negative impressions— the "part" and "counterpart" of the fossils. Of the animals of Liaoning (known collectively as the "Jehol biota"), Zhonghe Zhou, Paul Barrett, and Jason Hilton (2003) wrote: "The spectacular fossils of the Jehol Group have already provided many important insights into the evolution of birds, angiosperms, and mammals. Nevertheless, the rate of fossil discovery presently outstrips the rate of description, and detailed monographic treatment of all species from the biota is needed if the full potential of these deposits is to be realized." At the end of the article, the authors placed a "note added in proof," in which they wrote, "New material of the dromaeosaurid theropod *Microraptor* indicated that the animal possessed wing-like arrays of asymmetrical feathers on both its fore- and hind-limbs."

If possible, the "new material" reported by Chinese paleontologists in early 2003 was even more surprising than feathers on dinosaurs. In *Nature*, Xing Xu and five other authors (one of whom was Zhonghe Zhou) revealed that *Microraptor gui*, the new species of dromaeosaurid theropod that was mentioned in the above note, had *four wings*—or at least what appeared to be flight feathers on the hind limbs. It was placed in the genus *Microraptor* because it was similar in almost every respect to *M. zhaoianus*, except for the "prominent biceps tubercularity on radius, much shorter manual digit I, strongly curved pubis, and bowed tibia."

And its feathered hind limbs. Incredibly, in a *1915* paper, William Beebe had postulated (and illustrated) a hypothetical four-winged bird ancestor, which he called *Tetrapteryx* ("four wings"), and which looked remarkably like the reconstruction *M. gui* in the *Nature* paper. Beebe's postulations were based on his discovery of tiny "sprouting quills" on the upper part of the hind legs of various nestling birds, including doves, pigeons, jacanas, owls, and even *Archaeopteryx*. He wrote that "the line of feathers along the leg of a young bird reproduces on this diminutive, useless scale the glory that was once theirs," and proposed an evolutionary succession from *Tetrapteryx* to an unnamed creature with reduced hind wings, to *Archaeopteryx*, and finally to modern birds. He concluded his article by writing, "No fossil bird of the ages prior to *Archaeopteryx* may come to light, but the memory of *Tetrapteryx* lingers in every dove-cote." *Microraptor gui* is considerably younger than *Archaeopteryx*, and is therefore unlikely to have been its ancestor, but, wrote Richard Prum in his introduction to the article by Xu *et al.*, "*Microraptor* is a basal member—an early evolutionary branch—of the closest relatives of *Archaeopteryx* and other birds."

Xing Xu and his fellow authors suggested "that basal dromaeosaurid dinosaurs were four-winged animals and could probably glide, representing an intermediate stage toward the active, flapping-flight stage." The authors dismissed the possibility that this little dinosaur was a runner, because they had earlier concluded that *M. zhaoianus* was arboreal, and also "because such long feathers on the feet would be a hindrance for a small cursorial [running] animal." Besides, wrote Richard Prum, "dragging your wing feathers in the dirt would doubtless be disadvantageous." In a splashy article titled "Lord of the Wings" in the May 2003 *National Geographic,* Christopher Sloan says that the four-winged *M. gui* supports the theory that bird flight began with gliders that took off from trees, rather than the "ground up" theory, which holds that flight began with feathered creatures that flapped their "wings" until they became airborne. He doesn't think that this animal (*M. gui*) "ran or flapped well enough to take off. Its leg feathers would have tripped it up like a hurdler in a ball gown."

Four wings might take a little getting used to (for the paleontologists, not the dinosaurs), because there is nothing alive today that *has* four wings, let alone flies with them.* (There are, however, some hawks and eagles whose hind leg feathers, while not as long as their flight feathers, might confuse some future paleontologists into believing that they had four wings.) At a total length of about thirty inches—including the feathered tail which was longer than the head and body combined—*M. gui* had asymmetrical feather vanes on both forewings and hindwings; such vanes are an indication of aerodynamic flight in the wings of modern birds (Prum 2003). A total of six specimens of *M. gui* (named for Chinese paleontologist Gu Zhiwei) were examined, and all had feathers the length of the metatarsus. *Archaeopteryx*, which preceded *M. gui* by about 30 million years, has long been considered the first bird, but it now appears that while some of the earliest dino-birds (e.g., *Archaeopteryx*) had two wings, others had four; whether or not four wings led to two as Beebe suggested, the hindwings were lost, and the two-wing plan has perdured. If ever there was an example of the nonlinear—yes, even capricious—directions that evolution can take, the history of dino-birds is surely it.

Until the discovery of feathered dinosaurs, probably the most notorious of all fossils was *Archaeopteryx*. It appeared in the 150-million-year-old Solnhofen limestones, so it obviously predated the known "birds" by tens of millions of years; but because feathers were identified as the single identifying characteristic of birds, *Archaeopteryx* had to have been a bird. It was often hailed as one of the few known "missing links" between one major group of vertebrates and another, for it seemed to mark the transition from reptilian dinosaur to reptilian bird. As Edwin Colbert wrote in 1961, "Here was a truly intermediate form between the reptiles and the birds. The skeleton was essentially reptilian, but with some characters trending strongly toward the birds. The feathers, on the other hand, were typical bird feathers, and because of them, *Archaeopteryx* is classified as a bird—the earliest and most primitive

---

* The problem of how four-winged dinosaurs moved through the air is not unlike the problem of how four-flippered plesiosaurs swam through Cretaceous seas. Sea turtles are the only living reptiles that have four flippers, so we have to compare the four-flippered plesiosaurs to them; but while the turtles are slow swimmers, some of the plesiosaurs were aggressive predators, and had to chase down their prey. How they actually swam—whether they "flew," "rowed," or performed some combination of the two—has been a subject for much paleontological speculation.

member of the class." It had feathers to be sure, but it also had a mouthful of teeth and clawed talons on the leading edge of its wings, two characteristics that most modern birds lack.* All books on dinosaurs published before, say, 1990, showed their subjects as smooth or scaly-skinned reptiles. The exception, of course, was the feathered *Archaeopteryx,* but as Romer wrote in 1966, "Although the feathery covering of birds is in contrast with the horny scales which normally cover a reptile body, the difference is, in reality, not so great as it seems; feathers are comparable to such scales, although with a complex structure of barbs and barbules instead of a simple scale shape."

Like every other life-form that exists on earth today, birds have gone through a long and convoluted evolutionary history. They seem to have begun their journey as little dinosaurs, sprouted feathers—which may or may not have evolved from scales and may or may not have been used for flying—and then slowly metamorphosed into the forms we can recognize today. In the deep past, strange fuzzy dinosaurs roamed the earth, learning to climb, to run, to swim, and eventually, to achieve a mastery of flight unequaled by anything else in history. We don't know that much about the flying skills of the pterosaurs, but it is probably safe to say that they could not fly as well as falcons, swallows, or hummingbirds. Birds developed arms that became feathered wings, but bats borrowed from the wing design of the pterosaurs, with a membrane of skin that stretched between the finger bones.

The "ancient wing" (*Archaeopteryx*) is long gone, and all the unfeathered dinosaurs have been extinct for at least 65 million years. There are no four-winged birds any more, but the tradition of feathered reptiles, unearthed recently in China, is alive and well. It required 150 million years, and many failed designs, but our world is now populated by an abundance

---

* No living bird has teeth, but the hoatzin (*Opisthocomus hoazin*), a strange-looking, crow-sized bird of northern South American forests, has functional claws on its forelimbs when it hatches. Using all four limbs, the young hoatzins crawl around in trees until their wings develop and they can fly. In his 1961 *Birds of the World*, ornithologist Oliver Austin wrote of the hoatzin: "It is one of those peculiar relics that has remained relatively unchanged for ages in a narrow, restricted range, while its farther-ranging relatives moved away and disappeared in the process of evolution. The Hoatzin remains behind as a vague clue to possible characteristics of the preavian ancestors of modern birds."

of bird life that can be seen in and over fields, forests, mountains, deserts, icepacks, and cities.

Feathers are beautifully designed for flying, but in order to fly, birds have light, hollow bones; this makes for delicate and fragile beings. Canaries have long been used to identify bad air, and we have already compiled a long list of bird species that we have ushered along the road to extinction. Think of the millions of years that were required to create something as splendid as a passenger pigeon, or as imposing as a moa. How much evolutionary tinkering was needed to produce the Eskimo curlew, the Carolina parakeet, or the Hawaiian honeycreepers? Then think of the evolutionary nanosecond that it took for us to kill them all off, along with hundreds of other avian species. "We have learned," wrote Errol Fuller in *Extinct Birds* (2001), "that nothing is as dead as a dodo."

Your Extinct Ancestors

We are classified as primates, along with lemurs, tarsiers, lorises, monkeys, and apes, but exactly where we fit in this classification is not clear. There are those who would separate humans from apes on the grounds that we are quite different, with our hairless bodies, upright stance, large brains, power of speech, and other readily observable differences. But on the molecular level, the differences are not nearly as obvious. When your DNA differs from that of another mammal (the chimpanzee) by less than 2 percent, the distinctions are not so clear. As Matt Ridley wrote in *Genome* (2000):

> Apart from the fusion of chromosome 2, visible differences between chimp and human chromosomes are few and tiny. In thirteen chromosomes no visible differences of any kind exist. If you select at random any "paragraph" in the human genome, you will find very few "letters" that are different: on average, less than two in every hundred. We are, to a 98 percent approximation, chimpanzees. If that does not dent your self-esteem, consider that chimpanzees are only 97 percent gorillas. In other words, we are more chimpanzee-like than gorillas are.*

---

* Although the figure of 98.5 percent as the difference between the genomes of humans and chimpanzees has long been accepted as gospel, recent research has shown that the actual figure

All your hominid ancestors are extinct, but beyond that we know little about the origin and evolution of *Homo sapiens,* the only mammal able to read this book. We used to believe that the earliest hominid was "Lucy," whose proper name is *Australopithecus afarensis,* and whose fossil bones have been dated from an Ethiopian site about 4.4 million years old. Then they found *Ardipithecus ramidus,* who is believed to have walked upright 5.2 to 5.8 million years ago. In early 2001, a French team working in Kenya found a fossil hominid they named *Orrorin tugenensis,* who they believed also walked upright—6 million years ago.

Recent discoveries have pushed this date back even further. In Ethiopia, anthropologist Yohannes Haile-Selassie (2001) uncovered the fossil evidence of a hominid that was dated at 5.2 and 5.8 million years old. The bones and teeth resembled those of the hominid *Ardipithecus ramidus,* described in 1994 by White, Suwa, and Asfaw, but they were much older, and were therefore nominated as a new subspecies, *A. ramidus kadabba,* from the Afar word *kadabba* that means "basal family ancestor." In 2002, an even older fossil hominid was discovered, pushing back the earliest evidence of the human lineage another million years, to 6 and 7 million years ago (Brunet *et al.* 2002). To upset the traditional view that the first hominids evolved in Africa's Rift Valley, the skull of *Sahelanthropus* was found in Chad, some 1,500 miles away. Because it shows ape- and humanlike characteristics, *Sahelanthropus* is believed to represent the common ancestor of modern humans and chimpanzees.

Heading in the other direction—that is, toward modern times and us—we find a sort of treadmill with other upright hominids, including *Australopithecus africanus, A. aethiopicus,* and *A. robustus.* Suddenly, because all those Australopithecines got off, we begin to find fossils that anthropologists have identified as belonging to the genus *Homo.* That's you, of course, but also (maybe) your distant ancestors, *H. rudolfensis, H. ergaster,* and *H. erectus.* The actual evolutionary order of these species—if they really are species—is far from resolved, but anthropologists believe that *H. sapiens* arrived on the scene somewhere between 400,000 and 150,000 years ago. The oldest accurately dated *Homo* fossils were recently found at Herto in Ethiopia; these remains, of three humans,

---

is greater. Roy Britten of Cal Tech, who developed the measuring technique in the first place, found that the difference was actually closer to 95 percent (Coghlan 2002).

have been dated at 160,000 years old, pushing back the previous record by some 45,000 years, and lending strong support to the idea that our species originated in Africa (White, Asfaw *et al.* 2003). The last coexisting hominids were *H. neanderthalensis* and *H. sapiens*; but in the last 10,000 years or so, the Neanderthals disappeared, leaving no clues—fossil or otherwise—to explain their departure. As Ian Tattersall (2000) wrote, "*Homo sapiens* has had the earth to itself for the past 25,000 years or so, free and clear of competition from other members of the hominid family. This period has evidently been long enough for us to have developed a profound feeling that being alone in the world is an entirely natural and appropriate state of affairs."

We still have no idea why early hominids developed an upright, bipedal stance, but two anthropologists from George Washington University said they might *not* have—at least not at first. Brian Richmond and David Strait published a study in which they concluded that *Australopithecus anamensis* and *A. afarensis* (that's Lucy), "retain the specialized morphology for knuckle-walking." Our closest living relatives are the anthropoid apes—gorillas, chimpanzees, and bonobos—and they all "flex the tips of their fingers and bear their weight on the dorsal surface of their middle phalanges, permitting them to use their hands for terrestrial locomotion while retaining long fingers for climbing trees." Bipedaling or knuckling along, our ancestors are believed to have originated in Africa and then wandered off into Europe. The Neanderthals branched off as a different species, and eventually became extinct, leaving the whole kit and kaboodle to the Cro-Magnon people. They (that is, we) painted cave walls, built cities, raised pyramids, went on crusades, invented gunpowder, painted the Sistine ceiling, realized that the earth orbited the sun instead of vice versa, discovered America, developed the calculus, recognized the origin of species and the process of evolution, invented the telephone, the airplane, the automobile and the computer, dropped atomic bombs on Hiroshima and Nagasaki, discovered DNA, and walked on the Moon. On September 11, 2001, some of our fellow *sapiens* ushered in the twenty-first century of "civilization" by destroying the World Trade Center.

And to bathe *H. sapiens* in even more glory, we are doing a pretty good job of eliminating even more of our fellow mammals from our home planet. Listed among the permanently missing are the Atlantic gray whale, Steller's sea cow, the Tasmanian tiger, Caribbean monk seals, bluebucks, quaggas, sea mink, various wallabies and bandicoots, and many bats and rats—though not the ones that a lot of people would like to see eliminated, *Rattus norvegicus* and *R. rattus*. On our list of potential extirpations in near time include the Chinese River dolphin, the vaquita, the Ganges River dolphin, the Siberian tiger, the giant panda, the mountain gorilla, the black-footed ferret, the black rhino, Javan rhino, Sumatran rhino, northern right whale, and Cape hunting dog. In his book on fossil fishes, evolutionary biologist John Maisey said:

> Just as the Renaissance discovery of perspective changed the way reality was represented, so our modern understanding of the fossil record has reduced humankind to a single thread in the tapestry of the living world. The history of *Homo sapiens* is interconnected with the history of every other species, living and extinct, that has rented space on Earth. Uniquely, however, humankind has discovered its own origins. Like all species, we evolved from forgotten ancestors, we exist briefly, and we will be gone. This is perhaps the strongest message sent by the fossil record, and we would do well to listen.

The Pleistocene Extinctions

In 1906, Henry Fairfield Osborn wrote a piece for *American Naturalist* entitled "The Causes of Extinction in Mammalia," in which he wrote, "In my opinion, the most striking advance toward a complete theory of the causes of natural extinction has come from recent discoveries regarding the real nature of the animal diseases and how they are communicated. Only recently have we come thoroughly to understand that insects are the most active means of introducing and spreading fatal diseases over great geographical areas and on a vast scale." The "greatest destroyer of wild African quadrupeds" was rinderpest, which is fatal to wild buffalo, kudu, sable antelope, wildebeest, hartebeest, and 90 to 100 percent of infected domestic cattle. Osborn was writing about terrestrial mammals, and mentions neither dinosaurs nor marine reptiles, but recent discoveries of virulent marine microorganisms have extended the reach and scope of Osborn's observations to every habitat on the earth and to all time periods, past and present.

In *The Dinosaur Heresies,* Robert Bakker applies the idea to the disappearance of the dinosaurs: "As dinosaurs were snuffed out, at the end of the Cretaceous, the great sea lizards, and the snake-necked plesiosaurs were also dying out, as were a host of large and small invertebrates, from coral-like oysters to shelled squid and microscopic plankton." Like the writer of a detective story, Bakker lays out the *modus operandi* of what he calls "the great mass murderer": "The suspect: (1) kills on land and sea at the same time; (2) strikes hardest at large, fast-evolving families on land; (3) hits small animals less hard; (4) leaves large, cold-blooded animals untouched; (5) does not strike at freshwater swimmers—most of these creatures are cold-blooded anyway, so criterion (4) applies; (6) strikes plant-eaters more severely than plants."

After dismissing the "Death Star" proposed by Alvarez & Co. as the agent of destruction, Bakker introduces the "old and well-thought-out" theory that the "best answer for the extinction of the open water, deep-sea creatures is that the surface water becomes colder and more thoroughly mixed with deep water when the shallow seas drain off . . . and at the same time, mountain-building forces weaken so that there are fewer barriers dividing the terrestrial regions . . . [and] the net result is a more homogenized ecosystem where species can pass from one end of a continent to another, and from one continent to another."

"The probable culprit," he wrote, "is a natural agent so ordinary and earthbound that it seems totally devoid of glamour compared to the hypothesis of death-dealing cosmic collisions." It is disease spread by the introduction of parasites or foreign organisms for which populations have no resistance, and therefore it "will run amok." As an example he cites the cattle disease known as rinderpest, which was unwittingly imported into Africa in the late nineteenth century when Indian cattle were brought to that continent; it spread like wildfire through African animals, killing millions of ruminants, from native cattle to antelopes. Bakker concludes:

> The Late Cretaceous world contained all the prerequisites for this kind of disaster. The shallow oceans drained off and a series of extinctions ran through the saltwater world. A monumental immigration of Asian dinosaurs streamed into North America, while an equally grand migration of North American fauna moved into Asia. In every region touched by this global intermixture, disasters large and small would occur. A foreign predator might suddenly thrive unchecked, slaughtering virtually defenseless prey as its populations multiplied beyond anything possible in its home habitat. But the predator might suddenly disappear, victim of a disease for which it had no immunity. As species

intermixed from all corners of the globe, the result could have only been biogeographical chaos.

After the disappearance of the dinosaurs, the most celebrated extinction event is the disappearance of the large terrestrial mammals—the "charismatic megafauna"—of the late Pleistocene in North America. Some 13,000 years ago, there were mammoths, mastodons, saber-tooth cats, wooly rhinos, and ground sloths on the plains of North America, as well as American camels, lions, and horses. Incontrovertible evidence shows that they disappeared in a geological instant, and by 12,000 years ago, they were all gone. What happened to them?

The predominant theories ascribe their disappearance to hunting or climate change ("overkill" or "overchill"), but it is more reasonable to assume that a combination of factors was responsible. Björn Kurtén and Elaine Anderson wrote in 1980:

> Extinction, the end of a phyletic line without replacement, has occurred throughout the history of life on earth and is the ultimate destiny of every species. The extinction of the dinosaurs at the end of the Mesozoic and the "sudden" disappearance of the megafauna at the end of the Wisconsonian are the best-known examples of widespread extinction. Hundreds of causes have been suggested to explain extinction (Van Valen, 1969, lists 86 reasons for the late Pleistocene alone) yet, for the most part, extinction remains an enigma.

Those who do not subscribe to the theory that humans were responsible for the disappearance of large mammals in North America suggest that it can be attributed to climate change, such as the retreat of the ice sheet some 12,000 years ago, which caused a prolonged period of increased temperatures. As Ernest Lundelius and Russell Graham (1999) wrote, "The land grew warmer and drier, the glaciers withdrew and the seasons, once dampened by the ice sheets, became sharply defined. Changing habitat forced species that once coexisted to separate and find new areas that met their individual needs. The price of failure was extinction—and many failed." Donald Grayson, an archaeologist at the University of Washington, does not believe that human hunters could possibly have killed every last member of

every species of large mammal that became extinct during the last ice age.*
For one thing, he asks, where are the "kill sites," those locations that would
contain evidence of such mass slaughter? As he wrote in a 1995 paper entitled "Late Pleistocene Mammalian Extinctions in North America: Taxon-
omy, Chronology and Explanations,"

> The results of that search [for reliable radiometric age dates on
> the extinct mammals] strongly suggest that overkill could not
> have been the force that Martin has claimed. The differential ap-
> pearance of kill sites (only proboscideans, and within the pro-
> boscideans, almost only mammoth) and the strong hints that
> many of the taxa involved may have been on their way to extinc-
> tion, if not already gone, by 12,000 years ago imply a far lesser hu-
> man role in the extinction than the overkill model allows. The
> climatic models account not only for the extinctions, but for
> the histories of smaller mammals during the Pleistocene. With
> greater explanatory power, most scientists studying the extinc-
> tions issue accept climatic, not overkill, accounts, while recogniz-
> ing that far more precision is needed in these accounts. This does
> not mean that people played no role in causing the extinctions. A
> multivariate explanation may yet provide the best account of the
> extinctions. But no matter what the human role might have been,
> overkill was not the prime cause of the extinctions. That cause
> rather clearly lies in the massive climate change that marks the
> end of the Pleistocene in North America.

In *Guns, Germs and Steel* (1997), UCLA physiologist Jared Diamond
says that he can't believe in a climatic theory of megafaunal extinction:

---

* Killing every living representative of a species will certainly result in that species' extinction,
but it is probably not necessary. In *The Call of Distant Mammoths* (1997), Peter Ward wrote, "Any
species has an extinction threshold: a minimum population below which the species is unlikely
to survive. Small populations are vulnerable to extinction even if, on average, they are showing
an increase in numbers (births outstripping deaths), because chance events can destroy entire
populations. . . . Whales come immediately to mind. We have hunted many of the larger
baleen species nearly to extinction, but the well-publicized, eleventh-hour moratorium on most
commercial whaling seems to have saved most species from extinction. Or so we think. It is not
certain that we have really achieved this success. A millennium from now we will know better
how successful our 'Save the Whales' campaigns have really been."

The Americas' big animals had already survived the ends of 22 previous Ice Ages. Why did most of them pick the 23rd to expire in concert, in the presence of all those supposedly harmless humans? Why did they disappear in all habitats, not only in habitats that contracted but also in ones that greatly expanded at the end of the last Ice Age?

Throughout the world, the Pleistocene extinction events took place on completely different schedules. The North and South American extinctions took place about 13,000 years ago, but the giant birds, megakangaroos and giant varanid lizards had all disappeared from Australia some 50,000 years ago. What ties these timetables together? We do. Human beings arrived in Australia about 50,000 years ago and in North America about 37,000 years later. Paul Martin of the University of Arizona contends that human hunters crossed the Bering Land Bridge some 13,000 years ago and methodically hunted all the large mammals to extinction. In a 2003 article, K. Alden Peterson wrote,

> Throughout the Americas and Eurasia, the loss of mammals during this time can be correlated directly to size: 100 percent of mega-mammals (over 100 kilograms); 76 percent of large mammals (100 kg to 1000 kg); 41 percent of intermediate-sized mammals (5 kg to 100 kg); and 1.3 percent of small mammals (0.01 kg to 5 kg) became extinct. In North America the late Quaternary Extinction meant that of the 35 genera of mega and large mammals roaming the continent at the end of the Pleistocene, 29 disappeared from the fossil record by 10,000 years ago.

The "blitzkrieg" theory is not restricted to North America; its proponents believe that similar extirpations took place in Europe, South America, and Australia. Martin (Martin 1999.) believes that people actually did the killing; more specifically, that "countless animals must have died out at the hands of the hunters." John Alroy agrees. In *Extinctions in Near Time* (1999), Alroy wrote a chapter "Putting North America's End-Pleistocene Megafaunal Extinction in Context," in which he said that the "overkill hypothesis, at least in general terms, has been 'proven' as thoroughly as any historical hypothesis can be." Alroy's conclusion: "Long before the dawn of written history, human impacts were responsible for a fantastically destructive wave of extinctions around the globe. This message should be seen as a wake-up call instead of a mere omen of disaster. Although the fossil record proves that biodiversity is far more vulnerable than we would like to think, the conservation movement's past successes show us that much can be done to save our biological heritage."

In 1999, with Karen Sears, Alroy published an abstract entitled "End-Pleistocene Megafaunal Extinction in South America: Massive Overkill in the Tropics," which can be translated to mean that at the end of the Pleistocene, human beings killed off large numbers of mammals in South America, too, and sometimes wiped them out completely. Whatever the cause, the North and South American Pleistocene extinctions have been dated at roughly 13,000 years ago, while in Australia, the larger mammals, reptiles, and birds perished much earlier.

> At the March 2003 meeting of the Geological Society of America, Fordham University paleontologists Guy Robinson, David Burney, and Linda Pigott Burney presented evidence that humans arrived in North America so close to the collapse of the populations of mammoths, mastodons, ground sloths, and saber-tooth cats that they had to be the perpetrators of the extinctions. Using microscopic bits of charcoal as markers of the presence of humans, Robinson *et al.* concluded that the elimination of the herbivores led to an increase in fuel for fires, hence the tenfold increase in charcoal bits. At the same time, there was a tenfold *decrease* in spores of the fungus *Sporormiella*, which was an indicator of the presence of large herbivores—spores wash into ponds, lakes, and bogs from the dung of watering large animals. "Knocked down," wrote Richard Kerr in a summary of Robinson and the Burneys' presentation, "the megafauna finally went out under the weight of further hunting, the cold, the altered environment, or some other stress or stresses that would have otherwise not caused extinction."

As do their living descendants, ancestral Australian mammals differed markedly from most other mammals in that they were marsupials. (It is possible to differentiate fossil marsupials from fossil placentals by differences in the skull, but particularly by the presence of two extra bones, the epipubics, in the pelvic girdle.) And if you think today's kangaroos, koalas, wombats, and bandicoots are weird, wait 'til you meet the extinct ones. Like the North American megafauna, the Australian representatives were bigger— in some cases, *much* bigger. Yes, cave bears, cave lions, and Irish elk were larger than any bears, cats, or deer alive today, but in Pleistocene Australia, there was a kangaroo that stood over eight feet tall, a wombat as big as a tapir, and an echidna the size of a boar. There were also creatures that would be utterly unrecognizable today, such as the "marsupial lion" (*Thylacoleo carnifex*), as big as a modern lioness, with a catlike skull and large slicing carnassials or cheek teeth. With a large retractable thumb claw and powerful forelimbs this animal would have been a fearsome predator. A somewhat smaller catlike marsupial predator was *Wakaleo vanderleuri*, the size of a leopard and capable of leaping from trees onto its unsuspecting prey (Wroe 1999a). There was also a throng of herbivorous giant marsupials that looked like nothing alive today.

One of them, *Zygomaturus trilobus*, was a heavy-bodied, thick-legged creature that resembled a pygmy hippopotamus in size and build. It is thought to have lived in small herds around the wetter coastal margins of Australia around 19,000 years ago, occasionally extending its range along the watercourses into the center of the continent. As a ground-dweller, it moved on all four limbs, and probably fed by shoveling up clumps of reeds and sedges with its fork-like, lower incisor teeth. Although unrelated in chronology and phylogeny, some of these giant Australian marsupials probably resembled the desmostylians (extinct for 9 million years), which were also large, semiaquatic mammals with teeth adapted for eating grass and sedges.

Even larger than *Zygomaturus* was the rhinoceros-sized *Diprotodon optatum*, a two-ton behemoth; it was the largest marsupial ever to live in Australia. *Palorchestes azael* was originally described from a few teeth as a giant kangaroo, but as more fossils of this species were found, it was realized that it must have walked on four legs. (Modern kangaroos cannot actually "walk" in the traditional sense; they swing their hindlegs forward between their forelegs in a slow gait; for speed, they hop on their hindlegs only.) *Palorchestes* may also have had a small trunk, hence the name "marsupial tapir." With a long tongue, strong arms, and big claws, *Palorchestes* was a herbivore. It may have used its claws to tear bark from trees or to pull up small plants, and it may have probed with its long tongue to find food inside trees or holes.*

Australia was also the home of immense flightless birds, such as *Dromornis stirtoni*. At a weight of half a ton (500 kilos) and a height of nearly ten feet (three meters), it is generally considered the biggest bird that ever lived. These monsters had long necks and stubby, useless wings; although they had powerful legs, they are not thought to have been particularly fast runners. Built along the lines of ostriches or emus, the dromornithids are actually more closely related to geese (Murray and Megirian 1998). *Dromornis* had a huge beak and jaw capable of great force, and while early workers decided that the powerful beak could have been used for shearing through the tough stalks of plants, Stephen Wroe of the Australian Museum argued that a beak like that could not possibly have been used by a herbivore, and that *Dromornis* was more than likely a meat-eater. And so was *Bullockornis* ("ox-bird"), equally as formidable as *Dromornis*. Dubbed "the demon duck of doom," *Bullockornis* stood ten feet tall; its head was as large as that of a small horse. Its beak was a terrifying weapon, unlike those of the herbivorous moas. In a fictional reconstruction of the life of *Bullockornis*, Wroe has one of these "terror birds" driving a marsupial lion (*Thylacoleo*) from its kill: "Our lion has just been muscled off its kill by the most formidable bipedal carnivores since the extinction of the dinosaurs."

Among the large Australian animals that became extinct during the late Pleistocene was another large flightless bird known as the mihirung (*Genyornis newtoni*), whose name comes from the Tjapwuring people's *mihirung paringmal*, meaning "giant emu." Based on analysis of preserved eggshells, Miller *et al.* (1999) narrowed down the date of extinction of the mihirung to about 50,000 years ago. Climate changes at that time appear to have been moderate, and therefore unlikely to have been the cause of the mihirung's demise. The oldest widely accepted date of arrival of humans in Australia is

---

* Unlike the Pleistocene fauna of North America, some of the Australian species survived into modern times. The thylacine, or Tasmanian tiger, was a large marsupial predator that lived well into the twentieth century, becoming extinct only when bounty hunters killed the last of them in the 1930s. Some giant kangaroos became extinct, but D. R. Horton (1995) believes that the extant red kangaroo (*Macropus rufus*) qualifies for the category of "megafauna," as large males can weigh 200 pounds (ninety kilograms); it can still be found hopping around arid regions of Australia. The gigantic "elephant birds" and moas of New Zealand were all eliminated, but there are still emus and cassowaries.

Bullockornis

about 55,000 years ago, which would fit with a "human impact" explanation. The authors postulate that large-scale fires, started by human colonists, destroyed so much of the browsers' food supply that their populations were unable to sustain themselves. Populations of top predators, deprived of their prey, also collapsed. Herbivores with less restricted diets, such as the emu, were able to survive the loss of shrubby vegetation.

Most of the Australian land mammals weighing more than 100 kilograms (220 pounds) perished in the late Quaternary, but the timing and causes of these extinctions remain uncertain and controversial. In *The Future Eaters* (1994) Tim Flannery speculated that the megafaunal collapse was the result of overhunting by early human colonizers, combined with fire-stick farming practices that changed the ecology of the continent so dramatically that many larger marsupial species were driven to extinction. The blitzkrieg model was arguably fortified by the dating of the extinction of the giant mihirung to about 45,000 years ago in central Australia, though some claimed that it survived until later in other locations. The time frames range from 51,200 to 39,800 years ago.

In their 2001 analysis of burial sites across the continent, Richard Roberts and his colleagues reported much more specifically that the last Australian megafauna became extinct 46,000 years ago. Roberts *et al.* presented optical dates on sediment layers at a series of sites containing articulated megafauna remains. They argued that the megafauna went extinct within a short time period. In a response to the Roberts *et al.* paper, Judith Field and Richard Fullagar (2001) wrote that because human exploitation of animals inherently involves the disarticulation of the remains (taking them apart), the earlier paper on articulated remains had ignored archaeological sites which "provide evidence of human/megafauna coexistence."

In a direct critique of the Roberts *et al.* "blitzkrieg" scenario, Stephen Wroe and Judith Field (2001) wrote:

> Many media reports have presented these findings as holy writ. This is unfortunate because we believe that Dr. Roberts and his co-authors have misinterpreted the data and made claims that the results just don't support. . . . In fact, within the data they present may be the solution to at least one problem of significance. Curiously, in our view, the one important conclusion that could be drawn from this study was not.

Wroe and Field believe that Roberts *et al.* ignored sites such as Cuddie Springs, where megafaunal fossils and human cultural remains have been found together; these sites give much younger dates than 46,000 years ago. Besides, wrote Wroe and Field, "Even if we did accept that the dismissal of all younger sites was justified (and we don't), this data can only be interpreted as support for the argument that the megafauna persisted until at least 46 kya [thousand years ago]. There is a great difference between this and the authors' claim that the megafauna were extinct by 46 kya."

Instead, Wroe and Field conclude that climate change—the drying up of the Australian continent between 65,000 and 40,000 years ago—did not so much kill off the megafauna as it killed off their food. As Vickers-Rich and Rich wrote (1999), "What kills the last survivors is not lack of water, but lack of fodder. Under these conditions, the animals are 'tethered' to their water supply." An arid environment also makes fossils much harder to find because they may not form; water is "another factor that very strongly influences the likelihood of fossil deposits being formed" (Wroe and Field).

Because, say Wroe and Field, there were no humans in Australia until about 46,000 years ago, they could not have killed off the megafauna at that time. (Indeed, they maintain that some species of the megafauna may have disappeared before humans even arrived, including *Genyornis*, whose departure is dated at 50,000 years ago.) "The most significant conclusion that can be drawn from this study is that, if the dates produced by Dr. Roberts and his colleagues are anywhere near the mark, then 'blitzkrieg' was not the *modus operandi* for megafaunal extinction in Australia. Furthermore, the work of Roberts *et al.* also constitutes strong evidence for the persistence of Australian megafauna until at least 46 kya as well as further support for already corroborated ages of 36 to 27 kya for sites containing megafauna at Cuddie Springs. In no way can the study be used to argue that the megafauna disappeared by 46 kya, less still that humans were responsible."

In *The Future Eaters* (1994), Tim Flannery introduced a new, testable hypothesis for the extinction of the Australian megafauna. The hypothesis "centered around the concept that megafaunal extinction was rapid and continentwide, and was caused by the arrival of humans." Analysis of *Genyornis* eggshells at Lake Eyre gives 50,000 rcyrbp (radiocarbon years before present) as a reliable date for the extinction of this giant bird, which, Flannery says, coincides with the arrival date of the first humans in Australia. The implicit assumption is that people colonized the entire continent rapidly after the first landfall and eliminated the megafauna as they spread out. But, he says, "if it can be demonstrated that humans and megafauna overlapped [at Cuddie Springs] for a substantial period of time between 30,000 and 40,000 rcyrbp, then clearly the hypothesis is refuted."

Whether or not they were responsible for the rapid demise of the megafauna, humans have been in Australia for about 50,000 years. It is the "about" that causes the problem. Some authors, such as Roberts *et al.*, believe that the evidence supports the idea of an aboriginal invasion of Australia some 55,000 years ago; others (Wroe and Field 2001b) write that "it is not yet clear that humans were present in Australia earlier than 40,000 years ago." Flannery and Roberts (1999) write, "A human presence in Australia has been firmly dated at 20,000 rcyrbp in 1965, and over 30,000 by 1970. . . . By the 1980s, dates for an even earlier human presence were being presented [and] by 1994 . . . dates from Malakunanja and Nauwalabila rock-shelters in the Northern Territory suggested that humans had arrived in Australia by between 50,000 and 60,000 yrbp."

Human beings tend to complicate things. Few nowadays dispute the extraterrestrial body that slammed into Chicxulub 65 million years ago, but there were no people around to witness the event, so all conclusions have to be drawn from the geological and paleontological evidence. When *Homo sapiens* crossed the Bering Land Bridge some 13,000 years ago, they were heavily armed (in Pleistocene terms, anyway), and quite capable of killing the mammoths, mastodons, wooly rhinos, bison, and any other large herbivores that they found waiting for them. (The saber-toothed cats present a different problem. It is hard to imagine prehistoric hunters killing off every single lion, but if they killed off all the lions' prey, perhaps the cats starved to death en masse, and thus became extinct.) In North America evidence of killing exists, such as kill sites and weapons (arrow- and spearheads) found in conjunction with disarticulated bison and mammoth skeletons. By contrast, there are no known kill sites in Australia, and no evidence whatsoever to indicate that the early aboriginals (or the later ones, for that matter) had developed the weapons technology that would have enabled them to kill off even a few of the giant carnivorous birds, or the rhinoceros-sized *Diprotodon*. Still, because blitzkrieg is the most sensational of possible scenarios (Tim Flannery wrote an entire book about it), it is the most popular, and people like Stephen Wroe will have to labor even harder in the field or in the lab to get public (or even professional) acceptance of his disputations. Paul Martin, the developer of the "blitzkrieg" theory, has written that in order for the idea to be accepted, we must concede that the whole process of

Saber-tooth "Tiger"

extinction took place within 2,000 years. That requires what Wroe calls a "mega-ignorance of the megafauna," who would surely have learned to avoid hunters over a longer period.

Gittleman and Gompper (2001) wrote, "The most powerful illustration of how naïveté to danger may lead to elimination comes from the extinctions of the late Quaternary, during which more than half of the 167 genera of large land mammals became extinct, primarily because of the rapid and catastrophic effects of 'first contact' with colonizing hunters." This statement appeared in the same issue of *Science* (February 9, 2001), in which Berger, Swenson, and Persson suggested that the "current extinction of many of Earth's large terrestrial carnivores has left some extant prey species lacking knowledge about contemporary predators, a situation roughly parallel to that of 10,000–50,000 years ago, when naïve animals first encountered colonizing human hunters."

One creature unlikely to have been eliminated by Australian Pleistocene human hunters was the giant monitor lizard *Megalania prisca*. From the fossil evidence, it seems more likely that this lizard may have preyed on people, rather than vice versa. Built along the lines of the Komodo dragon (*Varanus komodoensis*) but twice as large, *Megalania* was the largest-known terrestrial lizard, reaching a known length of twenty-five feet (six meters); it may have weighed more than a ton (900 kilograms). A skull in the collection of the Museum of Victoria is nearly three feet long; it is armed with a set of daggerlike teeth almost an inch (two centimeters) long and curved with a serrated rear edge. These teeth could disembowel a large animal with ease. This one-ton meat-eater, the largest terrestrial carnivore since the tyrannosaurids, probably preyed on the larger kangaroos as well as other marsupials and big birds, but if it was anything like its Komodo island descendant, it was also a prodigious consumer of carrion.

Eurasia experienced a similar rash of Pleistocene extinctions, but not on the same scale as the North American or Australian events. In Europe, wrote Anthony Stuart (1999), "all of the animals of 1000 kg (1 metric tonne) or more became extinct." These include the cave bear (*Ursus spelaeus*), the spotted hyena (*Crocuta crocuta*), the straight-tusked elephant (*Paleoloxodon antiquus*), the wooly mammoth (*Mammuthus primigenius*), the narrow-nosed rhinoceros (*Stephanorhinus hemitoechius*), the wooly rhinoceros (*Coelodonta antiquitatis*), the hippopotamus (*Hippopotamus amphibius*), a form of giant bison (*Bison priscus*), and the giant deer known as the Irish elk, (*Megaloceros giganteus*.) In his 1999 review of the European megafaunal extinctions, Stuart eliminates hunting as the cause:

> It is difficult to see how hunting with upper Paleolithic technologies could have led to the extinction of so many species with wide geographical distributions as the wooly mammoth, which ranged across the northern half of North America and most of Northern Eurasia. Overkill would predict marked extinctions closely following the arrival of modern humans *Homo sapiens* . . . in a given region, or perhaps the first humans in regions colonized by earlier *Homo* species. However, in Europe major megafaunal extinctions correlate neither with the first arrival of *Homo* nor with the arrival of *Homo sapiens*.

There were great climatic fluctuations in Eurasia during the Pleistocene, with glaciers receding and open steppe areas being replaced by forests, but there were long enough periods of stability to mitigate against a rapid extinction of a given species. Stuart wrote, "All of this strongly indicates that climatic/environmental changes of the kind experienced in the Pleistocene would not themselves produce accelerated extinctions of megafauna (although they very probably drove the much more gradual 'background' extinctions of large and small mammals, and plants, which occurred through the European Pleistocene)."

In the seventeenth and eighteenth centuries, it was becoming increasingly apparent that many fossils represented organisms that were not known to be alive at that time anywhere on Earth. To scientists who believed in the divine creation of life, this posed a jarring philosophical problem: why would God allow any of the animals in His perfect creation to die out? Many simply denied the reality of extinction, and instead suggested that animals known only as fossils would one day be found alive in some unexplored part of the globe, or, as Thomas Molyneux wrote in 1697, "That no real species of living creatures is so utterly extinct, as to be lost entirely out of the World, since it was first created, is the opinion of many naturalists; and 'tis grounded on so good a principle of Providence taking care in general of all its animal productions, that it deserves our assent."

Commenting on the large number of gigantic deer antlers found in Ireland, Molyneux, a seventeenth-century British physician (and the first scientist to describe the Irish elk), wrote,

> By what means this Kind of Animal, formerly so common and numerous in this Country, should now become utterly lost and extinct, deserves our Consideration. . . . I know that some have been apt to imagine this like all other Animals might have been destroyed from off the Face of this Country by that Flood recorded in the Holy Scripture to have happened in the time of Noah, which I confess is a ready and short way to solve this Difficulty, but it does not satisfy me . . . it seems more likely that this kind of animal might become extinct here from a certain ill Constitution of Air in some of the past seasons long since the flood, which might occasion an Epidemick Disaster, if we may so call it, or Pestilential Murren, peculiarly to affect this sort of creature, so as to destroy at once great numbers of 'em, if not quite ruine the species.

The disappearance of the Irish elk (*Megaloceros giganteus*), which is neither exclusively Irish nor an elk, has always been one of the paradigmatic extinctions. The name "Irish" has stuck because most of the well-preserved fossils of this giant deer have been found in lake sediments and peat bogs in Ireland, but the fossil has also been recorded throughout Europe, northern Asia, and northern Africa, with a related form from China. The largest deer species ever, megaloceros (which simply means "giant deer") stood up to seven feet high at the shoulder, with antlers that could span up to twelve feet. In the past, it was believed to have become extinct in Ireland around 11,000 years ago, but recently this chronology has been questioned. It was also assumed that the immense antlers somehow contributed to its extinction, but exactly how this might have happened was unclear. A once-popular hypothesis held that the evolutionary mechanism was orthogenesis, in which change in organisms is due not to natural selection, but to internal directional trends within a lineage. The Irish elk was once considered a prime example of orthogenesis: it was thought that its lineage had started evolving on an irreversible trajectory toward larger and larger antlers. The Irish elk finally went extinct when the antlers became so large that the animals could no longer hold up their heads, or got entangled in the trees. (The Irish elk is usually recreated looking like a large fallow deer or a moose, but examination of French Paleolithic paintings, particularly in the caves at Cougnac, show that the great stags had a large muscular hump on their shoulders, like that of a brahma bull or a bison, which would certainly have been helpful in holding up the head and massive antlers.)

These same cave paintings have led some to postulate a very undeerlike coloration for the Irish elk. In a fascinating essay, R. Dale Guthrie of the University of Alaska shows the relevant Paleolithic drawings of the giant deer (which he calls the "shelk"), showing a large dark mane on the shoulder hump and various markings on the face, neck and body. "Paleolithic drawings," he notes, "suggest that both sexes had a relatively light 'ground color' to the body. Paleolithic artists represent shelk markings as a striking contrast to the body color. Without a light body these stripes would not even have been visible." Valerius Geist (1998) agrees and writes, "The cave paintings suggest body markings similar to those of other highly evolved cursorial deer." He then describes these markings:

> In the open, the stag would be fully exposed to the sun. To minimize solar heat gain, it was probably of light body color. This is supported by signs of external body markings found in the cave drawings. These show a dark dorsal stripe, a dark lateral line running from shoulder to haunch, a light throat patch flanked by dark stripes and a dark hump.

Early discussions suggested that the antlers grew so large and unmanageable that the poor creature could no longer walk through the forest; it became so sad and depressed that it lay down and became extinct. This seems an unlikely scenario, so Stephen J. Gould (1973) investigated the mystery, and concluded that megaloceros could not adapt to climatic changes that altered its habitat from "the sparsely wooded open country . . . to the subarctic tundra that followed in the next cold epoch or to the heavy forestation that developed after the final retreat of the ice

Irish Elk

sheet." In a 1999 discussion, Moen, Pastor, and Cohen compared the mineral content of the antlers of megaloceros with that of moose (*Alces alces*) antlers, which are quite large but still only about half the spread of megaloceros. They concluded that the minerals that are necessary for antler growth, particularly calcium and phosphorus, would not have been available to the Irish elk; the stags would have had great difficulty replenishing fat reserves depleted during the rut. They suggested "that the inability to balance these selection pressures in the face of rapid environmental change contributed to the extinction of the Irish elk 10,600 years ago."

Although humans drew the cave pictures and therefore coexisted with Irish elk, they left their descendants (us) no indication as to why or how the giant deer became extinct. Circumstantial evidence—the presence of man the hunter and shelk the prey—may implicate Paleolithic man in the great deer's demise, but we cannot prove it. Other successful predators, such as large-bodied wolves and bears, were present in the same place and time as the Irish elk, and they must have contributed to its downfall. (Geist also notes that "Ice Age giants such as these [the wolves and bears] featured relatively large brains and luxury organs, such as ornate antlers, horns, tusks, or hair coats, and large fat deposits. *Homo sapiens* is a classic Ice Age giant.") As the climate cooled, the range of the giant deer was pushed southward. "This move," wrote Geist, "came at a time of hardship for humans and probably greater keenness to acquire food. . . . It is not unlikely that hungry upper Paleolithic hunters killed and ate the last of the giant deer."

It appears that giant deer lived through the climatic changes. According to Gonzalez *et al.* (2000), the "survival of *M. giganteus* in the temperate, forested environment of northwestern Europe in the early Holocene allows the possibility that Mesolithic hunters could have been responsible for the giant deer's final demise." Once again, we are facing the problem of overchill vs. overkill, but there is also the possibility that the Irish elk may have fallen to the "Epidemick Disaster" or "Pestilential Murren" suggested by Molyneux.

The Pleistocene megafaunal extinctions are still unsolved mysteries. Were they caused by hunters, climate change, disease, some combination of the foregoing, or did all these populations of very large animals disappear for reasons we cannot understand? Whatever the cause (or causes) the same sort of thing happened in North America, South America, Eurasia, and Australia. In unexplainable contrast, however, virtually all the one-ton or larger Pleistocene mammals of sub-Saharan Africa and southeastern Asia survive to the present day. In Africa we can still see the largest of all land mammals, the African elephant, and the third largest, the white rhinoceros. Africa also has the black rhino. Asia has the second largest land animal, the Indian elephant, as well as three more species of rhinoceros: the Indian, the Sumatran, and the Javan.

Some of these large animals are presently endangered. If they were to go extinct, there would be no problem identifying the causative agent this time as *H. sapiens*. Since *H. sapiens* began in sub-Saharan Africa, he was certainly available to hunt all these megamammals, but for some reason they were spared. There were certainly Pleistocene extinctions in Africa—Martin (1985) lists various hyenas, cats, pigs, camels and giraffes—but alone among the large landmasses, Africa retains its terrestrial giants.

If we can't figure out why so many species disappeared, we are completely baffled by those that didn't.*

Most of the theories about the Pleistocene extinctions depend on comparing the timing of human arrival with the extinctions ("overkill"), or on analyzing climate and vegetation changes at the end of the last ice age that could have negatively impacted the mammalian megafauna ("overchill"). But Christopher Johnson of James Cook University in Queensland (2002) has taken a different approach. He has compared the biological traits of extinct species with those of survivors and found that

> large size was not directly related to risk of extinction, but rather, species with slow reproductive rates were at high risk regardless of their body size, a finding that rejects the "blitzkrieg" model of overkill in which extinctions were completed during brief intervals of selective hunting of large bodied prey. Second, species that survived despite having low reproductive rates typically occurred in closed habitats and were arboreal or nocturnal. Such traits would have reduced their exposure to direct interaction with people. Therefore, although this analysis rejects blitzkrieg as a general scenario for the mammal megafauna extinctions, it is consistent with extinctions being due to interaction with human populations.

Johnson found that the lower the species' reproduction rate, the higher the likelihood of extinction, regardless of body size. "It seems that what was lost during the Pleistocene," wrote Cardillo and Lister (2002) in a synopsis of Johnson's work, "was not so much the megafauna as the 'bradyfauna' [from the Greek *bradys*, meaning "slow"]—a whole way of life based on slow life-history." If human hunting caused the extinctions, then body size should be an important determinant. But not every Pleistocene extinction involved large creatures; any animals with low reproduction rates were

---

* Whales of course grow much larger than any terrestrial animals—the blue whale is longer and heavier than the largest known sauropod dinosaur—but most sea mammals were not accessible for hunting with Paleolithic technology. Steller's sea cow (*Hydrodamalis gigas*) was an aquatic mammal that lived just offshore in the Pacific during the North American Pleistocene, and although it was a clumsy swimmer and probably couldn't dive, it managed to survive until the middle of the eighteenth century, when the last individuals were killed by Russian fur trappers on the Commander Islands off the Kamchatka peninsula.

candidates, from tiny lemurs to giant mastodons. Johnson's findings complement the hunting theory, because even low levels of hunting (as suggested by Alroy 2001) of animals with low reproductive rates would throw the population into a downward curve from which it might not recover. Species with slow reproductive rates would also be vulnerable to climate change, because they would not be able to adapt as rapidly as those that achieve a more rapid generational turnover. Cardillo and Lister draw parallels with the current extinction crisis from Johnson's findings: "Today's extinctions have accelerated to an observable pace," they write. "Moreover, slow life-history is a strong predictor of current extinction risk in living mammals. Perhaps in another 50,000 years—or even sooner—we will be left only with those that live life in the fast lane."

Ross MacPhee and Preston Marx (mammalogist and virologist, respectively), believe that humans were responsible for the Pleistocene extinctions, but not with their spears and arrows. They suggest (1997) that "hyperdisease"—somehow connected with the arrival of the first humans—is the agent of destruction. "Conceivably," wrote MacPhee and Marx (1999), "something else came with these immigrants: a pathogen (a disease-causing organism) carried by the humans or their Ice Age entourage. Perhaps it was the lice that prowled their hair, the fleas that harried the [accompanying] rodents or wolf-dogs, or something that lived in the humans' gut. The animals of the New World had no immunity to this microbe and they were helpless against it. As HIV would do thousands of years later, the microbe 'jumped' from its original host among the newcomers to native species. It spread across North and South America, killing by the tens of thousands. For many species, it was a death sentence." In his 2001 book about the Siberian mammoth hunt, Richard Stone wrote, "A spine-chilling new idea appeared in 1997 when Ross MacPhee, the curator of mammals at the American Museum of Natural History in New York, argued that prehistoric hunters, or perhaps their dogs, carried a deadly microbe that tore through with the virulence of the flu and the lethal quality of the Ebola virus." In a 2000 study of emerging infectious diseases of wildlife, Daszak, Cunningham, and Hyatt wrote, "MacPhee and Marx implicate the introduction of infectious diseases in the striking loss of biodiversity after human colonization of continental landmasses and large islands over the last 40,000 years, including many of the Pleistocene megafaunal extinctions."

"MacPhee and Marx," wrote Richard Stone in *Science*, "defined a hyperdisease pathogen as one that strikes in all age groups, killing at least three out of every four individuals and capable of taking a heavy toll on several species simultaneously." There is no known pathogen that fits this description precisely, but there are some that historically have been so efficient at wiping out large numbers of animals—and even humans—that the idea cannot be dismissed out of hand. The influenza virus that killed upwards of 30 million people in 1918 "jumped" from nonhumans— perhaps from ducks or pigs—to a form that was unbelievably lethal to people. And as of now, something is attacking frogs and toads around the world with such virulence that some species, such as the golden toad of Costa Rica, have already been completely wiped out, in what MacPhee has described as "essentially an overnight collapse . . . that's what a hyperdisease would be like." MacPhee has taken his quest to the Taimyr Peninsula of northern Siberia, where he hopes to find RNA or DNA evidence in the bones of mammoths that perished there a mere 10,000 years ago.

In their chapter on prehistoric extinctions in *Extinctions in Near Time*, Martin and Steadman discuss the theory of MacPhee and Marx: "They take the high ground and mount a strong case for what appears to be a long shot, the relatively neglected model of highly lethal disease as the driving force in first contact extinctions." John Alroy, however, dismisses the "disease model" out of hand, writing that "There is no such thing as a pandemic disease that can drive virtually any large mammal species into extinction—or even cause high levels of mortality throughout populations belonging to different orders. Most of the well-known mammalian diseases are restricted to a single order, including rinderpest (Artiodactyla) and myxomatosis (Lagomorpha). The few diseases that can strike multiple orders (e.g., rabies, influenza) typically either have low transmission rates or produce low mortality rates."

Nobody seems to be able to account for the disappearance of wild horses. (The only remaining "primitive" wild horses are to be found in Mongolia, but they had to be saved from extinction, as discussed on pp. 303–7.) In *Evolution of the Horse* (1936), W. D. Matthew said, "It is also unknown why the various species which inhabited North and South America and Europe during the early part of the Age of Man should have become extinct, while those of Asia (horse and wild ass) and of Africa (wild ass and zebra) still survive. Man, since his appearance, has played an important part in the extermination of the larger animals; but there is nothing to show how far he is responsible for the disappearance of the native American species of horse." And George Gaylord Simpson (1951), wrote, "The extinction of the horse over the whole of North and South America, where they had roamed in vast herds during the Pleistocene, is one of the most mysterious episodes of animal history." Simpson asks, "Were the horses all carried off by some plague, perhaps a fly-borne epidemic such as sleeping sickness or some other deadly and new infection? This is a possibility and a tempting one, but it cannot be checked. If the last horses all died from the same epidemic, no evidence is provided by their fossil remains. If such an epidemic occurred, how did the pronghorn antelopes, the bison, and other animals that lived along with the horses escape it? . . . Thus there are objections, but disease cannot be ruled out as a possible cause or as a contributing factor."

Of the three hypotheses that attempt to explain the extinction episodes of the Pleistocene, only hyperdisease can possibly explain the disappearance of an entire mammalian species. Humans have certainly had an effect on animals that they have hunted, but in the last 500 years, while there have been many opportunities to drive a species to extinction, we have "succeeded" in only a couple of cases. With weapons whose effectiveness far exceeded those of the Stone Age hunters, we killed off all the Tasmanian tigers, quaggas, and Steller's sea cows; but while we behaved as if total annihilation was the goal, we failed with the bison, fur seals, and sea otters, and though we tried for the better part of two centuries, we could not eliminate a single species of whale. Glaciers move, well, glacially, and they usually allow enough time for a species to move elsewhere or adapt to changing temperatures. As MacPhee and Marx (1999) wrote, "We propose . . . that the first contact [with humans] did indeed cause the extinctions, but the means was an unusually lethal disease—a hyperdisease. In our model, a hyperdisease need not kill every individual of a species. When the mortality rate rises above 75 percent, populations tend to be so ravaged that random effects—including human hunting or environmental changes—might finish off the last survivors or prevent them from continuing their kind." In other words, once the disease has taken out most of the population, human or environmental effects might finish the job. Moreover, wrote the authors, "Extinction rates in affected areas consistently dropped off after an initial period of mass losses . . . a hyperdisease would rapidly kill off susceptible species, leaving behind those that were not vulnerable to the pathogen." And it also explains the disappearance of so many large mammals; "small mammals have higher reproductive rates and shorter gestation periods, so their populations would recover much more quickly after major die-offs than larger, slower-breeding species."

In John Alroy's scenario, extinction driven by human hunters was not only plausible, it was unavoidable. He postulates a human population that grew 2 percent annually; if each band of fifty people killed, say, fifteen or twenty large mammals a year, the mammals would have been eliminated within 1,000 years. MacPhee contests Alroy's assumptions and his numbers; he believes that humans probably killed no more than one or two a year, and spent the rest of the time gathering roots and tubers. He feels that the evidence isn't convincing that hunters with stone point technology could have wiped out the mammoths. "Of course," he says, "there are cases where projectile points have been found embedded in mammoth bones. But when you take a look at the number of instances, you can barely come up with a dozen for the relevant time period in North America—between 11,000 and 12,000 years ago. In other words, although people were clearly hunting, it is not a demonstration by that evidence alone that they were hunting on a scale that would have made any difference to the survival of the species." Moreover, some of the Pleistocene mammals were so widely distributed—such as the giant ground sloth, which was found from the Yukon to Mexico—that killing them in large numbers would have been almost impossible.

Species can bounce back from a base population of only a few dozen individuals. It becomes incrementally harder to kill off a remainder population, especially one with a wide geographic range, and humans had no incentive to try. Evidence shown that the extinction rate typically declined rapidly after the period of first contact. Why would the Native Americans, for example, have caused dozens of extinctions around 11,000 years ago and none thereafter? "What else in nature, besides direct human impact, could have such a dramatic effect in such a short time?" wondered MacPhee. "Why is it that again and again, all over the planet, I find the same story: that there's really nothing happening until people come, and then the animals go down." After reading a magazine article about the

Ebola virus outbreak, MacPhee was struck by a sudden insight: the only thing capable of causing extinctions of this type and scale was a highly lethal infectious disease.*

He teamed up with virologist Preston Marx of the Aaron Diamond AIDS Research Center at Rockefeller University in New York City, an expert in the ways by which emerging diseases can cross species boundaries. Together they came up with the hyperdisease hypothesis: as human populations expanded into new land masses during the Pleistocene era, they brought along one or more virulent pathogens (disease-causing agents) that wiped out native animals. "New" diseases, that is, ones encountered for the first time by a population, can indeed cause rapid crashes. Marx theorized that in order to effect mass extinctions, a "hyperdisease" would have to: (1) kill rapidly, causing death in over 75 percent of "new" hosts in a matter of days or weeks; (2) infect individuals in all age groups, wiping out a species' reproductive capacity; (3) infect many species without having much of an impact on human groups; and (4) have a vector (carrier), either human or associated with humans, to move it from place to place.

In *Mammoth*, Richard Stone mentions a finding that would tend to refute the human killer scenario, and support the possibility of disease:

In my opinion, intriguing circumstantial evidence for a plague can be found near Sevsk, a town 250 miles southwest of Moscow. During the late 1980s, scientists from the Institute of Paleontology in Moscow unearthed some 4,000 bones from thirty-six woolly mammoths, including seven full skeletons, from a sandy pit outside Sevsk. They found the remains of young and old alike—from a six-week-old baby, its tusks just beginning to protrude from its cheeks, to a fifty-year-old bull. There's no evidence that hunters felled these mammoths, which died about 13,680 years ago, nearly the end of the line for mammoths this far south in Russia. What calamity befell them remains a mystery; perhaps

---

* See pp. 238–44 for a discussion of a "highly lethal infectious disease"—Ebola, in fact—that is bringing the gorillas of central Africa dangerously close to extinction.

they all plunged into a sinkhole—or perhaps they were all killed by hyperdisease.

The killer plague that MacPhee envisioned might not necessarily have been a single disease; it could have been caused by different agents in different populations. Because of the pattern of extinction in the Americas, which MacPhee likens to a spreading wildfire, the pathogen would have to have been able to travel independently of humans. The fact that its transmission requires human contact at the start is one weakness of the theory. There aren't many good examples of human diseases that infect animals. However, tuberculosis (TB) can pass from humans to other mammals. Furthermore, the pathogen could have been carried by a parasite like a tick or flea, as are many infectious agents.

The hypothesis has to be tested by finding evidence of a unique pathogen in Pleistocene animals. Chronology is crucial because the evidence must be found in the last, terminal populations of species that came into contact with the hyperdisease. By definition, this period of overlap with humans must be a narrow window. Finding numerous samples of well-preserved mammal remains is the first challenge, and that is one of the reasons that MacPhee turned to Wrangel Island, where Russian scientists had shown that woolly mammoths (*Mammuthus primigenius*) survived 6,000 years longer than anywhere else (Vartanyan *et al.* 1995). Each specimen would have to be radiocarbon dated to make sure it coincides with the time of species extinction. Once the animal's DNA had been isolated by the latest molecular biology techniques, the final step is the search for "foreign" or "contaminant" DNA that could belong to a pathogen—the "smoking gun" of the hyperdisease theory.

The field of ancient molecular pathology is new, but MacPhee's team has recruited Alex Greenwood, a postdoctoral fellow in his department and an expert in the area of ancient DNA. Greenwood has been working on isolating DNA from the Wrangel Island material using polymerase chain reaction (PCR). PCR is the key technology of genetic analysis, allowing researchers to amplify tiny fragments of DNA so they can be studied more easily. The team hopes to find evidence of viral families that are already known to cause disease in all kinds of mammals. If there's a credible match, and if it turns up in a wide variety of remains from terminal populations, MacPhee and his colleagues will have a strong case.*

Alex Greenwood (quoted in Stone 2001) said, "The best-case scenario is that we find a correlation, a specific virus showing up at the time of extinction that cannot be found prior to that time." They plan to start their virus hunt by sweeping for suspects from broad classes such as the morbilliviruses, including the species-jumping canine distemper virus and rinderpest, which at the turn of the twentieth century ravaged wildebeest, hartebeest, and various other African ungulates after jumping from Asian cattle. "With a hyperdisease pathogen, you should have jillions of particles in the body," says MacPhee. But the odds of finding a pathogen would rise if it were in the form of a virus wrapped in a protective sheath called a capsid. "If it's something like flu or distemper," MacPhee says, "we stand a chance." As of 1999, MacPhee and Marx wrote that they "have identified no pathogen that could account for the extinctions, although we are looking. Our scenario requires something extremely nasty, but no disease known to science is capable of killing off an entire species, one after another, in what amounts to an instant in geologic time." It turned out to be much simpler than MacPhee and Marx thought. The DNA testing of Wrangel Island mammoths has been inconclusive, but the answer had been lying under our noses ever since the first mammoth bones were collected in the eighteenth century.

In August 2001, as if in response to the search by MacPhee, Marx, and Co., Bruce Rothschild and Mark Helbling published a paper with this startling title: "Documentation of Hyperdisease in the Late Pleistocene: Validation of an Early 20th Century Hypothesis." Here is the entire abstract from the *Journal of Vertebrate Paleontology:*

> Suggested in the early 20th century and selectively espoused over the past several decades, the hypothesis of disease-based large

---

* Most of the information on MacPhee and hyperdisease was taken from the American Museum of Natural History's 1998 extinction website, "Humans and Other Catastrophes" *http://www. amnh.org/science/biodiversity/extinction*), a lengthy interview in Scientific American's website, "Interview with Ross MacPhee" (www.sciam.com/interview/2001/10201macphee/index.html), and the 1999 book *Extinctions in Near Time,* which MacPhee edited. Richard Stone's *Mammoth* (2001) also contains plentiful material on MacPhee's investigations, as well as detailed discussions of cloning, re-establishing mammoths in Siberia, and the perils of making a television documentary.

mammal extinction has remained theoretical—until now. It is ironic that *Mammut americanum* provides actual demonstration of hyperdisease, given the historical importance of mastodons in the Washingtonian and Jeffersonian traditions and recognizing that it was the first fossil skeleton to be mounted in America.

Recognition of a unique pathologic alteration in a Hiscox site foot bone stimulated systematic macroscopic evaluation of North American Collections to establish the population frequency of the phenomenon. Such a pathologic zone of resorption, undermining the articular surface, was noted in at least one manus/pes bone in 22 (45 percent) of 49 *Mammut americanum* available for examination. Metacarpals (12) were most commonly affected, with less frequent involvement of metatarsals (6), phalanges (4 manus, 4 pes), carpals (6) and tarsals (5). Six (18 percent) of 33 with associated ribs had periosteal reaction on the pleural surface of several of those ribs. Such rib reaction was limited only to individuals with foot involvement.

The rib periosteal reaction is highly suggestive of tuberculosis and the metacarpal lesion is identical to that documented in *Bison* as pathognomonic for tuberculosis. Given the fractional frequency of bone involvement among contemporary animals affected with tuberculosis, it seems likely that the entire population was afflicted with what has been termed the "white plague." Recognizing that only a portion of animals infected by tuberculosis develop bone involvement, the high frequency of the pathology in *Mammut americanum* suggests that tuberculosis was not simply endemic, but actually pandemic, a hyperdisease.

Despite its name, *Mammut americanum* is a mastodon, characterized by its cheek teeth, which have blunt crests that were likened to breasts, hence *masto* ("breast"), and *odont* ("tooth"). The true mammoths are different animals, with very different, ridged teeth; the best known was the familiar "wooly mammoth" of Europe, Asia, and North America. Rothschild found no evidence of tuberculosis in mammoths. Thus the mastodon, the very species that Ross MacPhee examined in the Siberian Arctic (and on the PBS television program, *Raising the Mammoth*), turns out to be the one in which incontestable evidence of hyperdisease has been found.

Mammoth

Tuberculosis (TB) (which in humans used to be known as "consumption") is an infectious disease caused by the bacterium *Mycobacterium tuberculosis*. TB usually occurs as pneumonia, where the lungs are primarily involved, but the infection can also occur in the brain, bones, lymph nodes, intestines, or kidneys. Tuberculosis is usually communicated by the inhalation of droplets sprayed into the air (aerosols) from a cough or sneeze by an infected individual. The disease is characterized by the development of granular tumors in the infected tissues. Patients with tuberculosis of the lungs have a prolonged illness with fever, cough, night sweats, and weight loss, and may occasionally cough up blood. As the disease progresses, symptoms get worse and shortness of breath ensues. Untreated, the disease can be fatal. One billion people are infected with the tubercle bacillus; in 1997 there were 8 million new cases; of these, 3.5 million were considered highly contagious. According to Cosivi *et al.* (1998), "The annual global incidence is predicted to rise to 10.2 million by the year 2000, an increase of 36 percent from 1990. Southeast Asia, Western Pacific regions, and sub-Saharan Africa will account for 81 percent of these new cases. . . . During 1990 to 1999, an estimated 30 million will die of TB, with 9.7 percent of the cases attributable to HIV infection." People with HIV who become infected by the tuberculosis bacterium almost always die. *M. tuberculosis* is the paradigmatic drug-resistant bacterium, evolving rapidly in response to drugs that were employed in its treatment. There is now a worldwide epidemic of MDR-TB—multiple-drug-resistant tuberculosis.

There are many different strains of tuberculosis, capable of infecting different kinds of animals. In addition to humans, *M. tuberculosis* infects other primates, as well as cattle, dogs, swine, and parrots. *M. avium* infects birds, swine, and sheep. After *M. tuberculosis*, the most common species of mycobacterium is *M. bovis*, which infects cattle, dogs, and swine. Humans can contract tuberculosis from *M. bovis* by drinking raw milk, but with the introduction of pasteurization, this form of human tuberculosis has become very rare in industrialized countries. But "zoonotic" (animal to human) TB caused by *M. bovis* is a major health hazard in developing countries. As Cosivi *et al.* wrote, "*M. tuberculosis* will be largely responsible for the new TB cases and deaths, but an unknown, and potentially important, proportion will be caused by *M. bovis*." *M. bovis* has been shown to infect wild animals like European badgers and the African buffalo, but until recently *M. tuberculosis* was reported only in domestic or wildlife species living in close, prolonged contact with humans.

And now it has been shown (Alexander *et al.* 2002) that *M. tuberculosis*, the pathogen for human tuberculosis, can infect free-ranging wildlife. It has been found in South Africa in free-ranging banded mongooses, and in Botswana in a band of suricates, another kind of mongoose also known as the meerkat. Summing up their findings, Kathleen Alexander and her colleagues wrote, "Ecotourism brings large numbers of people to wildlife areas and provides both important economic benefits and an instrument for the conservation of biodiversity. However, susceptible wildlife populations may be negatively affected by the increased exposure to humans and their pathogens."

Through contact between wild and domestic animals, bovine tuberculosis too has now begun to appear in wildlife populations around the world where it does not naturally occur. It is one of the most infectious forms of TB, and it can be transmitted to other farm animals, as well as to bison and various deer species under certain conditions. According to the veterinary literature, tuberculosis has been diagnosed in snow leopards, badgers, baboons, sea lions, rhinoceroses, and Australian brushtail possums. Thousands of Cape buffalo in South Africa's Kruger Park have been infected by TB, originally spread by domestic cattle on the park's borders, and there is considerable evidence that the disease is spreading to other species. Reports of tuberculosis in wild carnivores are rare, but in 2000, Briones *et al.* reported bovine tuberculosis in the Iberian lynx (*Lynx pardina*), which is found only in isolated areas of Spain and Portugal and is already considered the most endangered feline in the world. Wild and captive elephants have tested positive for tuberculosis, and three captive elephants died of the disease, apparently transmitted by humans (Michalak *et al.* 1998). Given the pernicious nature of this widespread disease, it does not seem at all unlikely that it could have infected mammoths.

The Holocene epoch is the geologically brief interval of time encompassing the last 10,000 years, the period when the so-called charismatic megafauna of North America were somehow eliminated, Sabertooth cats, mammoths, mastodons, ground sloths, wooly rhinos, cave bears, and numerous other species disappeared at approximately the same time that humans crossed the Bering Land Bridge from Asia. One of the possible explanations for this mass extinction has to be the arrival of the most destructive predator in the history of the planet, but another possibility is disease, perhaps brought over by *Homo sapiens* or their dogs. But as S. David Webb (1995) wrote,

> In the broader scale of the late Cenozoic, however, man appears too late to be the major cause. The genus *Homo* evolved midway in the late Cenozoic glacial ages but arrived in the New World only for the last in a long string of extinctions. Hunting cultures surely played a role in this last great extinction pulse. Just as surely, however, man could not have figured in the preceding extinction pulses. Thus by elimination, *climatic deterioration* becomes the primary causal hypothesis for the late Cenozoic late mammal extinctions.

In recent years, we have been able to document the disappearance of a substantial number of creatures as it occurred; in many cases, we can identify the causative agent by looking in a mirror. We are the same species, indeed the very same creature, that lived 30,000 years ago, having added only a thin veneer of "civilization" to our *modus vivendi*. In an essay published in 1996, Stephen J. Gould argued that since Cro-Magnon people were just like us, there is no reason why their artists would not be capable of producing cave-wall masterpieces such as those at Lascaux and Chauvet. The painters are, after all, our ancestors, and do not differ in any substantive way from us—or from Michelangelo, Picasso, or Einstein, for that matter. Earlier students of Paleolithic cave painting, such as Abbé Breuil and André Leroi-Gourham, assumed some sort of a chronological progression from the primitive to the sophisticated; since the paintings were so old they had to be rudimentary. Gould wrote: "The Cro-Magnon cave painters are us— so why should their mental capacity differ from ours? We don't regard Plato or King Tut as dumb, even though they lived a long time ago."

Today on our home planet, some of us are terribly worried about the loss of biodiversity and the extinction of various life forms, especially now that we have been able to learn more about the history of past life on earth. Although there is fierce debate about the details, hominids are believed to have been around, in one form or another, for some seven million years. For most of that time, they have devoted their energy to the processes of eating and reproduction—staving off extinction as it were. The same can be said of any other animal of the past or present. But despite all this dedication to the fundamental processes of self- and species perpetuation, no strategy, as far as we know, has been particularly successful. Estimates of inclusive extinction run as high as 99 percent of all the creatures that have ever existed. To be sure, there are some living fossils that seem to have defeated extinction, such as cockroaches, horseshoe crabs, nautiluses, coelacanths, etc., but these examples are so rare and unusual that they can be cited only as glaring anomalies. All other creatures are long gone, and we don't know why. Since so few animals have fossilized, it is fair to add that we don't know very much about what used to live here; we have only scratched the surface.

The disappearance from North America in the past 20,000 years of the mammoths, mastodons, saber-toothed cats, dire wolves, ground sloths, Irish elk, etc., and a similar (although not simultaneous) reduction of large mammal populations in Australia points an accusatory finger at *Homo sapiens*, because in both cases, these extinctions corresponded to the arrival of humans. In his recent book about the ecological history of

North America, Australian Tim Flannery unequivocally identifies *H. sapiens* as the villain: "I believe that analysis of global extinction points towards humans rather than climate change as the cause of the demise of North American giants." He also assigns our species comparable blame in the "continentwide extinction" of Australian megafauna about 46,000 years ago (Roberts, Flannery *et al.* 2001). A recent study by John Alroy (2001) supports this humans-are-the-bad-guys scenario. Alroy ran a computer simulation of North American end-Pleistocene human and large herbivore dynamics, and found that

> The improved ecologically realistic model outlined here challenges the commonsense notion that no amount of overkill could have resulted in a true megafaunal mass extinction. The simulations demonstrate not merely that overkill scenarios are plausible, but that an anthropogenic extinction was unavoidable given the facts of ecology and the fossil record—even assuming that human predation was limited and non-selective. The overkill model thus serves as a parable of resource exploitation, providing a clear mechanism for a geologically instantaneous ecological catastrophe that was too gradual to be perceived by the people who unleashed it.

It may have happened the way Alroy's computer simulation plays it out, and indeed, his analysis is quite persuasive. But even the most plausible of scenarios are only guesses at to what actually happened 15,000 or 50,000 years ago. All of the aforementioned discussions include the possibility that there is a third explanation for the extinction of the mammoths and the saber-tooth cats, and that is hyperdisease, somehow transmitted by man or the animals that accompanied him. As Edward O. Wilson (1992) eloquently put it, "From prehistory to the present time, the mindless horsemen of the environmental apocalypse have been overkill, habitat destruction, introduction of animals such as rats and goats, and diseases carried by these exotic animals." Unless it was climate change (over which we had no control) that brought about the Pleistocene extinctions, it is difficult to escape at least part of the responsibility. At the same time, our ability to affect the *evolutionary* process (the flip side of the extinction process) is equally pervasive. We can now change climate (think of global warming), modify landscapes to an unprecedented extent, introduce alien animals to environments they never inhabited before, and watch helplessly as microbes of all sorts evolve to develop immunities to whatever antibiotics or pesticides we throw at them. In *The Beak of the Finch* (1994), Jonathan Weiner wrote:

> What we don't understand on either front is that the more pressure we put on our pests, the more we cause them to evolve around the pressure. The pressure is evolutionary pressure; what we fail to understand is evolution itself. . . . Precisely where we wish to control the environment most tightly and possess it most completely we are powerless to do so, besieged and beleaguered by resistance movements that seem to spring up faster the more we lop them off, like the heads of Hydra. The harder we fight those resistance movements, the harder and faster they evolve before our eyes—precisely because it is our effort to control them that is driving their evolution.

III
FINALE

Extinctions (and Nonextinctions) in Near Time (The Last 1,000 Years)

It is most difficult always to remember that the increase of every creature is constantly being checked by unperceived hostile agencies; and that these same unperceived agencies are amply sufficient to cause rarity, and finally extinction. So little is the subject understood, that I have heard surprise repeatedly expressed at such great monsters as the Mastodon and the more ancient Dinosaurians having become extinct; as if mere bodily strength gave victory in the battle of life.

—Charles Darwin,
The Origin of Species, 1859

We are now hell-bent on destroying rain forests, wetlands, plains, coral reefs, and almost all those places where it has been possible for plant and animal life to live. With our concentrated assault on the earth's pristine regions, from the poles to the tropics, from the highest mountains to the uninhabited steppes and deserts, down to the oceans' greatest depths, we have multiplied a thousandfold our pernicious influence on the ecology of our planet. In the process, we have driven many creatures over the precipice of extinction, and many more now crowd the edge of the cliff, waiting for the final shove. On extinction's current short list are mammals such as the Florida manatee, black-footed ferret, Mediterranean monk seal, Gulf of California porpoise, Chinese river dolphin, giant panda, mountain gorilla, African black rhinoceros, koala, Indian lion, cheetah, Tibetan antelope, giant river otter, Ethiopian wolf, Siberian and Southeast Asian tigers; snow leopard, and Javan and Sumatran rhinoceroses. Figuratively and literally, birds like the Spix's macaw and the Kauai honeycreeper (o'o) now hover near extinction. Fishermen will soon catch the last freshwater sawfish and barndoor skate. Every one of these unfortunate creatures has the ongoing misfortune to live in or around areas where people want to live or work. It is not hard to imagine who will win the battle between Florida waterfront developers and manatees.

To date, our most successful attempt to completely eliminate a large mammal species took place in the icy reaches of the Bering Sea, two and a half centuries ago. When Commander Vitus Bering's ship *St. Peter* was wrecked on a remote, rocky island in 1741, the surviving crew members found, in addition to bewhiskered sea otters, immense "sea cows," which were subsequently named for Bering's naturalist, Georg Wilhelm Steller. Bering died there of scurvy on what was later named Bering Island. With Copper Island, the group is known as the Commander Islands (*Komandorskiye Ostrova* in Russian. Upon Steller's return to Kamchatka on the Russian mainland, the existence of the sea otters, the sea cow, and the islands themselves became known. Russian sealers began to visit the Commander Islands for the fur seals, the sea otters, and the sea lions that also bear Steller's name, and also for meat and oil for their voyages.

We have no way of knowing how many sea cows existed when Bering landed on these chilly islands, but Leonhard Stejneger, Steller's biographer and a biologist himself, has estimated that there were about 1,500. They are now extinct, but there are enough contemporary illustrations and descriptions to give us ample information on what they looked like and how they lived. As far as we know, *Hydrodamalis gigas* was the only cold water sirenian; it was also, at a length of thirty feet and a weight of ten tons, the largest. The animal, which Russians called *morskaya korova* (marine cow) was an overstuffed sausage of a beast, with a small head, piggy eyes, and skin that was likened to the bark of a tree. It had a forked, horizontal tail like its relative the dugong (manatees have a rounded, paddle-like tail), and its forelegs were unique in the mammalian kingdom: they had no finger bones. The animal, which could probably not dive below the surface, pulled itself along the shallows on its stumps as it browsed on kelp. Instead of teeth, the mouth of the sea cow was equipped with horny plates that it rubbed together to grind plant matter into a pulp.

The skin of the sea cow—which Steller himself called "the manatee"—was so thick it could be used for the soles of shoes and for belts. But these massive beasts were not killed for their skin, but rather for the subcutaneous

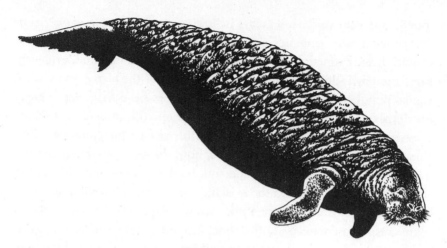

Steller's Sea Cow

fat, which could be as much as nine inches thick. In his 1745 description, Steller wrote (of the fat):

> It is glandulous, stiff, and white, but when exposed to the sun it becomes yellow like May butter. . . . Its odor and flavor are so agreeable that it can not easily be compared with the fat of any other sea beast. . . . Moreover, it can be kept a very long time, even in the hottest weather, without becoming rancid or strong. In flavor it approximates nearly the oil of sweet almonds and can be used for the same purposes as butter. In a lamp it burns clear, without smoke or smell. And indeed, its use in medicine is not to be despised, for it moves the bowels gently, producing no loss of appetite or nausea, even when drunk from the cup.

The sealers killed the huge, slow moving, oil-rich "manatees" with such insensitive profligacy that there were none left by 1768. It had taken only twenty-seven years for the Russian adventurers to eliminate the hapless sea cow from the face of the earth, but the sealers had no way of knowing that this was the last of them; they probably assumed that there were similar undiscovered islands with more sea cows. There were not.

Nowadays, most people attribute the extirpation of *Hydrodamalis gigas* to the rapacious Russian sealers who simply killed every one they found. But Paul Anderson (1995) believes that this simple explanation "may hide a more complex and interesting tale that bears both on the evolution and extinction of this giant sirenian." His suggestion, evidently first raised by Delphine Haley in 1980, puts part of the blame on the aboriginal populations who removed the sea otters from the inshore areas previously inhabited by the sea cows. Without the otters, sea urchins increased exponentially (sea otters feed on sea urchins, among other things), and consumed the algae throughout the sea cows' range depriving them of food. Thus, says Anderson, by the time Bering arrived, the sea cows were restricted to those islands like Copper and Bering where there had never been a human population, and were therefore easy to eliminate. "Sea cow evolution," concluded Anderson, "may have been dependant upon otter predation on urchins, and sea cow extinction may have been hastened by otter declines. The moral may be that extinction is rarely 'simple.'"

The arrival of white men into the Aleutians signaled big trouble for the native wildlife. When Bering and Steller were shipwrecked on the barren Commander Islands, they and their fellow survivors had to subsist on whatever they could catch. They began with sea otters and fur seals, and initiated what soon became the total elimination of the hapless sea cow. But they also found a large flightless bird, ripe for the plucking. It was the spectacled cormorant (*Phalacrocorax perspicillatus*), an ungainly bird who, most unfortunately for its future, had wings so small it could not fly. Like the sea cow, it was ridiculously easy to kill, and it tasted delicious; the combination almost guaranteed its demise. In his 1936 biography of Steller, Stejneger described it:

> The flightless spectacled cormorant is another of Steller's sensational discoveries—sensational not only because its wings were too small to carry its gigantic body, but chiefly because—like the sea cow—it is known only from Bering Island and was exterminated by ruthless hunting. Steller, the only naturalist to see the bird alive, and his comrades were able to vary their fresh meat diet of sea-otter and seal with roasts of this stupid bird, which was as large as a goose, weighing 12 to 14 pounds, "so that one single bird was sufficient for three starving men," Ordinarily cormorants are not considered particularly savory eating, but Steller avers that

Flightless Cormorant

when properly prepared according to the method employed by the Kamtchadals, namely by burying it encased—feathers and all—in a big lump of clay and baking it in a heated pit, it was a palatable and juicy morsel.

The Russian-American sealing company transported Aleuts to the islands in 1826 to kill seals and otters all year round. Although the sea cows had been extinct for fifty-eight years when they arrived, there were still flightless cormorants for the Aleuts to eat; by 1850, they had eaten every last one.

Steller's sea cow and Steller's flightless cormorant are gone, every one of them killed and eaten by fur trappers in the North Pacific, who, in the process, came perilously close to eliminating the chief object of their hunt, the richly furred sea otter. In Africa, hunters of "bushmeat" (primates used for food) are threatening everything from gorillas and chimpanzees to various monkeys, and in the year 2000, scientists declared Miss Waldron's red colobus monkey (*Procolobus badius waldroni*) extinct. This marked the first time that a primate species has been hunted to death, although the *2000 IUCN Red List of Threatened Species* cautions: "While available evidence suggests that this taxon is probably extinct, it does not fulfill the IUCN criteria for extinction, which is: "there is no reasonable doubt that its last individual has died."

Of the monk seals (*Monachus*), Karl Kenyon (1981) wrote, "Perhaps because of their primitiveness, the monk seals seem far more sensitive than other phocids to the intrusion of man into their environment. In recent years, and concomitant with the rapid spread of human activity to even the most remote and isolated areas, all three species have shown alarming population declines. One species, the Caribbean monk seal, probably became extinct during the 1950s."

When Christopher Columbus arrived at the island of Alta Vela, south of Haiti, on his second voyage in 1494, he saw a group of eight "sea wolves" on the beach. A shore party killed all eight of them, and as Peter Knudtson (1977) wrote, "thus ended, in a prophetically bloody manner, the first recorded encounter between Europeans and the sea mammal now known as the Caribbean monk seal, *Monachus tropicalis*." Their placid and unaggressive nature made them easy to kill, and there was already a fishery for these animals in 1675, when William Dampier visited the Bay of Campeche off the western Yucatán:

> . . . there being such plenty of Fowls and Seals (especially of the latter) that the Spaniards do often come hither to make Oyl of their fat; upon which account it has been visited by English-men of *Jamaica*, particularly by Capt. *Long;* who, having command of a small Bark, came hither purposely to make Seal-Oyl, and anchored on the North side of one of the sandy Islands, the most convenient Place for his design.

In the Smithsonian Institution's archives, Peter Adam and Gabriela Garcia (2003) found the heretofore unpublished field notes of biologist

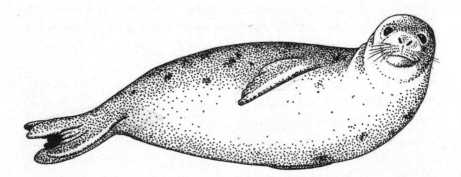

Caribbean Monk Seal

Edward W. Nelson, who had observed a herd of wild monk seals in the Triangle Islands (Arracifés Triangulos) also off the Yucatán in the Bay of Campeche in 1900. Nelson wrote:

> We found these seals much less numerous than they were reported to be by men at Campeche who have visited the Triangles to kill them for oil during the past few years. The man from whom we hired our schooner has made two sealing expeditions to the Triangles and under his directions hundreds of the seals have been killed with clubs. . . . In this way the great majority of the existing seals of this species have been destroyed within the last ten years. . . . They lie directly against one another, sometimes half overlying one another and thus reminding one of a mass of hogs. . . . They lie ashore in this manner for days at a time in a lethargic sleep with the sun shining on their backs with all the fierce heat of the tropical spring in this region. . . . Part of this sluggish carelessness is due to their not having been hunted sufficiently to arouse them but at the same time their stupidity seemed one of their most characteristic features. . . . The males have a hoarse, guttural roar which they utter at intervals while lying on shore and more rarely when in the water. After a number had been shot from a "pod" on shore the survivors always floundered into the water in a wild panic and most of them disappeared but several always remained for some time swimming back and forth near shore raising their heads high out of [the] water and watching us curiously as we were skinning their companions. On shore their motions are almost exactly like those of a large maggot but once in the water they are very graceful.

According to Joel Allen, who reproduced many of the nineteenth-century descriptions of this species in his 1880 *History of the North American Pinnipeds,* "the habitat of the West Indian Seal [Caribbean Monk Seal] extends from the northern coast of Yucatan northward to the southern point of Florida, eastward to the Bahamas and Jamaica, and southward along the Central American coast to about latitude 12°. Although known to have been once abundant at some of these localities, it appears to have now well-nigh reached extinction, and is doubtless to be found at only a few of the least frequented islets in various portions of the area above indicated." Monk seals are the only seals that live in warm waters year-round; this fondness for tropical waters meant that they were likely to inhabit the very locations where human beings from colder climates wanted to spend their vacations. In the latter years of their existence, the "Pedro seals" found themselves on beaches that developers wanted for hotels and condominiums, and this shy and inoffensive creature proved no match for land sharks.

All species of monk seals look very much alike; if one were to somehow transplant a Hawaiian monk seal to, say, the eastern Mediterranean, nobody would be able to tell the difference. So it is only the caption that identifies the forlorn-looking monk seal in a photograph in the New York Aquarium's 1937 *Guide Book* as a "West Indian Seal." This lonely creature, whose species was described by the Aquarium's director, Charles H. Townsend, as "now approaching extinction," was probably one of the last of its kind. In his 1942 study of extinct and vanishing mammals of the Western Hemisphere, Glover Allen penned this plaintive cry for the Caribbean monk seal:

> While conclusive information is at present unobtainable, it nevertheless seems very likely that there may be a few seals still resorting at the Triangle Keys, and possibly, on rare occasions individuals may turn up elsewhere, but clearly the species was of restricted habitat, and within historic times has been brought nearly to the verge of extinction. It would appear to be a simple matter for the British Government to pass protective regulations for the preservation of any that might still exist in the Bahamas and for the Mexican Government to prohibit their killing on the islands of the Yucatan so that they might breed up to numbers placing them less close to the danger line.

No such luck. In 1973, under the auspices of the U.S. Fish and Wildlife Service, biologist Karl Kenyon carried out a 4,000-mile aerial survey of the Caribbean basin, searching for monk seals. He visited almost every island where they had been seen in the past, and interviewed fishermen, sailors, and anyone else who might provide information on the missing animal.

He concluded that they were gone, and wrote wistfully, "Man has now dominated the environment." The final sighting of a wild Caribbean monk seal occurred in 1952, off the Serranilla Bank, 250 miles southwest of Jamaica (Kenyon 1977).

The "horse antelopes" (genus *Hippotragus*) are among Africa's most impressive animals. There are three distinct species and one subspecies. One of the three species is extinct, and the "giant sable" subspecies—perhaps Africa's most spectacular antelope—is dangerously close.

The roan antelope (*Hippotragus equinus*) is a large, light-colored, reddish gray animal, once found throughout the savannas and woodlands surrounding the West African rain forest, from Gambia east to western Ethiopia, south to southwestern Tanzania, west to the Angolan coast, southwest to northern Namibia, and southeast to Swaziland. The somewhat smaller sable antelope (*Hippotragus niger*) occupied similar country in Central and southeast Africa, from southeastern Kenya south through eastern Tanzania and Mozambique to Swaziland and the Transvaal, and west through Zimbabwe, Malawi, Zambia, and southern Zaire to northeastern Namibia and eastern Angola (Klein 1974). An isolated population in central Angola that contained very large, dark individuals with massive horns was distinguished as the giant sable, or *Hippotragus niger variani*.

The giant sables are compact, powerful animals, with thick necks enhanced by upstanding manes. Males and females look alike until about the age of three, when the males become darker and develop bigger horns. A full-grown bull can weigh over 600 pounds, and stand almost five feet high at the shoulder. The coat is short and glossy like that of a horse, and it even comes in horselike colors: the bulls are chocolate brown to black, the females and young, sorrel to chestnut. All sables have a white belly and a white rump patch. Giant sables can be differentiated from their smaller relatives by the grackle black coat color of the adult males, and by the lack of the usual white facial stripe. Where not persecuted, sables are not excessively wary; they will often run a short distance when startled, only to stop and look back. When closely pursued, however, they can run as fast as thirty-five miles per hour for a considerable distance. Juveniles are preyed upon by leopards and hyenas, but only lions will attack the impressively armed adults. Both sexes

Giant Sable Antelope

have horns (as is the case with most antelopes) but the ridged, sickle-shaped horns of sable bulls are among the most spectacular adornments in the animal kingdom, arching over the animal's back to a record length of sixty-four inches (Mochi and Carter 1971). The giant sable rivals the greater kudu as Africa's most handsome antelope, and its great sweeping horns made it one of the most popular trophies for African big game hunters.

In *Green Hills of Africa*, a chronicle of a monthlong safari in East Africa in 1933, Ernest Hemingway describes his first sighting of sable antelope:

Below us across the open space where the gully we could not see opened onto the head of the valley, sable started to pass at a running stampede. . . . They all looked like the one I had shot and I was trying to pick a big one. They all looked about the same and they were crowding running and then came the bull. Even in the shadow he was a dead black, and shiny as he hit the sun, and his horns swept up high, then back, and huge and dark, in two great curves nearly touching the middle of his back. He was a bull all right. God, what a bull.

Hemingway shoots the bull but does not kill him ("I had gotten excited and shot at the whole animal instead of the right place and I was ashamed"). The safari follows the wounded animal for hours by spatters of blood on the grass, but they lose him. "I was thinking about the bull and wishing to God I had never hit him," wrote Hemingway. "Now I had wounded him and lost him. I believe he went right on travelling and went out of that country. He never showed any tendency to circle back. To-night he would die and the hyenas would eat him."

Thirty years ago there might have been as many as 1,000 of the big, graceful antelopes that Angolans call *palanca preta gigante*. The last authoritative sighting of a giant sable was made by the American antelope specialist Richard Estes, who studied the coal black bulls and chestnut-colored females from 1969 to 1970, the only study ever conducted of giant sables in the wild. During a year in the field, Estes estimated there were about 1,000 giant sables, found only in Cangandala National Park, the Luando Nature Reserve, and the corridor between these two parks. But between Estes's published observations and the recent past, a violent, take-no-prisoners war has raged throughout Angola, which has not been beneficial to the country's wildlife—especially those that could be butchered and eaten by jungle fighters.

In 1975, Angola declared its independence from Portugal. For the next twenty-seven years, various factions fought bitterly for control of the tropical, heavily forested country in southwestern Africa. Countless animals were hunted for food by the rebels, many of whom spent decades living in the bush. The huge herds of elephants were shot out by both government forces and the União Nacional para a Independência Total de Angola (UNITA), which sold ivory from the tusks to fund its war efforts. Rhinos and Cape buffalo also vanished, many slaughtered for food or used as target practice by soldiers thundering over the game parks in helicopters. Land mines inflicted heavy casualties as well. Today, antelope, monkeys, and even domestic animals are a rarity in war-ravaged Angola, one of the poorest nations in the world. It is also one of the most dangerous places on earth, and until very recently, biologists were unwilling to risk their lives in hopes of catching a glimpse of a rare antelope.

In *A Certain Curve of Horn,* journalist John Frederick Walker lays out the story of the discovery of the giant sable, first described by British railway engineer Frank Varian (hence *variani*) in 1916. Much of the book is devoted to Angola's turbulent political history with various guerrillas fighting the Portuguese for independence: Cuba and South Africa committing troops and materiel to the battle, and UNITA leader Jonas Savimbi (who was killed in February 2002) waging a massively destructive war in the very area inhabited by the last of the giant sables. Walker participated in several expeditions in Angola, traveling (separately) with Richard Estes and Wouter van Hoven of the University of Pretoria, the two biologists who had invested so much time and energy in finding and studying the great black antelopes. At the conclusion of his book, Walker writes that he never saw a giant sable antelope:

> For years now I'd staged scenes of this anticipated final encounter in my head, rewriting them after every setback and incorporating every new whiff of hope. This was supposed to have been the last leg of the journey, the realization of a long-held dream; somehow we would fly over the warlord's forest and find the giant sable. It should have happened, and could have happened, and almost did, but for the war, and because few journeys end the way you think they were meant to end.

A book can be completed as much as a year before it is actually published, and *A Certain Curve of Horn* was probably in production well before it appeared in October 2002. Two months before the publication date, five giant sables, all bulls, were located in Cangandala National Park by van Hoven and other biologists. They had failed to spot any sables from helicopters, so they tracked the animals for twenty miles through the bush on foot, spotting the animals in grassy woodland before they bolted. The discovery obviously pleased the biologists—and Walker, who was on that expedition too—but it was also met with jubilation throughout Angola, because the antelope with the great sweeping horns appears on everything from postage stamps and passports to banknotes and the tailfins of TAAG, the Angolan national airline. National soccer players are referred to as the *Palancas Negras*, or black antelopes. Even UNITA used the giant sable as its symbol.

Now that the war is over, Angola is facing desperate shortages of food, housing, and medicine, and the government does not have the money to staff the country's game preserves with guards. To protect the giant sable, the Kissama Foundation, an Angolan wildlife conservation group headed by van Hoven, is trying to raise international money to begin policing the reserves. The foundation is also raising money for "Operation Noah's Ark" to restock Angola's once-teeming wildlife reserves with animals from Botswana and South Africa, and also to build tourist camps where visitors might be able to see the *palanca preta gigante*, in part to convince local residents that wild animals can be a valuable source of income.

Although the giant sable came close to extinction—and is certainly not out of danger yet—the closely related bluebuck (*Hippotragus leucophaeus*) is gone. It was the first large African mammal to become extinct in historical times.

Shortly after the last ice age, about 10,000 years ago, the bluebuck was probably common in the far south of Africa, which was largely covered with grassy plains. For reasons not clearly understood, their numbers began dropping about 2,000 years ago. Various factors have been suggested, including the change of grassland into bush and forest when the climate became warmer, and the introduction of domesticated livestock, particularly sheep, at about that time. The bluebuck may have suffered from habitat degradation by the livestock, or persecution by the herders because it competed with stock. It may also have been unusually susceptible to some of the epizootic diseases that stock carry.

In any case, the 2,000-years-ago date of blue antelope decline, if confirmed at more archaeological sites, would suggest that local Stone Age people, not climate, brought the species to the verge of extinction. European firearms then applied the final blow (Klein 1974). When the Dutch settled the Cape Colony in the seventeenth century, the bluebuck* was already rare. They called them *bluebook, bloubok,* or *blauwbok,* because of

---

* Not to be confused with the large Indian antelope, *Boselaphus tragocamelus,* known as the nilgai, but sometimes called "blue bull," or "blue buck."

Bluebuck

the dark blue gray color of their coats. The early travelers found the bluebuck only in relatively well watered grassy country, suggesting that it had to drink regularly, like the other horse antelopes, the roans and sables. Bluebucks lived in small herds of up to twenty individuals, and were primarily grass-eaters or grazers that sometimes fed on the same pastures as sheep.

The African blue antelope was clearly on its way to extinction when European naturalists and hunters finally discovered it. The first European to record it was probably Peter Kolb, a German who traveled extensively through the southwestern and southern Cape Province between 1705 and 1712. Probably because its horns resemble those of the European ibex, Kolb mistook it for a wild goat, and even portrayed it with a goatlike beard, which it did not have. It was subsequently noted by other eighteenth century travelers, who encountered it just east of the Hottentots Holland Mountains, in the region of Swellendam, Caledon, and Bredasdorp. In 1774, the Swedish naturalist Carl Thunberg reported that it had become extremely rare. According to the German zoologist Martin Lichtenstein, the last bluebuck was killed in 1800.

One of the keystone species for African extinctions has always been the quagga, a partly striped, horselike animal whose name is derived from its barking call. Its original scientific name was *Equus quagga*, but as we shall

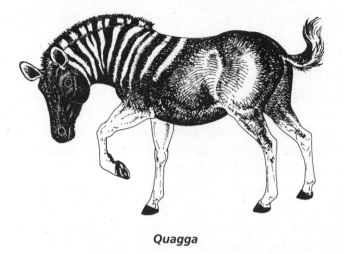

Quagga

see, there are good reasons to call it *E. burchelli.** Well into the nineteenth century, it was plentiful throughout southern South Africa; while it was being heavily hunted for meat and trophies, and its habitat appropriated for farms and ranches, the event of its extinction in the wild went almost unnoticed. The last wild quagga was killed in 1878. Once there were herds of quaggas in South Africa, and then there weren't. For a while, quaggas were kept in zoos, but the last captive specimen died in Amsterdam's Artis Magistra Zoo on August 12, 1883. It seemed as if the quagga, like the dodo, was gone forever.

The quagga looked like a brownish zebra with only the front half striped in darker brown. It had a white belly, white legs, and a white tail. It had no distinct markings on the hind quarters and only vague mottling on its back. The quagga seemed half zebra and half something else, but what? A wild ass perhaps? Lutz and Heinz Heck, who "backbred" the tarpan (see p. 303) and re-created the extinct giant ox known as the aurochs, believed that the quagga could also be re-created. In 1971, Reinhold Rau, a taxidermist at the South African Museum in Cape Town, visited

---

* There are even better reasons to call the plains zebra *Equus quagga*. Because the quagga was the first plains zebra to be named, its name takes precedence, and all plains zebras become *E. quagga* instead of *E. burchelli*. In Nowak's 1991 *Mammals of the World*, we read: "Since *E. quagga* of South Africa seems to represent a continuation of this trend [the disappearance of stripes on the hindquarters] there is increasing recognition that it is conspecific with *E. burchelli*, in which case, the proper name for the resulting species would be, by nomenclatural priority, *E. quagga*."

museums in Europe to examine most of the preserved quagga specimens; he decided that a program could be implemented to rebreed the animal by crossing specimens of the plains zebra that showed less evidence of striping on the hindquarters (Rau 1978). At that time, most zoologists considered the quagga a separate species, and therefore saw no merit in breeding zebras that sort of looked like quaggas. Rau considered the quagga to be a subspecies of the plains zebra (*Equus burchelli*), and did not abandon his program. He was vindicated in 1984, when a group of scientists undertook molecular studies on dried flesh and blood from an old museum specimen, and found that the protein and DNA fragments matched those of the plains zebra. Therefore, the quagga was a subspecies of the zebra, not a separate species at all. This meant, of course, that because the plains zebra was not extinct, the quagga wasn't either.

Dr. J. F. Warning, a retired veterinarian, had been associated with horse and cattle breeding in Germany and Namibia for more than fifty years. He was also a friend of Lutz Heck and spent much time with him during the latter's stay in Namibia. He contacted Rau in the latter part of 1985, and together they formed a committee in the South African Museum with the goal of rebreeding the quagga. In March 1987 nine selected zebras were captured at Etosha National Park. The first foal was born on December 9, 1988. In successive years, further selected breeding stock was taken from Etosha and Zululand, and the offspring are beginning to look remarkably like quaggas. The first foal of the second offspring generation (the "F2 generation") was born in February 1997. Reproductive maturity is reached only at age two or three in mares and four to five in stallions, so the project takes time. As of 1992, when six of the offspring were moved nearer Cape Town onto land which had sufficient natural grazing, they were still being called zebras. By April 1999, Quagga Project zebras were living in six localities. It is expected that this selective breeding will, with successive generations, reduce the high degree of individual variation, both in color and in the extent of striping, that is characteristic of the southern plains zebra. Eventually individuals should emerge whose coat pattern closely resembles that of the extinct quagga.

There are still those who have reservations about the project, or who oppose it altogether. They claim that creating something that looks like a quagga is not the same thing as creating a quagga. But if the definition of the quagga rests on its well-described morphological characteristics, and if an animal can be obtained that possesses these characters, then by definition, it will be a quagga. The quagga and other plains zebras belong to the same species; the quagga should be considered merely a different population. Since the characteristics used to identify the quagga all concern its coat pattern, it follows that if selective breeding can produce an animal with those characteristics, it can with full justification be called a quagga. Because it would possess the same coat-pattern genes as the original quagga, it would not be a "look-alike," but the real thing.

The Heck brothers began their recreation of extinct species with the aurochs. Now considered the ancestor of all domestic cattle breeds, the aurochs (*Bos primigenius*) was a huge ox that stood more than six feet high at the shoulder. Representations of the aurochs can be seen in wall paintings at Chauvet (painted 31,000 years ago) and at Lascaux (17,000 years ago).

  Although they are usually referred to as "bulls," the animals revered by the Minoans of Crete in 1500 were clearly aurochs. Their sheer size, along with the powerful neck muscles and forward-curving horns, identify them as *Bos primigenius*. Virtually every depiction of a "bull" in Minoan art—the famous "bull-dancer" fresco, scenes on various cups (rhytons) and sculptures, not to mention the great symbolic horns that decorate the palace at Knossos—all cry out "aurochs." The bull-dancer fresco shows a piebald brown and white animal, though for the most part the wild aurochs bull was black and the cow dark brown, both with a light stripe running down the middle of the back. The bulls were very much larger than the cows, with longer, differently shaped horns (Clutton-Brock 1989.) The Minoans disappeared around 1500 B.C. Fourteen centuries later, in his *Commentaries on the Gallic Wars*, Julius Caesar described the great black bulls he saw in the north:

> They are a little below the elephant in size, and of the appearance, color, and shape of a bull. Their strength and speed are extraordinary; they spare neither man nor wild beast which they have espied. These the Germans take with much pains in pits and kill them. The young men harden themselves with this exercise, and practice themselves in this kind of hunting, and those who have slain the greatest number of them, having produced the

horns in public, to serve as evidence, receive great praise. But not even when taken very young can they be rendered familiar to men and tamed. The size, shape, and appearance of their horns differ much from the horns of our oxen. These they anxiously seek after, and bind at the tips with silver, and use as cups at their most sumptuous entertainments.

"Soon after the *Commentaries* became known," wrote J. R. Conrad (1957), "aurochs bulls began to be sought eagerly for use in the arena. One of the most successful of the classical bullfighters is named by Martial as *Karpophorus*, whose strength and whose sword are described as prevailing best against the black bulls of the Hercynian Forest of the north [the Black Forest of Germany]." The bull of the Minoan frescoes and Caesar's *Commentaries* is not an ordinary cow or ox. Just smaller than an elephant, it is a huge, aggressive beast, probably responsible for innumerable references to bull fighting, bull worship, and bull hunting in antiquity. Although we know very little about the culture of the Minoans of Crete, the predominance of aurochs images at Knossos, Phaistos, and Zakros strongly suggests a bull-worshipping society. In Konrad Gesner's 1585 *Historia Animalium*, the first illustrated book on zoology, a huge *uro* (aurochs) is shown trying to get at a man who is stabbing it from behind a tree. The animal's shoulder is higher than the man's head.

Although the aurochs served as the baseline breeding stock for all known species of domestic cattle, from draft oxen and longhorns to beef cattle, zebus, and dairy cattle, the pure, wild form of the great bulls were dying out centuries ago. By the fourteenth century, they survived only in East Prussia, Lithuania, and Poland; they would have disappeared completely if it weren't for a royal decree in Poland that protected them under threat of death. They lasted longest in the Jaktorowski Royal Forest in Mazowsze, where the local villagers acted as gamekeepers. Exempted from taxes, their only required task was to look after the last herd of aurochs, but even this was not enough to save them. In 1564, there were eight bulls, twenty-two mature cows, and five calves. By 1602 there were three bulls and one cow. Eighteen years later there was only a single cow, which died a natural death in 1627.

To imagine an aurochs, think of a Spanish fighting bull—and then double its size. In the 1920s, Heinz and Lutz Heck, the sons of Ludwig

Aurochs

Heck, director of the Berlin Zoo, attempted to breed the aurochs back into existence from the domestic cattle that were their descendants. At the Tierpark Hellabrunn (the Munich Zoo), Heinz mixed and matched Polodian steppe cattle, Scottish Highland cattle, Alpines, Corsicans, and even Frisians, large dairy cattle from the Netherlands. Lutz, having taken over the reins at Berlin when his father retired in 1931, chose wild Corsican oxen, Spanish fighting bulls, and the wild black cattle of the Camargue. The results were the "recreated aurochs" or "Heck cattle", which bear a physical resemblance to what is known about the wild aurochs, but the Hecks never managed to replicate the size. A re-created aurochs bull is not much larger than the bull of most breeds of domestic cattle, while wild aurochs bulls are believed to have been almost as large as the one shown in the Knossos fresco. Lutz's Berlin breed failed, so today's animals are all descended from the Hellabrunn breed. These cattle are now quite uniform in conformation and have been largely uninfluenced by outside blood in recent years. The total population is about 150 animals.

Creatures of the air have not been able to fly away from their relentless pursuers; gone forever are the Labrador duck, Carolina parakeet, Eskimo curlew, heath hen, passenger pigeon, and ivory-billed woodpecker. And that is only North America. Other victims include a host of Hawaiian honeycreepers killed off for their red and yellow feathers by native Hawaiians, probably before Captain Cook paid his last, unfortunate visit to the islands; elsewhere, the Rodrigues solitaire, Pigeon Hollandaise, Bonin night heron, Guadalupe storm petrel, Tahitian sandpiper, painted vulture, Wake Island rail, pink-headed duck, mysterious starling, and hundreds of others suffered the same fate, including altogether too many parrots, parakeets, and macaws.

Grounded, flightless birds were easy prey: the giant moas of New Zealand are history; the dodo, the very symbol of extinction, was eliminated on the island of Mauritius by the seventeenth century, not because it could be eaten (it was said to taste terrible), but because it was too clumsy and flightless to escape. The same fate befell the great auk, although it was said to have been tastier than the hapless dodo. Clumsiness and inability to escape also rang the knell for the giant tortoises; the last remaining populations are found in the Galápagos and the Seychelles, where they are considered endangered. Also to be found only in lists of extinct mammals are the quagga, the Tasmanian tiger, and the Caribbean monk seal.

Until the arrival of Polynesian colonists in New Zealand around 700 years ago, there were eleven species of moas there, gigantic flightless birds that stood up to ten feet tall and weighed more than 500 pounds. They were gone long before the first *pakeha* (Maori for white man) arrived in the person of Captain Cook in 1769; because there is ample evidence for their recent existence, people have long wondered what happened to them. Early theories held that the first Polynesians (who became the indigenous Maori) killed them off slowly. Until recently, it was supposed that the extinction of the moas was a gradual process, taking place over a period of about six hundred years following the arrival of the Polynesian ancestors of modern Maoris. Now it is thought that the moas succumbed much more quickly. Reassessment of archaeological evidence suggests that the first settlement of New Zealand by Polynesian colonists took place in the late thirteenth century, not the tenth or eleventh century as previously thought. It also appears that moas had already become scarce, if not extinct, by the end of the fourteenth century.

To simulate the decline of moas, researchers inferred moa life history parameters from fossil evidence and by comparison with other long-lived

Moa

birds. They determined the size of the initial population of human colonists from an analysis of mitochondrial DNA diversity in the existing Maori population. Even with a generous estimate of the original size of the moa population, and conservative exploitation scenarios, the simulation suggests that moas were effectively extinct not long after human colonization. The most likely scenario predicts moa extinction after only fifty years. Using radiocarbon dating and a technique called the Leslie Population Matrix Model, Richard Holdaway and Chris Jacomb (2000) were able to show that instead of taking several centuries, the elimination of the moas took no more than 160 years. They wrote in a *Science* article:

> The simulations showed that the moas, like most long-lived birds, were very vulnerable to any increase in adult mortality. When subjected to even a low level of human predation, moas required a disproportionate (and impossible) increase in recruitment to maintain their numbers. . . . The elimination of the moa by Polynesians was the fastest recorded megafaunal extinction, matched only by the 'Blitzkrieg' model for North American Late Pleistocene extinctions.

In an accompanying essay (also in *Science*), Jared Diamond wrote, "anyone who has hiked over New Zealand's incredibly rugged terrain is staggered by the suggestion that a few Maoris could have quickly found and killed every single individual of those dozen moa species, with a total initial population estimated at 160,000 birds." Diamond says, "Yes, this was a blitzkrieg; yes, a few people could and did kill every moa." On the other hand, commenting in the *New York Times* (Stevens 2000) on the articles in *Science* discussing the moa extinctions, Ross MacPhee of the American Museum of Natural History "expressed skepticism that New Zealand's early hunters could have tracked down every member of every moa species, especially the smaller ones and those that might have lived in hard-to-get-at places in the islands' rugged mountains."

Also found in mountain forests, the honeyeaters (family Meliphagidae) constitute a large family—perhaps as many as 160 species—of forest and brushland birds that feed on nectar and nectar-feeding insects with their long, down-curved bills. They are particularly numerous in Australia, but they have managed to cross vast areas of open ocean to colonize islands from the Moluccas to Papua New Guinea, the Solomons, Micronesia, Fiji, and Samoa. They managed to make it to New Zealand (1,000 miles from Australia), and one small group somehow established itself in the Hawaiian Islands, thousands of miles from any other landmass. Many Australasian honeycreepers are brightly colored, but the Hawaiian immigrants were black. Like Darwin's finches, they colonized various islands, developing minor variations that eventually led to four different species, all known locally as o'o (and scientifically as *Moho*). Each species inhabited a different Hawaiian island. There was the Kauai o'o (*M. braccatus*) the Oahu o'o (*M. apicalis*), the Molokai o'o (*M. bishopi*), and the Big Island o'o (*M. nobilis*). They were all beautiful black or metallic brown birds, about the size of a slender robin, with long, slightly down-curved bills, long wings, and long, pointed tails. Varying slightly from species to species, all the Hawaiian o'o birds had tufts of crocus yellow feathers that glowed in stark contrast to their glossy black plumage. These yellow feathers proved to be the birds' undoing.

The Hawaiian Islands were uninhabited by humans until about 1,600 years ago, when Polynesian voyagers landed there, probably setting out from the Marquesas. But the islands were the home of a fully developed culture when Captain James Cook arrived in 1778 and bestowed the name "Sandwich Islands" on them. (Cook was killed there the following year.) Hawaiian royalty favored elaborate red and yellow feather capes, preferably

Hawaiian Honeycreeper (O'o)

made with the feathers of either the o'o, a little orange red bird called i'iwi (*Vestiaria coccinea*), or the crimson 'apapane (*Himatione sanguinea*). These colorful feathers were also used in covering ceremonial wands (*kalihis*), and elaborate crested helmets. On exhibition at the Bishop Museum in Honolulu is a cape worn by King Kamehameha I, which is composed of 450,000 feathers taken from an estimated 80,000 black and yellow honeycreepers (*Drepanis pacifica*), known to the Hawaiians as *mamo,* and to us as extinct. The birds were captured by an expert called a poe hahai manu, who mixed an adhesive paste made from the sap of the breadfruit tree, smeared it on tree limbs, and then picked off the trapped birds. It is said that after the yellow feathers were plucked, the birds were released to regrow their feathers, but this presupposes an ecological sensitivity not always evident in the early Hawaiians. In any event, the i'iwi and the 'apapane are still to be found at high altitudes on many of the Hawaiian islands, but with the possible exception of some stragglers on Kauai, no more o'os are to be found in Hawaii. The last mamo was collected in 1892.

The dubious title of "most endangered bird in the world" may belong to the Hawaiian honeycreeper known as po'ouli ("black face," *Melamprosops phaeosoma*). As of February 2003, there were three known specimens of this bird, and the U.S. Fish and Wildlife Service sent biologists to the forests of eastern Maui, the last refuge of this five-inch-long black and white bird. Avian diseases such as malaria and poxvirus were probably responsible for the devastating declines of perching birds in Hawaii during the 1980s. Destruction of habitat by pigs, goats, and other introduced ungulates has had devastating impacts on all native habitats in Hawaii. Feral pigs destroy understory vegetation, spread alien weeds, and create mosquito breeding areas by their rooting and wallowing in wet forests. For birds such as the po'ouli, which specialize in foraging in the understory, disturbance by pigs has been a major threat. Other introduced predators such as the black and Polynesian rat, the small Indian mongoose, and the feral house cat are all known to be predators of Hawaii's native birds. Three remaining po'oulis have been spotted in the Hanawi Natural Area Reserve and adjacent portions of Haleakala National Park and the Ko'olau Forest Reserve on Maui. They live in three distinct home ranges and do not appear to interact with one another. Scientists believe there is one male and two females; there may be a chance to save this species if a pair can be captured and induced to breed in captivity.

It has been politically convenient to assign the degradation of the Hawaiian ecosystem to the newly arrived Europeans and spare the noble Polynesians, but when Storrs Olson and Helen James (1995) looked at the fossil record for four of the six major Hawaiian Islands (Kauai, Oahu, Molokai and Maui), they found that "the destruction of the greater part of the avifauna took place well before Cook's arrival." (None of the remains they found were fossils in the traditional sense; they were simply the unmineralized old bones of birds that had become extinct within the past 1,000 years.) Evidently, the islands had been populated by various species of flightless geese, ibises, and rails; three species of long-legged owls; an eagle; a hawk; two species of crows; four petrels; and assorted honeycreepers and finches. As many as forty-four species of birds became extinct in Hawaii before they could be recorded by ornithologists. Much of the evidence was discovered in archaeological sites, indicating that the early Hawaiians had killed and eaten the birds. (Evidence of rats in these sites exonerates the Europeans from the charge of introducing these destructive creatures to the Hawaiian Islands; the rats obviously came with the voyaging Polynesians). It is likely that the early Hawaiians killed many of the defenseless, flightless geese and ibises, and plucked nesting petrels from their underground burrows, but because "they could hardly have been hunted to extinction by any means available to the Polynesians, some other mechanism may have been at work," wrote Alson and James. What was this mechanism? The Polynesians cleared the lowlands for their large, aquatic taro plantations, thus eliminating the lowland forest habitat of the various birds. Endemic island birds (especially, but not exclusively flightless ones) are prime candidates for extinction if their habitat is disrupted, or, as in the case of the Hawaiian Islands, eliminated completely.

As humans occupied and farmed more and more of the Hawaiian lowlands, they eliminated those birds that could only live there, while driving other species higher and higher into the mountains. Nowadays Brazilian cardinals hop on the sidewalks; flocks of red-billed leiothrix (Pekin robins), and orange-cheeked waxbills from Africa pop in and out of the trees. Not only exotic species were introduced; massive damage has been done by commonplace aliens. Rats, pigs, goats, dogs, cats, snakes, and mongooses have fed on birds and birds' eggs, contributing to the downfall of native Hawaiian birds. No single factor is responsible for the crash of these bird species, but habitat loss, introduced avian diseases, and the deliberate taking of certain species have combined to make the Hawaiian Islands a showplace for island extinctions.*

The Hawaiian Islands have the dubious distinction of being the site of some of the world's worst ecological devastations. Since humans arrived sixteen centuries ago, the islands' specially adapted native plants and animals have suffered mightily. Lowell Dingus and Timothy Rowe (1997) succinctly summarize the situation:

> On the Hawaiian Islands, 60 native bird species have become extinct since the arrival of the Polynesians between 1,500 and 2,000 years ago. Roughly one-third of those remaining (20 to 25 species) disappeared in the two centuries following the arrival of the Europeans, and two-thirds of the surviving species are now endangered. . . . Before discovering the great Holocene extinction of Pacific birds, biologists and biogeographers thought that today's distribution patterns for birds were natural. Now they are

---

* David Quammen's 1996 *The Song of the Dodo,* subtitled "Island Biogeography in an Age of Extinctions," is a brilliant and detailed discussion of the history and causes of island extinctions, which occur at a rate that far exceeds those on continental landmasses. I recommend this book unequivocally to those interested in learning more about the sorry history of mankind's pernicious influence on the island wildlife of the world.

beginning to reinterpret the patterns and to recognize human effects. . . . While some of the range losses for living species might be restored through conservation efforts, it is increasingly apparent that we are centuries too late to preserve any true reflection of the original Pacific avifauna.

At the northwestern extension of the 1,500-mile-long Hawaiian chain are the islands of Kure, Midway, Pearl and Hermes Reef, Lisianski, and Laysan. The islands are largely unoccupied by people, serving mostly as home to sea birds and the remaining population of Hawaiian monk seals. Although the islands might have been visited by Hawaiian voyagers, they were not officially discovered by Europeans or Americans until 1828. In 1857, Captain John Paty annexed Laysan Island to the Kingdom of Hawaii. As with all island wildlife, the endemic species of the Leeward Islands have been threatened by the sequential specters of diminution and extinction, usually engineered by human visitors. A little red bird known as the Laysan honeyeater closely resembled its bigger Hawaiian Islands counterpart (*Himatione sanguinea*), but while the Hawaiian bird was a vivid vermilion color, the Laysan Island version (*H. sanguinea freethi*), was much paler, shading toward pink. In a wrongheaded commercial venture, rabbits were introduced to the island in 1903, and they proceeded to gobble up most of the island's sparse foliage, leaving the honeyeaters nowhere to nest. In 1923 a sandstorm obliterated the last of the Laysan honeyeaters.

Rails are strong-billed, secretive brown birds that range in size from sparrow to chicken. Their ability to scuttle invisibly through cattails and marsh grasses is said to be responsible for the phrase "thin as a rail," although the rails of railroads are none too fat either. They can fly more than adequately (some species migrate great distances), but they prefer to hide in the grass from predators. On certain islands where natural predators were absent, the birds lost the power of flight, and their wings were reduced to useless appendages. This evolutionary strategy backfired terribly when a new predator arrived on the islands, one that specialized in catching flightless birds. And this predator brought with him even more predators, in the form of rats, goats, pigs, cats, and dogs. In Greenway's 1958 *Extinct and Vanishing Birds of the World,* here are the flightless rails and their status:

Auckland Island Rail (*Rallus muelleri*): Extinct
Wake Island Rail (*Rallus wakensis*): Extinct
Red-billed Rail of Tahiti (*Rallus pacificus*): Extinct
Diffenbach's Rail (*Rallus diffenbachii*): Extinct
Chatham Island Rail (*Rallus modestus*): Extinct
Jamaican Wood Rail (*Amaurolimnas concolor*): Extinct
Fiji Barred-wing Rail (*Rallina poeciloptera*): Extinct
Laysan Island Rail (*Porzulana palmeri*): Extinct
Hawaiian Rail (*Penniula sandwichensis*): Extinct
Kosrae Island Crake (*Aphanolimnas monasa*): Extinct
Iwo Jima Rail (*Poliolimnas cinereus brevipes*): Extinct
Samoan Wood Rail (*Pareudiastes pacificus*): Extinct
Tristan Island Gallinule (*Gallinula nesiotis nesiotis*): Extinct

To this list, David Day (1989) added the Macquarie Island rail (*Gallirallus philippensis macquariensis*), the Mauritian banded rail (*Aphanopteryx bonasia*), and Leguat's gelinote (*A. leguati*), all of which are also extinct.

The Wake Island rail was abundant on that central Pacific island until 1941, when Japanese troops overran an American seaplane base there. U.S. planes repeatedly bombed the island (and the rails) until the Japanese surrender in 1945. When Japanese troops were cut off from supplies, they ran down and ate every last Wake Island rail. Laysan Island, a two-mile-long raised coral reef, was home to the Laysan rail. Flightless and no larger than a sparrow, it was said to scuttle with "mouselike speed." In 1902, Walter Fisher wrote, "The rails are everywhere on Laysan in great numbers. Nearly every bunch of grass seemed to harbor a pair." When Alfred Bailey visited Laysan in 1912, "the rails were abundant, and remarkably tame . . . they were in and out of our dwelling place, capturing moths and flies." When rabbits were introduced to Laysan, they ate the foliage that provided cover for the little rails. Because the birds were fast disappearing, some were collected and transported to the rabbitless Eastern Island of the Midway Atoll, where they flourished. In 1943, a US Navy landing craft drifted ashore on Eastern, bringing an invasion of rats, the only invasive predators that can be mentioned in the same breath as human beings. But because the men were responsible for the rats, the destruction of the Laysan Island rail can be laid directly at the feet of *Homo sapiens*.

Laysan Island Rail

The final chapter of *The Mistaken Extinction: Dinosaur Evolution and the Origin of Birds,* by Dingus and Rowe (1998), is entitled "The Real Great Dinosaur Extinction." It is a lengthy discussion of human-induced bird extinctions in recent history. The chapter begins with a discussion of the dodo, the giant flightless pigeon that couldn't get out of the way of sailors who landed on its home island of Mauritius in the seventeenth century, and was gone by the eighteenth. Then we read about the moas of New Zealand, which inhabited those isolated islands until the first Maori adventurers arrived there about a thousand years ago and killed all the giant birds. As Dingus and Rowe put it, "the diet of the Maoris must have consisted predominantly of moas." Fast forward to the North Atlantic, where we encounter the great auk, a yard-high, upright bird that, like the dodo and the moas, couldn't fly, and was thus easily slaughtered by hunters, who simply walked up to the birds on the outer islands of the Hebrides, the Orkneys, St. Kilda, Newfoundland, and Iceland, killed the hapless birds, and ate them.

It is rare that the entire history of an extinct species can be found in a single work, but Errol Fuller, a young Englishman, has evidently read everything ever written about the great auk, and then published a 448-page, coffee-table-sized volume that incorporates practically everything known about *Alca impennis.* This monumental book is titled *The Great Auk.* Fuller, (I assume he wrote the flap copy—the book is self-published) describes it accurately as "an astonishingly comprehensive record of a species that is gone forever." It includes chapters on appearance, lifestyle, extermination and how that came about, a complete catalog (with illustrations) of the known stuffed specimens, and, in keeping with his description, a catalog of every known great auk egg—also with illustrations. Finally, there is a ten-page bibliography, which probably includes every mention ever made of this bird since it was first seen standing around on barren, rocky North Atlantic islands.

Crossing the Atlantic from Iceland, the tenth century Vikings were probably the first Europeans to see this great, upright bird. It looked like today's living razorbill, but where the razorbill (*Alca torda*) stands about sixteen inches tall and can fly, the great auk was more than thirty inches tall, and with its tiny wings, couldn't fly a stroke. It was this helplessness on land that made the *geirfugl* (hence "garefowl," its first English name) so vulnerable to hunting; practically everybody who landed on the North Atlantic islands that the great auk called home whacked a few on the head for provisions. The first records of people killing great auks are from 1497, when French ships arrived at the prodigious cod-fishing grounds of New-foundland. (The French called the birds *pingouins,* the name that would later be applied to the upstanding, flightless, black and white birds of the Southern Hemisphere.) In 1535, Jacques Cartier visited an "Island of Birds" (probably Funk Island) off Newfoundland, where his crews crammed auks into barrels and collected as many eggs as they could carry. It is said that an Icelander named Latra filled a boat with great auks at Gunnbjorn Rocks on the eastern coast of Greenland in 1590 (Greenway 1958). The hapless birds nested on the various islands in the western North Atlantic around Newfoundland, including the Magdalens in the Gulf of St. Lawrence, and as far south as the Gulf of Maine and Massachusetts Bay. In the eastern Atlantic, they nested at various islands off Europe, espe-cially St. Kilda (forty miles west of the Outer Hebrides of Scotland), and several tiny islands off Iceland. Wherever they were found, they were harvested.

One of the centers of nesting grounds was Funk Island, so-called be-cause of the overpowering stench ("funk") of the tons of dung that cov-ered every square foot of the half-mile-long island, some forty miles off northeast Newfoundland, about 130 miles north of St. John's. (Others have attributed the name to the smell of roasting and rotting birds.) Fuller says that the first sailors to encounter this speck of land may have found 10,000 breeding pairs there:

Great Auk

Such a convenient larder of fresh meat was bound to be plundered, especially by ships that had just made the long voyage across the Atlantic. For something like 300 years the larder held up, despite the fearful onslaught on the birds. Eaten or used for bait, boiled down for their fat or stripped of their feathers, the Great Auks of Funk Island were exploited in every way human ingenuity could devise. To facilitate removal of their feathers individuals were scalded in vats of boiling water and the fires to heat the cauldrons were fueled by fat from already butchered birds.

By 1810, Funk Island was the only rookery where any great auks still nested, and sailors returned every spring until they had killed every one.

In Icelandic, the great auk is still known as *geirfugl,* and there are several Icelandic islands known as *Geirfuglasker*. Several of these islands disappeared beneath the sea in 1830, during one of Iceland's frequent volcanic cataclysms, but off the Reykjanes Peninsula at the southwestern tip, there are still three islands known collectively as Fuglasker: Geirfugladrangur, Eldeyjardrangur, and Eldey. There may have been other great auks on some of the more remote outposts, but legend has it that the last living garefowl was killed on Eldey on June 3, 1844, when three fishermen found a breeding pair with a single egg. Jon Brandsson and Sigurdur Islefsson strangled the birds, and Ketil Ketilsson smashed the egg with his boot.

⋆   ⋆   ⋆

Although its very name now suggests extinction, the poor dodo was probably named because of its lumbering size, its awkwardness, and what was perceived as its stupidity. After the ornithological definition, the *Oxford English Dictionary* further defines "dodo" as "An old-fashioned, stupid, or inactive person or institution." There weren't that many opportunities to test the dodo's IQ, of course, because this giant, landbound pigeon was relegated to the rolls of the extinct less than 200 years after it was discovered by Europeans. In 1507, the Portuguese navigator Alfonso Albuquerque discovered the Mascarene Islands (Mauritius, Réunion, and Rodrigues) off the island of Madagascar in the Indian Ocean. When he and his men landed on Mauritius, they encountered these clumsy, turkey-sized birds foraging for fallen fruits and nuts in the underbrush. They killed them, but why—besides the fact that it was ridiculously easy—is not evident, since their flesh was deemed tough and unpalatable. Concurrent with human depredations were the violations perpetrated by introduced monkeys, pigs, rats, cats and dogs. It is impossible to determine when the last one died, but the last recorded sighting of a living dodo was made in 1662, and the best guess for its final waddle on earth is 1681.

The Mascarene Islands hold a very special place in the history of bird extinctions, and not only because the dodo once lived there. Rodrigues had its own endemic giant pigeon, known as the Rodrigues solitaire (or solitary), which was "as large as a swan but without any tail or wings" (Fuller 2001). From the few contemporaneous illustrations that survive, we see that the solitaire (*Pezophaps solitaria*) had a much more pigeonlike visage than the dodo, lacking the great hooked beak. Also unlike the dodo, the solitaires were said to taste good, a characteristic that doubtlessly led to their demise. On Réunion there is said to have been a white dodo, which some authors recognize as a valid species, while others opine that it may have been an apocryphal description, and may not have existed at all. Real or imaginary, the white dodo is long gone from Réunion, as are many species of Mascarene parrots, discussed on pp. 165–66.

Now known as *Raphus cucullatus,* the dodo was named either by the Portuguese or, more likely, by the Dutch sailors that followed them to

Dodo

Mauritius. According to Fuller (2001), one school of thought holds that the name is a clumsy onomatopoetic rendering of its call, but most now believe it is a corruption of a word that meant "stupid" or "slug-gard." Greenway (1958) wrote that "Dutch sailors called them *dod-aarse* ["lump-arse"]." He also wrote that the Portuguese term *duodo*, which means "simpleton," may never have actually been applied to the bird, be-cause "no reference to dodo birds has been found in Portuguese." The Dutch sometimes referred to it as *walck-vogel,* which means "disgusting bird," which may have been a reference to its unpalatability. Dodos may have weighed as much as fifty pounds, and they stood about three feet tall. They had a round body, a small head with a powerfully hooked bill, stout yellow legs, stubby, useless wings, and a tuft of curly feathers high on the rear end. The skin around the eyes was bare, but the body plumage was ash gray, lighter on the belly. We know what they looked like because some bones and parts have been preserved, and there are numerous contemporary illustrations.* David Quammen named his 1996 book on island biogeography *The Song of the Dodo*. He wrote, "the song of the dodo, if it had one, has been lost to human memory."

For some reason, perhaps its awkward, silly appearance, the dodo looks like something that was *supposed* to be extinct. (Greenway wrote that "the bird was extraordinarily gross.") It seemed unable to fly, unable to defend itself, unable to endure. But that is ridiculous; no animal—at least no animal more than any other—is *supposed* to be extinct; with us, they are all, as Henry Beston wrote, "fellow prisoners of the splendour and travail of the Earth." No matter how ridiculous its looks or behavior, every animal deserves the right to live out the full term of its existence. "The death of the dodo," wrote Stephen J. Gould in 1996, "really doesn't make sense in moral terms and didn't have to occur. If we own this contingency of actual events, we might even learn to prevent the recurrence of undesired results. For the Preacher of Ecclesiastes wrote: " 'I returned, and saw under the sun, that the race is not to the swift, nor the battle to the strong . . . but time and chance happeneth to them all.' " Errol Fuller concludes his discussion this way:

The dodo was one of the most fantastic birds ever to have lived. Even during the century in which it came to notice—only to become extinct—the species aroused enormous interest in Europe. Had dodos survived for a few more decades, colonies might have established themselves in European parks and gardens; they were probably hardy enough creatures. Today, dodos might be as common as peacocks in ornamental gardens the world over! Instead, all that remains are a few bones and pieces of skin, a collection of pictures of varying quality, and a series of written descriptions enormously expressive of the age in which they

---

* Recent research has challenged the traditional depiction of the dodo as being so fat and clumsy that it couldn't run without dragging its belly on the ground. Measurements of the skeleton at Oxford University, plus the hundreds of bones amassed in London's Natural History Museum and the Cambridge Zoology Museum, have been used to calculate how much weight the bird could have carried. Researchers concluded that the fat dodo of legend would have been too heavy for its skeleton to support and would have collapsed. The new reconstruction of the dodo is much thinner and looks much more like the first contemporary drawings of the bird.

were conceived yet curiously inadequate in the information they convey.

The legacy of the dodo endures, not only in *Alice in Wonderland,* but in the forests of Mauritius as well. While studying endangered birds on the island in the late 1970s, Stanley Temple noted that the few surviving examples of the tambalacoque tree (*Calvaria major*) were all more than 300 years old. The trees produced apparently fertile seeds, but they didn't germinate and there were no saplings to be seen. Because the age of the trees coincided with the time since the extinction of the dodo, Temple (1977) suggested that there was "obligate mutualism" between the dodo and the Calvaria tree:

> In response to intense exploitation of its fruits by dodos, *Calvaria* evolved an extremely thick endocarp as a protection for its seeds; seeds surrounded by thin-walled pits could withstand ingestion by dodos, but the seeds within were unable to germinate without being abraded and scarified in the gizzard of a dodo.

The dodos ate the Calvaria seeds, scarified them with their gizzard stones, and then passed them, allowing them to take root and grow. Because there were no more dodos around to help the trees germinate, the Calvaria—now known colloquially as the "dodo tree"—was also en route to extinction, a powerful example of the unexpected connections that can be drawn between living things—or living and extinct things. The trees were saved when it was discovered that the same effect could be accomplished by feeding the pits to turkeys. New seedlings have germinated, and the species appears to have been saved, though the seedlings have not yet produced seeds of their own. The dodo tree is valued on Mauritius for its timber, so the foresters now scrape the pits by hand in order to get them to sprout, rather than feeding them to turkeys. Temple's idea, wrote David Quammen, "was embraced by other biologists as a textbook example of mutualism and a cautionary case of the ramifications of species extinction."

However, in an 1978 essay he called *Nature's Odd Couples,* Gould expressed his doubts about Temple's "obligate mutualism." After pointing out that the theory was not original with Temple but had been introduced in 1941 (without elaboration) by botanists named Vaughan and Wiehe, he published a response to Temple's *Science* article, written by A.W. Owadally of the Mauritian Forest Service. Owadally said Temple was wrong about the connection between the tambalacoque tree and the dodo; in fact, they didn't inhabit the same regions of Mauritius. David Quammen reopened the arguments in *The Song of the Dodo,* claiming that as a bird ecologist, Temple really didn't understand the trees; furthermore, he wrote, there are actually hundreds of younger trees that "did manage to germinate in the post-dodo era." Quammen concluded his discussion by quoting Carl Jones, an ornithologist on Mauritius: "Of all the Mascarene ecologists, no one accepts Temple's version." In his typically comprehensive book on the dodo, Errol Fuller (2002) also dismisses Temple's idea as "a highly unlikely proposition [that] gained something of a following despite its sheer illogicality. Young tambalacoques have been sprouting throughout the twentieth century and they have certainly been doing this without help from dodos."

A bluish green bird commonly known as Newton's parakeet (*Psittacula exsul*) once lived on the island of Rodrigues, but it was hunted to extinction by the end of the nineteenth century. The Mascarene parrot (*Mascarinus mascarinus*) was larger and more colorful, but it is just as extinct. *M. mascarinus* had a lilac-colored head, a black face, and a massive, bright red bill. The rest of its plumage was purplish brown, and it had a long, dark tail. The last of the Mascarene parrots died in the collection of the King of Bavaria in 1834.

Then there was the Rodrigues parrot, *Necropsitticatus rodericanus,* whose scientific name translates as "dead parrot of Rodrigues." Few descriptions of this dead parrot have survived, but it is believed to have been as large as a cockatoo and uniformly green in color. Known only from bones and a sketch made in 1603, the "broad-billed parrot" (*Lophopsittacus mauritianus*) is one of the most enigmatic of the extinct Mascarene parrots. The bones include a huge lower mandible, and the sketch, which comes from a manuscript kept by a Dutch East India Company admiral named Wolphart Harmanszoon, shows a huge, dark bird with a crest of feathers rising above the upper mandible. Studies of the osteological material by Holyoak (1971) suggest that the short wings and great size of this parrot rendered it flightless. In Errol Fuller's *Extinct Birds,* there is a painting by Julian Hume, showing the birds eating fruits from the forest floor. J. M. Forshaw does not illustrate this species in his *Parrots of the World,* because "there are no specimens. For an illustration based on Harman-szoon's sketch, see plate 7 in *Extinct Birds* by Lord Rothschild, 1907." The broad-billed parrot, which is not believed to have lived past 1680, thus joins the selective list of big, flightless, extinct birds from Mauritius.

Probably because they are so amenable as cage birds and pets, parrots have long been the favorite objects of collectors and hobbyists. "Yet," wrote Fuller, "human interest and favour do not seem to have done them much good. In terms of extinctions, parrot species have suffered badly and many of those that still survive are seriously threatened." Many parrot species have been successfully raised in captivity, but there are several species that became extinct before anyone realized that they could—or ought to be—saved. In northeastern Australia, the bird known as the "paradise parrot" (*Psephotus pulcherrimus*) was considered only somewhat rare, but it seems to have disappeared around 1927. "Since then," wrote Flannery, "a handful of unsubstantiated sightings have been made."

Macaws are large, colorful parrots that adapt readily to captivity, and are therefore particularly popular with collectors. They come mostly in the primary colors, often in dazzling combinations. There are green macaws; blue macaws; blue and yellow macaws; and even red, blue, green, and yellow macaws. The Cuban red macaw added orange to its suit of many colors, making it one of the brightest of all the macaws. Found only in Cuba, *Ara tricolor* was not large as macaws go, but its brilliant coloration made it particularly desirable as a cage bird. In the nineteenth century, the standard method of collecting birds like these was to fell the trees in which they nested in hopes of capturing undamaged fledglings, but this method also destoyed nests and eggs, and greatly contributed to the species' downfall. Flannery wrote, "Despite the fact that its flesh was reputed to be evil smelling and bad to taste, the Cuban red macaw was hunted for meat and its nests were raided for pets." The last wild Cuban macaw was shot in the Zapata Swamp in 1864, but a few probably lived on in captivity.

Like the macaws, amazon parrots are colorful and make good pets. They are "the best known of all New World parrots . . . medium-sized to large, stocky birds with strong heavy bills and short, slightly rounded tails" (Forshaw 1973). Forshaw lists fifty-five species, most of which are thriving, in the wild and also in captivity. Many species are found on the mainland of Central and South America, but others occur only on certain Caribbean islands, which, for some, meant incipient extinction. For his 1974 book, *To the Brink of Extinction,* E. R. Ricciuti went to its home island looking for the Puerto Rican parrot, *Amazona vittata.* Deep in the Luquillo National Forest he saw a small flock of the green birds with red foreheads and blue primary wing feathers. He wrote, "I was lucky even to glimpse the foot-long parrots, however, for the Puerto Rican parrot is a dying species. Soon it may vanish from the Earth as surely as the five parrots disappeared from my sight on that gloomy day in April 1968." *A. vittata* nests in hollow trees, making it susceptible to the ubiquitous rats that probably disembarked with the first Spanish explorers. An aggressive bird known as the pearly-eyed thrasher (*Margarops fuscata*) competes for nest sites with the Puerto Rican parrot, and may even destroy their chicks and eggs. Throughout the forest, the palo colorado trees in which the parrots nest are being cut for charcoal.

From 1968 to 1971, Dr. Cameron Kepler lived in the Luquillo Forest expressly to monitor the population of Puerto Rican parrots. At the conclusion of his study, he estimated that there were no more than fifteen of the birds left. As Kepler tried to capture some of them, The U.S. Department of the Interior obtained two birds that had been held in the Mayaguez Zoo for eighteen years, and brought them to the Patuxent Wildlife Research Center in Laurel, Maryland. Five more were held in an aviary in Puerto Rico, all in the hope that they could be captive-bred to forestall the looming inevitability of extinction. At the time of Columbus, this parrot's total population may have exceeded 100,000 individuals. In the 1950s, the population was estimated at 200 birds, and in 1975, it reached an all-time low of thirteen. By August 1989, the population count of the wild flock found a minimum of forty-seven birds. Hurricane Hugo hit eastern Puerto Rico on September 18, 1989, severely impacting the Luquillo Forest. Currently, there are about twenty-four to twenty-six parrots in the wild, including four breeding pairs, and fifty-six in captivity at the Luquillo Aviary. "Like the Puerto Rican parrot in years past," wrote Ricciuti, "the other Amazon parrots of the Caribbean are threatened by growing human populations and the destruction of their habitats."

The massive, ongoing devastation of the Amazon rain forest has had—and will continue to have—lasting effects on all life in that immense region, and, by extension, on all life everywhere. The Amazon basin, which is larger than all of Europe, supports millions of life forms, from jaguars and tapirs through thousands of endemic birds, to myriad insect species, most of which have not even been discovered. The people who live in the rain forest are also impacted by the ruination of their habitat, and we haven't even mentioned the destruction of the plant life itself. According to a recent study, the rain forest is being destroyed at a rate of 6,000 square miles per year—an area roughly equivalent to the state of Connecticut. Most of the destruction comes from farmers who need to clear the land in order to plant soybeans, most of which are sold to China. The most economical way to clear an area of jungle is to burn it. Burning the Brazilian jungle accounts for some 400 million tons of greenhouse gases entering the atmosphere every year. Land that is not burned outright is logged, and the lumber dragged on makeshift muddy roads, contributing even more to the degradation of Amazonia. The vast extent of this jungle has protected it and its wildlife up to now, but human activities affect every living thing that occupies its own special niche in the complex web that makes up the rain forest. How can an isolated, endemic parrot species stand up to the power of Chinese soybean markets or Japanese logging interests?

In the eighteenth century, the island of Martinique in the West Indies was heavily cleared for agriculture. One of the early casualties was the Martinique amazon, *Amazona martinica,* which was not seen again after a certain Monsieur Labat recorded its existence in 1722. A very similar form—perhaps the same species—inhabited the nearby island of Guadeloupe, and *Amazona vittata* was found only on Puerto Rico and its small neighboring island, Culebra. All these big green amazons are extinct or close to it. Other amazons, such as the St. Lucia amazon (*Amazona versicolor*), the red-necked amazon (*A. arausiaca*) from the island of Dominica, and the spectacular St. Vincent amazon (*A. guildingii*) were close to extinction until dedicated conservationists realized how precarious their status was, and initiated measures on each island to save them. In all cases, the campaign was led by an American named Paul Butler, who made the islanders aware of their invaluable natural heritage. As Tony Juniper (2002) wrote, "The parrots of St. Vincent and Dominica bottomed out after hundreds of years of decline and began to recover. Today their numbers continue to rise as they head out of immediate danger. For these little island nations, the great gorgeous parrots of the rain forests are among their most precious national treasures."

★   ★   ★

Discovered by Captain Cook in 1774 and named for the Duke of Norfolk, Norfolk Island is a thousand miles east of Sydney. It was settled as a penal colony in 1788, immediately after the First Fleet arrived in Australia with its cargo of convicts. When the first prisoners set foot on the thirty-four-square-mile island, they found among the local inhabitants long-billed, gray and gold parrots known as kakas (*Nestor productus*). The parrots were as guileless as the prisoners were hungry, and in a short while, the parrot population was drastically reduced. Flannery (2001) quotes from the journal of an American sailor named Jacob Nagle who stayed on the island from March 1790 to February 1791:

> There was but few land birds on the island eccepting quail, a few parrots, parokets that fed on wild red peper, and some wild pigeons of the same colour as our tame pigeons, but we reduced them a great deal before we left the island.

Those that were not killed for the pot lost their habitat when it was leveled for cultivation and lumber. (The indigenous Norfolk Island pine was used extensively for shipbuilding, especially for masts, since its straight trunk can grow to 200 feet high.) The original Norfolk Island prisons were abandoned in 1814, but reopened in 1826 to function as the harshest penal settlement in Australia, described as "punishment short of death." In 1855, the prison was closed once more and the last of the convicts removed to Tasmania. It is hard to imagine anyone or anything suffering worse than the Norfolk Island convicts, who were said to prefer hanging to the conditions of their incarceration; but the kakas, who had committed no crimes whatsoever, were not so lucky. In 1851, the last of them died in a cage in London.

We usually associate parrots with tropical forests (or cages), but once upon a time, there was a brightly colored parrot that lived in the United States. It was known to scientists as *Conuropsis carolinensis,* and to everybody else as the Carolina parakeet. It was a bright green bird with a yellow head and an orange forehead, about twice as big as a budgie. It was found throughout the Mississippi-Missouri drainage area, along the Gulf coast of Florida and Alabama, in the Carolinas, and as far to the northeast as New York.* When common, these birds appeared in flocks of two or three hundred. They were omnivorous, eating fruits, seeds, nuts, blossoms, leaf buds, and, as Greenway notes, "they were said to be notoriously fond of the fruits and seeds of cultivated plants as well, and were capable of destroying orchards and were said to have done so in a 'wanton and mischievous manner.' " They liked to pick all the apples off a tree and drop them to the ground, but never ate the fruit. Conflict with farmers is a fast ticket to oblivion, and by the end of the nineteenth century most of the hapless parrots had been shot. That they sought refuge in numbers made them that much easier to shoot.

It cannot be said that hunters and farmers killed all the Carolina parakeets, however. The elimination of a species is usually more complicated than that. Many disparate elements must be factored in, and in the case of *Conuropsis carolinensis,* habitat degradation also played a major role. Daniel McKinley (1980) identified honeybees as contributing agents of the parakeet's demise. The bees, which are not native to North America, were brought over by early settlers. By force of numbers they commandeered the hollow trees that the birds needed. "As a primary competitor for nesting and roosting sites," McKinley wrote, "the honeybee barnstormed across forested eastern North America in colonial and federal times, easily outdistancing the light-footed pioneers." The parakeets last stronghold was in the swamps of the Southeast, the same Florida and Louisiana habitats that harbored the last of the ivory-billed woodpeckers. The last Carolina parakeet died in the Cincinnati Zoo in February 21, 1918. If the same keeper at the zoo's bird house had been on duty on both September 1, 1914, and February 21, 1918, he would have been present at the sorrowful moments when two important American birds became extinct.

---

* Many species of parrots have established small breeding colonies in areas where they were not originally native, as a result of escaped or released cage birds. According to Sibley (2000), "There are no longer any native parrots in North America, but the imported species illustrated here are seen regularly. Some have established feral populations, especially in southern cities where exotic plantings create suitable habitat . . . more than 65 species have been recorded in Florida alone." The green-bodied, gray-faced monk parakeet (*Myiopsitta monachus*), originally from the scrublands of southeastern South America, has established colonies in Puerto Rico, Connecticut, Florida, Texas, Illinois, Oregon, and even New York, where there is a boisterous nesting colony in Central Park.

All the news about parrots is not bad. Along the eastern Amazon River system, scientists have collected several small green parrots that they have provisionally identified as immature vulturine parrots (*Pionopsitta vulturina*), whose name refers to their featherless heads. While the head of the standard vulturine parrot is black, the heads of the new ones are bald, with orange skin. At first they were thought to be immatures, but the orange-headed parrots turned out to be sexually mature. Furthermore, they did not mingle with the black-headed ones. In *The Auk* (August 2002) Renato Gaban-Lima and Marcos Raposo, graduate students at the University of São Paulo, Brazil, published their description of a new species, *P. aurantiocephala*. But parrot experts are worried that the newly discovered species might already be in trouble, as the home forests of the middle Tapajós and lower Madeira rivers are fast falling to loggers and ranchers. Evolution giveth and extinction taketh away.

With the extinction rate rocketing along, it is always gratifying to read of a saved, rescued, or rediscovered species. The *New York Times* of September 3, 2002, ran a color photograph of a parrot with the headline, "Parrot Turns Up After 91 Years." Far from the rain forest, high on the slopes of a volcano in the Colombian Andes, researchers found a long-lost parrot. Last seen in 1911 (on the same mountain) by American Museum of Natural History collectors, the Fuertes or indigo-winged parrot (*Hapalopsittica fuertesi*) was never photographed—until July 28, 2002. On that date, researchers Jorge Velasquez and Alonso Quevedo were astonished when a flock of fourteen of these foot-long birds spiraled down from the trees to alight near them. They made notes and took photographs and video footage to document their remarkable find. Known locally (but, evidently, not too well) as *loro multicolor* ("multicolored parrot"), the Fuertes parrot is green, with indigo wing feathers, red shoulders, a blue crown, and a wash of chestnut at the base of the pale ivory bill. Because the *loro multicolor* had not been seen for so long, it was thought to be extinct. A rule of thumb, not necessarily valid, is to consider a species extinct if it has not been seen for fifty years, but such a rule does not take into account inaccessible Colombian cloud forests.

The Mexican island of Guadalupe is in the North Pacific, 140 miles off Baja California, and 180 miles southwest of San Diego. The island is about twenty miles long by six miles wide, with a 5,000-foot-high volcanic ridge running down its entire length. Nineteenth-century California whalers in-troduced goats and feral cats to the island; the goats destroyed almost all the native vegetation, while the cats completely eliminated the Guadalupe storm petrel (*Oceanodroma macrodactyla*), a bird that nested and bred nowhere else.

Also endemic to Guadalupe was a large bird of prey called the quelili or calalie (*Polyborus lutosus*), which was believed by the goat herders that had settled the island to be taking the young goats. The calalie was a species of caracara, a hawklike bird that looks more formidable than it is, and is actually a carrion-eater. The introduction of goats was probably beneficial to the carrion-eating calalies, and in 1876 the birds were "abun-dant on every part of the island" (Grossman and Hamlet 1964). By 1906, however, the birds had been so efficiently eradicated that not one re-mained alive. Other avian and nonavian species have certainly been elimi-nated by man, but the calalie is the only bird whose extirpation was *premeditated*. The northern elephant seal (*Mirounga angustirostris*) was also thought to have been wiped out, killed for its oil, but a few remaining an-imals were found on Guadalupe Island in 1890. From this small herd, the elephant seal has recolonized its old breeding grounds on various islands and even on the California coast. Guadalupe Island is now unoccupied, except by the elephant seals and approximately 50,000 goats.

The Labrador duck (*Camptorhnychus labradorius*) is the only species of wa-terfowl to have become extinct in North America. The drake's head and neck were white with a black stripe on the crown and a narrow black collar.

Labrador Duck

The back, tail, and underparts were black, and the wings were white. Because of its striking markings, it was sometimes known as the "skunk duck" or "pie duck." Audubon's painting shows a black and white male and a brownish female, of which he wrote, "The female has not, I believe, been hitherto figured." It had a peculiar bill with numerous lamellae (filterplates), and it may have become extinct because it was somehow deprived of its food source—whatever that was. Audubon wrote, "at times it comes ashore and searches in the manner of the spoon-bill duck. Its usual fare consists of small shellfish, fry, and various kinds of seaweeds, along with which it swallows much sand and gravel." The Labrador duck inhabited eastern Canada and the United States, and may not have lived in Labrador at all. Although it was hunted, its flesh tasted strongly of fish, and it was never a popular food item. It does not seem to have ever been plentiful, and was recorded as rare by 1844. Although the actual cause is not known, it seems likely that hunting of the remaining population was partially responsible for its demise. The last one was shot on Long Island on December 12, 1878.

Once upon a time in recent history, the North American passenger pigeon (*Ectopistes migratorius*), was the most numerous bird on earth. In the eastern United States it numbered in the billions, outnumbering all other species of North American birds combined. Early American naturalist Alexander Wilson witnessed a flock between Kentucky and Indiana in 1800 that he estimated at a mile in width and 240 miles long. Wilson estimated that this flock contained 2.2 billion birds—and then said that was an underestimate. Larger and more graceful that the common street pigeon, *Ectopistes* was a soft gray on its head and back, and rufous pink on the breast, fading to white on the underparts. The bill was black, the feet red, and the eyes orange. Their flocks darkened the skies and extended for miles. Their nesting sites covered hundreds of square miles of forest, with dozens of birds in each tree. As the forests were cleared and converted to farmland, the pigeons' habitat began to disappear, but by far the major contributor to the decline was man. Many passenger pigeons were shot for the pot, but untold millions were shot for "sport." In one competition a participant had to kill 30,000 birds just to be considered for a prize.

In the sandy scrub-oak barrens of south central Wisconsin in April 1871,

the largest nesting assemblage of passenger pigeons ever recorded occurred: an estimated 136 million birds blanketed an area of 750 square miles.
By that time, the newly erected telegraph lines allowed hunters to commu-
nicate with each other, so everyone with a gun, stick, or bow and arrow
converged on Wisconsin to take advantage of the harvest. Thousands of
hunters slaughtered millions of birds, which were sold and shipped out by
railroad cars at a price of fifteen to twenty-five cents a dozen (Matthiessen
1959). Seven years later, at Petoskey, Michigan, the last great congregation of
passenger pigeons assembled, covering more than 200 square miles. The
population plummeted rapidly. The last legitimate sighting of a wild passenger pigeon was in 1900 in Ohio. This bird was shot; its remains are in the
Ohio State Museum. A few individuals lingered on in captivity into the early
years of the twentieth century. By 1909, the Cincinnati Zoological Gardens
had the three remaining birds, two males and a female. By 1910 only
"Martha" was left, named after the wife of George Washington. Martha
died in a cage at the Cincinnati Zoo at 5:00 P.M. on September 1, 1914.

Much changed in America from the time Audubon painted the passenger pigeon's portrait around 1824 until the last one died ninety years
later. Covered wagons lumbered westward, and the railroads steamed and
chugged back and forth across the continent. Gold was discovered in California, petroleum in Pennsylvania. Wars were fought, Indians displaced,
and the Alamo fell. *Moby-Dick, The Scarlet Letter, Uncle Tom's Cabin*, and
Huckleberry Finn were written. Abraham Lincoln wrote the Emancipation
Proclamation and was shot in Ford's Theater. The Confederacy seceded,
and lost the Civil War. General Custer and 264 of his men died at the Little Big Horn. America bought Alaska from Russia for $7.2 million. The
Great Fire destroyed Chicago and the Great Earthquake destroyed San
Francisco. Thomas Edison invented the kinescope and Henry Ford introduced the Model T. All of these events and inventions affected America in
one way or another, for better or for worse, but it might be argued that
nothing has had as much of an effect as the disappearance of the passenger pigeon. Flocks hundreds of miles long containing billions of birds
once darkened the skies; they roared like thunder as they passed overhead. Where once was heard the cacophonous percussion of billions of
wings, we are left with an echoing, hollow silence.

★   ★   ★

Passenger Pigeon

The prairie chicken (*Tympanuchus cupido*) was once found in great numbers on the grassy plains of the American West from the Mississippi to the Rockies. Not a chicken but a member of the grouse family (Tetraonidae), the bird is a large, hefty creature, heavily barred in brown, cinnamon and buff. During the breeding season, the males gather on open ground, stamp their feet, puff up the orange sacs on each side of their necks, and raise hornlike plumes behind the head, all designed to attract the attention of the females. The area chosen by the males for these demonstrations (and the demonstrations themselves) is known as a "lek." The males resonate their neck sacs to make a deep, hollow booming noise that has been likened to the sound you can make when you blow across the top of an empty bottle. Christopher Cokinos (2000) describes the sound made by prairie chickens in Kansas:

> I heard a slow, low *o-o-o* repeating and repeating, textured with pauses and, I thought, slight inflections. Naturalists once called the booming of the males "tooting," which connotes perhaps too high a pitch. Others have translated the call, somewhat loosely, to "old muldoon," which could be the name of a bar. Neither "tooting" nor "old muldoon" captures the hypnotic moaning quality of several birds booming and droning at once. . . . One writer said of droning heath hens, "It sounds as if a lot of lonesome little night winds had taken to crying whoo-oo-o in ragtime, mingled with whistles of syncopated measure."

Farmers and settlers destroyed their nesting places and breeding territories, and hunters shot the plump birds for the pot, so by the beginning of the twentieth century, the droning prairie chicken was an endangered species.

Another subspecies of prairie chicken is Attwater's, once counted in the millions on the coastal grasslands that extend behind the beaches of the Gulf of Mexico. Their range stretched from Corpus Christi, Texas, north to the Bayou Teche area in Louisiana and inland for about seventy-five miles. Acre by acre, coastal prairies diminished as cities and towns sprouted up, industries grew and expanded, and farmers plowed up native grasslands for crops or pasture. Like other grouselike birds (ptarmigan, ruffed grouse), prairie chickens were considered game birds, and those that were not crowded off their breeding and nesting grounds were shot by hunters. Smaller and darker than the prairie chicken and the heath hen, *Tympanuchus cupido attwateri* is a victim—like so many birds—of human encroachment. In 1937, when the first serious study of Attwater's was launched, the bird was extinct in Louisiana and reduced to about 8,700 individuals in Texas. Thirty years later, when it was listed as endangered, there were fewer than 1,100 birds left. By 1992, the population had sunk to 456 birds scattered across five Texas counties—a decline of 95 percent since 1937. In collaboration with the Nature Conservancy, the World Wildlife Fund financed the purchase of 3,450 acres near Eagle Lake in

Colorado County, Texas, as a preserve for this endangered bird. In the nick of time, Attwater's prairie chicken had been given at least one place to call home. The land was transferred to the U.S. Fish and Wildlife Service in 1972; today the Attwater Prairie Chicken National Wildlife Refuge incorporates 10,000 acres, more than twice its original allocation.

But even with government protection, things are not going well for Attwater's. Captive breeding programs have had little success. In the refuge itself, the beleaguered birds are unprotected from snakes and birds of prey, while fire ants attack new hatchlings. In a 2002 *National Geographic* article, prophetically entitled, "Down to a Handful," Douglas Chadwick wrote,

> It isn't that the public doesn't care. The problem is that the public hardly knows that Attwater's prairie-chickens exist. Funding for field research and captive breeding therefore stays scarce. Plans to reconnect fragments of coastal grassland habitat gather little momentum. A unique, irreplaceable creation and its elaborate behavior, special adaptations, particular store of genes, and the daybreak beauty of its fashions are fading out largely for want of attention.

East of the Appalachians, the prairie chicken was known as the heath hen, and awarded the subspecies appellation of *T. cupido cupido*. Prior to the American Revolution, these birds were found in the eastern United States from Maine to Virginia. They were shot for food throughout their eastern range, and by the middle of the nineteenth century they were gone from New England. All, that is, except the population on the small offshore Massachusetts island of Martha's Vineyard. In 1907 there were fifty heath hens left on the island; the following year, a 1,600-acre sanctuary was established for their protection. The sanctuary seemed to be successful, the original fifty birds reproduced rapidly, and by 1915 there were 800 heath hens on the island that locals call "the Vineyard." Because the entire population was confined to a single island, the species was terribly vulnerable; in 1916 an islandwide fire wiped out many of the birds, their eggs, and much of the habitat they used for breeding. The following winter was unusually harsh, and an influx of goshawks, predatory birds that fed on the heath hens, hurt the population even more. Finally, many of the remaining heath hens fell victim to a poultry disease brought to the island by domestic turkeys. There were only thirteen heath hens left by 1927 and most of them were males. Of the last of these birds, Arthur Gross, the author of the definitive study of heath hens, wrote, "It is truly remarkable that this lone bird, subject to all the vicissitudes of the weather, to disease, and to natural enemies, has been able to live in solitude for such a long time." In April 1931, Gross trapped and banded the lone bird and then released him. "Booming Ben" was spotted once, a year later, and was never seen again. The heath hen of Martha's Vineyard passed into the eternity of extinction some time in the spring of 1932.

In a time that is (sometimes) more ecologically aware than, say, the 1920s, efforts are currently underway to bring prairie chickens to Martha's Vineyard. In *Hope is the Thing with Feathers*, Christopher Cokinos discusses the plans of islanders Tom Chase and John Toepfer to import the chickens to the Vineyard and "establish a viable population of the birds, as a high-visibility consequence of the restoration of fire and scrub oak." It might require a massive public relations campaign to convince the islanders that the return of heath hens is in their best interest, and it might take a couple of thousand years before the prairie chicken became a heath hen, but if the birds adapted to Martha's Vineyard once, they could probably do it again.

Americans seemed to have a particular appetite for common shore birds, and in the nineteenth century, as the passenger pigeons were declining, the hunters trained their shotguns on the golden plover and the Eskimo curlew. Both species were plentiful; at one time their numbers contributed to the phenomenon of darkening skies over the American prairies. Somehow, the golden plover survived the carnage, but the Eskimo curlew was not so fortunate. Also known as "prairie pigeon" or "doughbird," *Numenius borealis* was a small, brownish bird, only twelve or thirteen inches in length, that resembled its somewhat larger relative the Hudsonian godwit. Twice every year, the curlews made prodigious, elliptical migrations. During the late summer, they would leave their breeding grounds in northwestern Canada and cross the entire North American continent to Labrador and Newfoundland, and then head south over the open Atlantic to the pampas of Argentina. In February of the following year, they began the return journey, but took a completely different route, crossing South and Central America until they progressed northward across Texas and Louisiana and through the Mississippi Valley to their northwest Canada breeding grounds. They were shot at everywhere. Like passenger pigeons on the American Great Plains, they were heaped onto wagons and brought to market. Heedless hunters shot as many as they could, exponentially reducing their numbers. It is possible that the curlews, like the passenger pigeons, required large flocks to stimulate breeding; once the huge numbers were gone, the species was doomed.

The Eskimo curlew is the bird of hope. Even though the last confirmed curlew was shot in Nebraska in 1915, and another may have been shot in Argentina a decade later, hopeful reports of sightings are occasionally heard. Out-of-focus photographs have been used to try to disprove that the bird is extinct; for example, a picture taken in Texas in 1962 appears to show an Eskimo curlew at a distance. In the collection of the Academy of Natural Sciences of Philadelphia there is the skin of a supposed Eskimo curlew that was shot in Barbados in 1963. After looking at the skin, Scott Weidensaul (1999) wrote, "Was the curlew, like the pigeon and the parakeet, a thing of darkening memory? Or were there still a few out there, somewhere, making the great arc from tundra to pampas each year?"

In 1954, Fred Bodsworth wrote a sad and beautiful little book called *The Last of the Curlews,* in which a single male bird, unaware of his eponymous status, instinctively replicates the migration of his species, beginning

Eskimo Curlew

in the Arctic, crossing the Gulf of St. Lawrence, and then passing over the Caribbean. The lone survivor flies dutifully over the Amazon jungles and lands in Patagonia, where he encounters the last remaining *female* Eskimo curlew, who has just completed the same migration:

> He had never seen a member of his own species before. Probably the female had not either. Both had searched two continents without consciously knowing what to look for. Yet when chance at last threw them together, the instinct of generations past when the Eskimo curlew was one of America's most abundant birds made the recognition sure and immediate.

The pair begins the northward leg of the migration, flying over the Andes and crossing the Yucatán before crossing the Gulf of Mexico and beginning the final leg of the journey that will take them up the Mississippi Flyway, thence to the breeding grounds in the north. As they cross the Canadian prairie, the female is shot and killed by a farmer. The male flies on:

> The snow water ponds and the cobblestone bar and the dwarfed willows that stood beside the S-twist of the tundra river were unchanged. The curlew was tired from the long flight. But when a golden plover flew close to the territory's boundary he darted madly to the attack. The Arctic summer would be short. The territory must be held in readiness for the female his instinct told him soon would come.

Extinction is not always so precipitous; we are usually aware that a species is in decline, and continue to hope that rumors of its existence in some inaccessible backwater will prove to be true. Before the Caribbean monk seal was "officially" declared extinct in 1956, biologists canvassed the entire region, checking out every rumor, regardless of how baseless. But America's most pointless extinction was the great ivory-billed woodpecker (*Campephilus principalis*), which was not hunted at all except for museum collections—what else would anyone want with a giant woodpecker? It lost its habitat, and thus its food source, when logging interests cut down the old growth swamp forests of the lower Mississippi basin, the Gulf Coast, and the southern Atlantic seaboard. Instead of a sad field guide description of an extinct bird, read this celebration of the ivorybill in Scott Weidensaul's *The Ghost with Trembling Wings* (2002):

> The ivory-billed woodpecker was nature's exclamation point, the personification of pizzazz—a full-throated yell of a bird with a scarlet crest, black-and-white wings like semaphore flags, and a walloping honker of a bill the color of old bone. It was big—as large as a duck or a crow, with a 3-foot wingspan—and noisy too, like someone blowing into a megaphone with the mouthpiece of a clarinet, blasting out single nasal notes that could be heard a half mile away. The ivorybill was a bird with *impact*.

Ivorybills inhabited the swamplands of the American Southeast, in isolated tracts in South Carolina, Louisiana and Florida. As the trees that harbored its favorite food—the larvae of the long-horned beetle—were chopped down, the great woodpeckers had nothing to eat, no place to nest, no place to live. In April 1935, ornithologists from Cornell University and the American Museum of Natural History headed for the Singer Tract, land owned by the Singer Sewing Machine Company in Louisiana, where they saw, recorded, and photographed ivorybills nesting in a tree. As Arthur Allen (1937) wrote,

Ivory-billed Woodpecker

The brilliant scarlet crest of the male, the gleaming yellow eye, the enormous ivory-white bill, the glossy black plumage with the snowy-white lines from the head meeting the glistening white of the wings, are as vividly pictured in my mind as if I were still sitting on that narrow board in that tree-top. . . .

James Tanner, the author of the definitive book on ivorybills in 1942, estimated that there were two dozen ivorybills alive in North America in 1939, six of which were on the 80,000-acre Singer Tract and the rest in Big Cypress Swamp and the Apalachicola River bottoms in Florida. He saw his last ivorybill in the Singer Tract in December 1941.

In recent years, there have been rumored sightings of ivorybills—the largest and loudest of the American woodpeckers and not easily mistaken for other birds—but none have been confirmed. Scott Weidensaul, whose book chronicles the search for lost species, wrote:

By the early 1940s, the situation was critical, and despite the overriding war effort, to which virtually everything else in America was considered to be of secondary importance, four federal agencies, the Audubon Society and the governors of four southern states, including Louisiana, appealed to Chicago Mill to sell the timber rights to some of the remaining forest as ivorybill sanctuary. Then timber barons declined the proffered $200,000, however ("We are just money grubbers," Chicago Mill's chairman cheerfully informed the coalition), and the rest of the Singer Tract was chopped down. The last time anyone saw an ivorybill on the Tensas River was April 1944.

A Cuban version of the ivorybill, known as *Campephilus principalis bairdii*, was last seen in the 1950s. There was an even larger model, the Mexican imperial woodpecker, whose high-altitude forest habitat was also demolished, and the birds with it.

Although hope lingers on, the imperial woodpecker of Mexico is believed to be extinct. Last spotted in 1956, the world's largest woodpecker now joins the ivorybill on the official oblivion list. A 2003 expedition to the Sierra Madre Occidental region of Durango, where the last *C. imperialis* was seen, found no signs of its existence. Now that most of its high-forest trees are gone, this noisy black and white bird with a three-foot wingspan would not be that hard to spot—if it was available for spotting.

The charismatic ivorybill, great auk, moas, Labrador duck, passenger pigeon, and Carolina parakeet were not the only recent avian victims. In *Vanished Species* (1989), David Day lists another 164 bird species that have become extinct in the past 400 years. Many of them are familiar to us, but others, like the Iwo Jima rail and the "mysterious starling," are known only to ornithologists or students of extinction. Some of these birds lived in dense forests or on inaccessible islands, but we managed to find them and eliminate them. (Small populations that live on islands are the most susceptible to extinction.)

Death (and Extinction)
by Disease

The golden toad (*Bufo periglenes*, "brilliant toad") was first described in 1963 by University of Miami biologist Jay Savage, working in the Monteverde cloud forest in northwestern Costa Rica. The males are bright orange, while the females are olive to black with large scarlet spots. Use of the present tense—i.e., males *are* bright orange—is probably incorrect: no golden toad has been seen since 1989, and they are believed to be extinct. It is the first animal in recent times known to have been driven to extinction by a lethal disease.

For amphibian populations already stressed by pollution, habitat loss, or other factors, the introduction of a new disease can be deadly. Scientists suggest that the mass disappearance of frogs in the Australian rain forest may be due to an epidemic of an exotic disease, caused by the fungus *Chytridiomycosis*. It is not clear what caused the outbreak of the fungus, but it might be related to climate change, particularly the phenomenon known as global warming. As Pounds *et al.* wrote in 1999, "Recent warming has caused changes in species distribution and abundance, but the extent of the effects is unclear.... Twenty species of anurans [frogs and toads] ... including the locally endemic golden toad, disappeared following synchronous population crashes in 1987. Our results [in the highland forests of Monteverde] indicate that these crashes probably belong to a constellation of demographic changes that have altered communities of birds, reptiles and amphibians in the area and are linked to recent warming." Amphibian die-offs around the globe are of great concern because amphibians are good barometers of significant environmental changes that may otherwise go undetected by humans at first.

Often compared to canaries in coal mines, frogs are a definite indicator of environmental degradation. Over the last fifty years, frog populations throughout the world have decreased drastically. Amphibian malformations have been reported in forty-four U.S. states since 1996. These deformities include extra legs, extra eyes, misshapen or incompletely formed limbs, limb formation in anatomically inappropriate places, missing limbs, and missing eyes. While anatomical deformities are a natural phenomenon within virtually all species, the widespread occurrence of these deformities, and the relatively high frequency of malformations within local populations (as high as 60 percent in some areas), is of great concern. The rate of deformities far exceeds what might be considered "natural," and scientists are currently investigating several factors, including parasites, contaminants, and ultraviolet (UV) radiation related to the thinning of the ozone layer. Because most frogs lay their eggs which are covered only by a thin gelatinous layer in shallow water, they are particularly prone to the dangers of UV radiation; frog eggs in containers with no UV protection failed to hatch, while those in which the UV was blocked had a much higher survival rate. Also, reports of frog loss seem to come more often from higher altitudes and latitudes than from lower altitudes and the tropics, apparently because these areas are closer to the sun and UV rays.

Populations of frogs, toads, and salamanders are declining or disappearing the world over. In the late 1980s the scientific community first took notice of this alarming trend, which continues today. A 1997 article in the *New York Times*, entitled "Protozoan Attacks Frogs," told of how biologist Karen Lips investigated the death of hundreds of frogs in the Panamanian rain forest. She said that a lethal protozoan was sweeping across Central America in a "death wave" that rapidly moved through the mountains, from one range to the next. First noticed in the Monteverde cloud forest in Costa Rica, the protozoan has since moved south and east into Panama and may have traveled north into Nicaragua. In July, a tourist found dead and dying frogs on an island in the middle of Lake Nicaragua, but none of the animals were preserved for autopsy.

In a finding that only serves to deepen the mystery, Australian biologists last year reported that they believe a similar "death wave" killed amphibians in the lush mountain forests of Queensland in the 1980s and 1990s. In the Australian rain forest, at least fourteen species of stream-dwelling frogs have disappeared or declined sharply (by more than 90 percent) during the past fifteen years. These precipitous declines are widespread across a large area of rain forest, suggesting that habitat destruction by itself is not a likely cause. Indeed, the golden toad was used as a public relations symbol in the United States to raise money to purchase and protect its habitat, creating the Monteverde Cloud Forest Preserve in 1972. Although this action seemed to secure the toad's future, it is now apparent that setting aside habitat is not enough to save most endangered species.

In Arizona, the Chiricahua leopard frog (*Rana chiricahuaensis*) has disappeared from 80 percent of its former habitat. New research is linking a chytrid fungus—which has been implicated as a probable cause of major amphibian die-offs in pristine areas around the globe—as a factor in the declines of some of Arizona's frog populations, according to recent studies by researchers with the U.S. Geological Survey, the U.S. Fish and Wildlife Service, the Arizona Game and Fish Department, and the University of Arizona. The chytrid fungus (*Batrachochytrium dendrobatidis*) in amphibians was first identified in 1998 by an international team of scientists from Australia, the United States, and Great Britain. They discovered that the fungus had most likely been responsible for large, previously unexplained amphibian die-offs in pristine areas of Panama and Australia. The disease is now being studied in detail, in order to understand its origin, incidence, distribution, and methods of control. Recent deaths of endangered boreal toads (*Bufo boreas*) in Rocky Mountain National Park have been linked to a chytrid fungus and this is the second such diagnosis in Colorado, according to scientists with the U.S. Geological Survey, who have been monitoring two populations of boreal toads in Rocky Mountain National Park for the past nine years. One of these two park populations of toads declined precipitously in 1996, and the second showed dramatic declines in the summer of 2000. In an article published in *New Scientist* in April 2002, Stephanie Pain wrote that *Batrachochytrium dendrobatidis* is on the way to killing *all* of New Zealand's native frog species. The fungal disease was first found in

New Zealand's frogs in 1999, and in some areas, more than 80 percent of the frogs are gone already. She quotes herpetologist Ben Waldman: "Once one population is gone, extinction is just over the horizon."

A study published in 2001 by Kiesecker, Blaustein, and Belden proposed that climate change—for which read: global warming—is a key factor in the decline of amphibian populations, particularly *Bufo boreas*. Warmer weather means more evaporation, leading to reduced water depth, which in turn means that embryos are exposed to greater UV-B (middle-wave ultraviolet) radiation, opening the door to a lethal infection of the fungus *Saprolegnia ferax*. Kiesecker and his colleagues found that it was not ozone depletion that increased the exposure to UV-B, but rather general warming conditions, most likely caused by a thirty-year rise in the surface temperature of the Pacific. This oceanic warming, a function of the El Niño/Southern Oscillation phenomenon, has affected biological communities from coral reefs to cloud forests, and can also be correlated with the mortality of golden toads in Costa Rica. Kiesecker *et al.* point out that extreme climate events can be connected with disease outbreaks, and J. Alan Pounds, in his synopsis of the Kiesecker article, concludes, "Today there is little doubt that both phenomena—amphibian declines and global warming—are real. If there is indeed a link between the two . . . there is clearly a need for a rapid transition to cleaner energy sources if we are to avoid a staggering loss of biodiversity."

As if to counterbalance the worldwide disappearance of frogs, researchers in Sri Lanka have found more than a hundred new species in a 750-square-kilometer (290 square mile) patch of rain forest on an island that has already lost 95 percent of its rain forest habitat (Meegaskumbura *et al.* 2002). The Sri Lankan frogs included examples of two groups of tree frogs, those that lay their eggs in foam nests and those that lay their eggs on the ground. The discovery means that we now know of at least 118 species of tree frogs where we only knew of eighteen before. Does that mean that the problem of the disappearing frogs has been solved? Of course not. It only means that there are that many more frog species to worry about. Still, the discovery of unexpected species is a tiny flicker of hope in a world made increasingly darker by extinctions.

In the January 21, 2000, issue of *Science*, Peter Daszak, Andrew Cunningham, and Alex Hyatt published a frightening study in which they pointed out that "many wildlife species are reservoirs of pathogens that threaten domestic animal and human health; [and] second, wildlife EIDs [emerging infectious diseases] pose a substantial threat to the conservation of global biodiversity." Among the EIDs they list as infecting wildlife and humans are Marburg and Ebola viruses (two filoviruses that infect humans and nonhuman primates); plague (a bacterium that affects a wide range of mammalian hosts including humans); CDV (canine distemper virus, a morbillivirus that infects "a wide range of carnivores"); and the neurotropic velogenic Newcastle disease 2 virus (a paramoxyvirus that is carried by cormorants, pelicans, gulls, and poultry).

Rinderpest is "the paradigm for the introduction, spread, and impact of virulent exotic pathogens on wildlife populations. This highly pathogenic morbillivirus disease, enzootic to Asia, was introduced into Africa in 1889. The panzootic front traveled 5000 km in 10 years, reaching the Cape of Good Hope in 1897, extirpating more than 90 percent of Kenya's buffalo population and causing secondary effects on predator populations and local extinctions of the tsetse fly.* Populations of some species remain depleted and the persistence of rinderpest in eastern Africa continues to threaten bovid populations."

Although the existence of crossover pathogens has been known for some time, they have now gained more attention, because they pose a threat to biodiversity and even to ourselves. In a study published in May 2000, Osterhaus *et al.* identified the virus influenza B in a harbor seal (*Phoca vitulina*). They wrote: "Herpes-, morbili-, calci-, and poxviruses have been identified as causes of significant morbidity and mortality among pinniped species. In addition, influenza A viruses of avian origin . . . have caused outbreaks of influenza among seals." They examined twelve juvenile harbor seals with respiratory problems that had been brought in to the Seal Rehabilitation and Research Center in Pieterburen, the Netherlands; they found that the seals didn't have phocine herpesvirus (PHV), or phocine distemper virus (PDV), but "to our surprise the presence of influenza B

---

* At first, the extinction of the tsetse fly (genus *Glossina*) might seem a blessing, because this fly has been the scourge of human habitation in Africa as long as humans have lived there. A number of the twenty-one species can transmit the trypanosomes that cause Gambian or Rhodesian sleeping sickness (trypanosomiasis) in humans, and they can also transmit nagana and other diseases of wild and domestic animals. But if the tsetse fly is eliminated, human herders can move in with their cattle, and the grasslands that were previously the domain of Cape buffalo, wildebeest, zebras, etc., will be overgrazed at the expense of the wild animals.

virus was identified by reverse transcription polymerase chain reaction (RT-PCR)." They wrote: "Our data not only highlight the fact that influenza B virus infections can emerge in seal populations but also show that seals may constitute an animal reservoir from which humans may be exposed to influenza B viruses that have circulated in the past."

Certainly no virus has ever received the intensity of study of HIV (human immunodeficiency virus, the AIDS virus). While its actual origins are still unknown, HIV is believed to have originated in Africa in chimpanzees or other primates, and somehow—with or without human assistance—made the jump to humans. As science turns its inquiring and investigative eyes to current EIDs, those of the past—in what Martin Rudwick refers to as "deep time"—are also brought into focus, albeit indistinctly. We do not know what happened to that vast menagerie of earthly and marine creatures that are now extinct, but there are so many of them, and in some cases they vanished so abruptly, that disease cannot be discounted as a possible cause. Moreover, our window into this subject is so limited—MacPhee and Marx are having trouble extracting DNA from 3,500-year-old mammoths—that we can only speculate as to what might have happened 100 or 200 million years ago.

Is there a connection between mammoths, measles, morbilliviruses, monk seals, and ancient marine reptiles? We know that the marine reptiles—the ichthyosaurs, plesiosaurs, and mosasaurs—are extinct, and something must have been responsible for (or helped along) their disappearance. That they could not adapt to changing conditions is obvious, but after that, we are in the dark. There are many situations—the extinction of the nonavian dinosaurs comes immediately to mind—where we have either too little information or too many conflicting theories to be able to produce a conclusive answer—if there really is one. Some suspect that large land mammals of Pleistocene North America may have been wiped out by a "hyperdisease," probably in concert with hunting and climate change. There are lethal marine viruses in the ocean today, and they have probably been there since the oceans formed. Morbilliviruses, such as the ones that cause measles, rinderpest, and CDV, have shown up in the carcasses of seals, sea lions, and dolphins. Pathogens like the AIDS virus and CDV are known to be able to jump from one species to another, mutate, and then become species-specific. It will be considerably more difficult to identify a submicroscopic pathogen in a marine reptile fossil that is

100 million years old, than it would be to identify 3,700-year-old mammoth DNA, and indeed, it may be impossible.* Nevertheless, it is worth a thought that throughout the history of animal life on this planet, mutating lethal viruses may have contributed to—or even been the primary cause of—the extinction of marine animals.

---

* In 1999, researchers from Japan and Chile (Li *et al.*) identified the virus known as HTLV-1 in mummies in northern Chile. In the bone marrow of a woman who had been dead for 1,500 years, they found two intact DNA strings that closely matched the sequence of HTLV-1 (human t-cell lymphotropic virus type 1), a virus that causes leukemia in modern Chileans and Japanese. (It is a retrovirus that suppresses the immune system, of the same family as HIV.) In addition to showing that Japanese and Chileans are genetically related—probably because the Mongoloid ancestors of the South Americans crossed the Bering Land Bridge from Asia to North America between 25,000 and 12,000 years ago—these findings have also shown that HTLV-1, which is an airborne virus, could have been carried across the Bering Land Bridge and transmitted to uninfected human populations.

Threatened Species,
or
Under the Gun

In *The Origin of Species,* Charles Darwin wrote, "Rarity is the attribute of a vast number of species of all classes in all countries. If we ask ourselves why this or that species is rare, we can answer that something is unfavourable in its conditions of life: but what that something is we can hardly ever tell." When he wrote his seminal study of evolution in 1859, Darwin knew that there were species that were already extinct (dinosaur fossils were beginning to emerge and fossils of huge marine reptiles were turning up all over Europe), but in the late nineteenth century, the dodo was long gone, Steller's sea cow disappeared a century after the dodo, and the last of the great auks had been killed in 1844. While humans were certainly responsible, the list was still fairly short. All these creatures were awkward and clumsy, not "fit" enough to survive. Because they appeared to be destined for extinction anyway, we were probably doing them and the world a favor by helping them along. In later years, Darwin's innocent remarks would be contradicted forcibly and painfully: the agent of rarity and extinction was about to take center stage. *Homo destructivus* was about to begin his campaign to rid the world of wild animals.

Similar to the Caribbean monk seal in everything except geography, the Mediterranean monk seal (*Monachus monachus*) is considered one of the most endangered of all large mammals. There are two widely separated populations: one on the Atlantic coast of North Africa and the other in Greece's Aegean and Ionian Seas. Both regions have been invaded and colonized by humans to such an extent that there is hardly any room left for the seals. In addition, they are entangled in fishing gear and drowned, or killed outright by fishermen who believe that the seals are stealing their fish. There may be fewer than 1,500 of these animals left alive. In his review of the 1988 North Sea seal epidemic, Joseph Harwood wrote, "Two Portuguese divers recently reported the first sighting of a common seal in Madeira during July 1986. Given the propensity for wandering that seals infected with PDV [phocine distemper virus] have shown, there is a real risk that monk seals could be exposed to the virus over the next few years."

In May 1997, a catastrophic epidemic struck the largest remaining group of Mediterranean monk seals, which were living in sea caves on the Mauritania coast of West Africa. Of the 270 seals known to inhabit these caves, only 109 survived. The killer was originally identified as a morbil-livirus similar to PDV-1 virus, that killed so many of Lake Baikal's seals; later analysis showed that it might have been a poisonous dinoflagellate, similar to the one that causes red tides. Whatever it was, it has reduced the breeding population to 77 or fewer, and in a 1999 report, Forcada, Hammond, and Aguilar wrote, "This number may not be enough to main-tain genetic variability, and overcome the effect of demographic stochas-ticity [randomness]." It is certainly a tragedy that harbor seals and various other marine mammals are being decimated here and there, but there are other populations of those species that are not affected. With *Monachus*, however, any intrusion into the only surviving population threatens its very existence, and pushes it closer to the brink of extinction. It was not a microorganism, but rather the encroachment of the macroorganism known as *Homo sapiens* that was the cause of the monk seal's early popu-lation reductions. Further invasions—no matter who or what the cause— might serve to eliminate the species.

Of the two other species of monk seals, only the Hawaiian (*M. schauinslandi*) is not in immediate danger of disappearing from the face of the earth (the Caribbean monk seal is already extinct). Although it was heavily hunted by whalers and sealers in the nineteenth century, and disturbed again when U.S. forces occupied Laysan and Midway Islands in the northwestern Hawaiian chain during World War II, the Hawaiian monk seal has managed to survive. No estimates are available for how many of these stocky, round-faced seals were on the islands when the first Europeans arrived with Captain James Cook, but there were probably more than the 1,400 that are there now. They were originally found on the main Hawaiian Islands, including Oahu, Kauai, and Hawaii, but now live only on the outer islands, such as Laysan, Midway, Pearl and Hermes Atoll, French Frigate Shoals, and Lisianski. They are fully protected from everything but tiger sharks.

Appending "Steller's" as a prefix to the common name of an animal is tantamount to a death sentence. Steller's sea cow and his cormorant are extinct and his sea lion is in decline (for reasons unknown). Only his jay, eider, and sea eagle remain unthreatened. (The sea otter was not discovered or named by Steller; it was known from Kamchatka long before Steller's adventures on the Commander Islands.) There is, however, another bird discovered by Steller, that also had the name of the great naturalist added to its own, and it too has barely managed to survive.

During the ill-starred 1741 voyage, a large flying bird was sighted, and although Steller thought it was a kind of large gull, it was actually a North Pacific albatross. (The albatrosses of the Southern Hemisphere had been known since navigators rounded the Cape of Good Hope on their way to India; in 1697, William Dampier had identified "algatrosses" as "very large long-winged fowl." The only other North Pacific albatrosses are the black-footed and the Laysan, which are dark-backed, while the back of Steller's is white.) During Steller's time, this bird—now known as Steller's (or the short-tailed) albatross (*Diomedea albatrus*)— probably ranged along the northern Chinese coast, Taiwan, Kamchatka, the Bering Sea ice edge, and the Pacific coast of North America, perhaps as far south as Baja California. They were said to follow whaling vessels in order to feed on scraps; it was probably their North Pacific habitat that served to identify them, not their blue-tipped pink bill. The name "short-tailed," is misleading, according to Tickell (2000): "Laysan and black-footed albatrosses also have short tails, and the Galápagos albatross has the shortest tail of all albatrosses."

The short-tailed albatross is the largest of the North Pacific albatrosses, and also the rarest. Indeed, it is one of the rarest birds in the world, with a total population that numbers somewhere around 200. Until its

Short-tailed Albatross

decline, *D. albatrus* was known or suspected to breed in the Ryukus, the Pescadores, and the Daitos, but then as now, its main area of concentration was Torishima Island in the southern Izus, some 360 miles south of Tokyo. Torishima Island is now its *only* breeding ground. In 1887, the South Seas Trading Company of Japan put fifty laborers ashore, whose sole function was to kill the *aho-dori,* or "fool birds." The white body feathers were used to stuff pillows and quilts; the wing and tail feathers made dandy quill pens, and some of the larger feathers, known as "eagle feathers," appeared on women's hats in Europe and North America. The fat was useful as food, and the dried meat was turned into fertilizer. Albatrosses are awkward and clumsy on land, and because they show no fear of predators while sitting on the nest, they were clubbed to death as they sat. The Japanese ornithologist Yoshimaro Yamashina estimated that 5 million of these birds had been killed between 1887 and 1902, when a volcano erupted and killed all 129 humans on the island. After five years, the island was resettled, and the new homesteaders picked up where their predecessors had left off, killing 3,000 birds between December 1932 and January 1933. By April 1933, there were fewer than one hundred birds left on Torishima. The volcano erupted again in 1939, and then again in 1941, destroying the old mooring cove. The breeding grounds of the short-tailed albatross were buried under thirty feet of ash and lava.

During World War II, the Japanese built an aircraft observation post on Izu, and the observers reported that a couple of *aho-dori* were breeding there again. In 1946, when Oliver Austin was Head of Wildlife for the Allied powers occupying Japan, he tried to visit Torishima, but "available transportation never coincided with freedom from other duties during the breeding season," and he never landed there. Then in 1949, he sailed around the Bonin Islands whaling grounds on a Japanese whale catcher, and "saw no albatrosses whatever." This led him to conclude that Steller's albatross was extinct, and he wrote, "Despite the most careful watch, I saw no albatrosses close to or on any of the islands, and no sign whatever of any bird that could conceivably be construed as Steller's albatross during the entire voyage."

But when Lance Tickell visited the island in 1973, he estimated that there had been ten breeding pairs in 1953, and that there were fifty-seven pairs when he managed to get there aboard HMS *Brighton,* a British warship that had been visiting Japan. A rare bird nesting on a live volcano seems like a recipe for disaster. But, as Tickell points out, "Steller's albatross is an oceanic bird and its habitat is not Torishima so much as a wide expanse of ocean not influenced by one small island volcano. No albatrosses are on the island from June to September, and throughout the rest of the year, a substantial proportion of the population is always at sea, even at the height of breeding. . . . As long as the island was not blown apart completely, an eruption would merely change the topography; much of Torishima is presently bare ash." Tickell noted the presence of at least one feral cat on the island, and wrote, "Even one cat is a potential menace, and we do not know if there are any more. . . . However hopeful we might be for this exceptional bird," he concluded, "there is obviously no room for complacency." Because part of the North Pacific range of the short-tailed albatross takes it over the northwest Hawaiian Islands, in November 1998, the U.S. Fish and Wildlife Service proposed to list the species as "endangered" within the United States.*

---

* I was about as far from the United States as you can get when I saw a Steller's albatross. In the summer of 2001, on an American Museum of Natural History cruise aboard the clipper *Odyssey,* I sailed from Hokkaido in northern Japan through the Kurile Islands to Kamchatka. One of the other lecturers on board was Peter Harrison, probably the world's foremost authority on sea birds. Peter had no sooner finished a lecture on his visit to Torishima to see the only nesting colony of short-tailed albatrosses, when one of these magnificent birds was spotted flying right by the ship. For me, this sighting was one of the highlights of a career devoted to looking for sea creatures.

★ ★ ★

There are five species of living rhinoceroses. Probably not for long.

The genus *Rhinoceros* ("nose horn") includes the great Indian rhino, *Rhinoceros unicornis*, and the one-horned Javan rhino (*R. sondaicus*). The remaining species, the Sumatran rhino (*Didermoceros sumatrensis*), the white rhino (*Ceratotherium simum*), and the black rhino (*Diceros bicornis*) all have two horns. The term "horns," of course, refers to the growths that protrude from the bridge of the animal's nose, and are composed of fibrous keratin with no firm attachment to the bones of the skull. (The horns of antelopes and cattle form over a bony core, while the antlers of deer, which have no bony core, are shed and regrown every year.) Both male and female rhinoceroses have horns, like antelopes but unlike deer, whose antlers are restricted to males. (Caribou, which are deer, are the exception; males and females both have antlers.)

Rhinoceroses are classified among the perissodactyls (hoofed animals with an odd number of toes) along with horses, asses, zebras (three toes, only one of which is functional) and tapirs (three toes). They are thus differentiated from the artiodactyls (even-toed ungulates), the large group that includes pigs, hippos, llamas and camels, deer, giraffes, antelopes, cattle, goats, and sheep.

The largest of the Asian rhinoceroses, the Indian rhino, is a huge, imposing creature with a thick, wrinkled hide arranged in several loose folds, and studded with tubercles which make the animal look as if it is riveted together. An adult male may measure fourteen feet from nose to tail, stand six feet high at the shoulder, and weigh more than two tons. At one time, this retiring, solitary species was widespread throughout the grassland jungles of northern India, but human encroachment into the rhino's territory has greatly reduced the available habitat, and hunting has also taken its toll. Indian royalty and European adventurers shot the inoffensive animals for sport (usually from the backs of elephants), and from 1871 to 1907, the Indian government put a bounty on rhinos that were raiding tea plantations in Assam and Bengal, resulting in a further wholesale slaughter. The species was close to extinction early in beginning of the twentieth century, when sport and bounty hunting were halted, and some protected areas established.

At one time, these behemoths were found throughout the alluvial

Indian Rhinoceros

plains of Pakistan, northern India, Bangladesh, and Bhutan, but the population has steadily and steeply declined in the past 400 years; what remains is considered endangered and is largely restricted to preserves. Throughout the nineteenth century, the great rhinos were hunted for sport by Europeans and Asians. In West Bengal and Assam, from 1871 to 1907 the Maharajah of Cooch Behar shot 207, "almost single-handedly sending the Greater One-Horned Rhinoceros to its doom," wrote Vivek Menon. "From the middle of the nineteenth century to the year 1900," wrote Martin and Martin (1982), "there was such a slaughter of Indian rhinos by sportsmen that their numbers were reduced to only a few hundred." To supply the insatiable thirst for tea in Britain, planters in Assam appropriated vast tracts for tea plantations, eliminating the resident rhinos in the process. Although the Indian government established Kaziranga as a forest reserve in 1908 and officially abolished rhino hunting in 1910, poaching has remained a persistent threat to the rhino population, and as the price of rhino horn continues to escalate, the number of remaining rhinos—even in protected areas—continues to fall.

In a 1996 Traffic report* called "Under Siege: Poaching and Protection

---

*Traffic is the wildlife trade monitoring program of the World Wide Fund for Nature (WWF) and the World Conservation Union (IUCN), established to monitor trade in wild plants and animals. It works in close cooperation with the Secretariat of the Convention on International Trade in Endangered Species of Wild Flora and Fauna (CITES). Reports such as this are commissioned regularly in an attempt to identify and understand specific conservation problems.

of the Greater One-Horned Rhinoceros in India," Menon tracks the history of poaching since the official ban on rhinoceros hunting was enacted in 1910. There were so many rhinos in India by the 1960s that poachers found it easy to find and kill them: "Between 1968 and 1972 . . . Jaldapara and Gorumara lost 32 rhinoceroses to poaching, or a third of the rhinoceroses present in these protected areas; Kaziranga lost 53 rhinoceroses between 1965 and 1969, the first large number lost to poachers, roughly; 15 percent of its population average for the decade." In the 1970s, the figures were lower, but the poachers continued to kill rhinos, especially in protected areas like Jaldapara, Gorumara, and Kaziranga, but the 1980s "marked a resurgence in poaching," and the introduction of two new methods, electrocution and poisoning. From 1980 to 1984, 251 (out of a known total of 1,125) rhinos were killed by poachers, including 61 in the sanctuary of Laokhawa and 125 in Kaziranga. In the next five years, another 232 rhinos were killed for their horns, bringing the total for the 1980s to 483. In the 1990s, the poaching rates, which had redoubled in the 1980s, accelerated again. From 1990 to 1993, 209 rhinos were killed, "equivalent to 13.8 percent of the remaining population. With the exception of the 'black year' of 1983, when the country lost 93 rhinoceroses, never before had more than 60 rhinoceroses been killed in one year, yet in 1992 and 1993, 66 rhinoceroses were poached annually. The most seriously affected reserve of this half-decade was undoubtedly Manas National Park, which officially lost 41 rhinoceroses (68 percent of its population)."

Most of the poached rhino horns make their way to China or Taiwan, where they are employed by practitioners of traditional Chinese medicine (TCM). Westerners reject the TCM notion that rhinoceros horn can reduce fevers, but it has been demonstrated that it can indeed reduce fever in rats, though the horn of the saiga antelope was just as effective (Menon 1996). (Aspirin and ibuprofen, for which no animals have to die, would probably work just as well.) While not prescribed as an aphrodisiac in China, rhino horn was used that way in Gujurat, India, until the price rose so high that Indians could not afford it. Tibetan medicine, somewhat different from Chinese, also employs rhino horn, although the Dalai Lama has expressly condemned the illegal killing of animals. Menon's report concludes:

Despite being considerably better off than the other Asian rhinoceros species and also showing an increase in numbers according to official statistics, the species is facing a very real threat throughout its range. With the exception of Kaziranga in India and Chitwan in Nepal, all other reserves are particularly vulnerable to poaching or have already been affected severely by poaching. Kaziranga today acts as a single reserve for more than half the world population of the species and therefore will need to be paid particular attention with reference to antipoaching strategies. Lessons from Laokhawa in 1983 and Manas 1987-1995 have shown that a single population can be particularly vulnerable to poaching at times of civil unrest and that numbers can quickly fall to levels where a population's long-term viability is doubtful. . . . This study concludes that the poached rhinoceros horn is used to a small extent in the domestic market in India, for Tibetan medicinal and other uses, but that the main demand is medicinal use in the markets of East Asia and Southeast Asia, which are reached using new routes wherever established ones have been put under enforcement vigil. Further, despite measures to prevent the sale of rhinoceros horn medicine in many East Asian countries, demand appears to remain sufficient to endow the horn of Greater One-horned Rhinoceroses with a value which is high enough to continue to induce poachers to take calculated risks to obtain it, even where areas are well protected.

Powdered rhino horn is extraordinarily valuable. The wholesale value of Asian rhino horn increased from $35 per kilo in 1972 to $9,000 per kilo in the mid-1980s. By the time the horn has been shaved or powdered the price leaps to an astonishing $20,000 to $30,000 per kilo (Nowak 1991). This made rhino horn about twice as valuable as gold, which was selling for $13,000 per kilo in 1990.

By 1968, the rhino population in the Chitwan Valley of Nepal had fallen to about one hundred animals. In 1973, the Nepalese government carved out the 600-square-mile Royal Chitwan National Park and posted a special "rhino patrol" to the region. Accounting for about 1,000 animals,

Kazaringa National Park is the last stronghold of the Indian rhino in India; there are some 375 rhinos in the Chitwan park, about 100 miles southeast of Kathmandu in Nepal. From 1986 to 1991, thirty-eight rhinos were relocated from Chitwan to the Royal Bardiya National Park, in the more remote southwest part of Nepal. Esmond Bradley Martin and Lucy Vigne (1996) called the protection of Nepal's rhinos "one of the great success stories in the world for the rhinoceros." According to the International Rhino Foundation (www.rhinos-irf.org), "With strict protection from Indian and Nepalese wildlife authorities, Indian Rhino numbers have recovered from under 200 earlier in the twentieth century to around 2,400. However . . . the success is precarious without continued and increased support for conservation efforts in India and Nepal."

Since a state of emergency was declared in Nepal in December 2001, bloody violence has been taking place throughout the country; by late 2002, it had become evident that the rhinos in Nepal's national parks were once again under the gun. From May to November 2002, twenty-nine Indian rhinos were found dead in Chitwan and four in Royal Bardiya. To fight the Maoist insurgents, security guards originally detailed to protect the rhinos were removed from their posts in the national parks, leaving the rhinos at the mercy of poachers. The hides and horns were removed for use in traditional Chinese medicine. In Chitwan, two tiger caracasses were also found, stripped of many body parts.

The Javan rhinoceros was long regarded as belonging to the same species as the Indian, but it has a proportionately smaller head and lacks the "rivets" that characterize its northerly cousin; it is now known as *Rhinoceros sondaicus*. The horns are also smaller, but no less valuable. In 1970, Simon and Géroudet wrote (somewhat hyperbolically), that "It is impossible to shake the firmly held belief, prevalent throughout the East, in the infallibility of rhino horn as a powerful aphrodisiac; and most Asians will go to any length and pay almost any price to obtain it. Not only the horn, but every part of the animal is utilized by Chinese pharmacists, including the blood, bones, various organs, flesh, and hide; even the urine is considered efficacious." The human population explosion in Myanmar (Burma), Thailand, Malaysia, Laos, Cambodia, Vietnam, and the Indonesian islands of Sumatra and Java has just about put an end to the Javan rhino. It is now considered the rarest of the rhino species—and one of the rarest large mammals in the world. There are fewer than sixty Javan rhinos surviving,

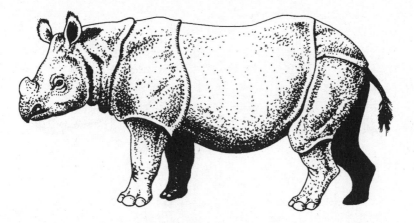

Javan Rhinoceros

in only two known locations, one in Indonesia and the other in Vietnam. Individuals have been poached from these small populations in recent years, and much more intensive protection is needed if this species is to survive.

The Javan and the Sumatran rhino compete for the dubious distinction of being the most endangered of their kind. The Sumatran is the smallest of the living rhinos, and also the strangest looking. While all other species are almost hairless, *Dicerorhinus sumatrensis* is covered with a coat of fur, densest in the juveniles, darker and more bristly in adults.* In the other two-horned rhinos, the forward horn is the larger; not so for the Sumatran rhino, which has an almost inconspicuous front horn and a larger rear one. Any rhino horn is valued in traditional Asian medicine, even the small and stumpy appendages of the Sumatran rhino, which has been slaughtered for its horns and other body parts. In the mid-1950s, Sumatran rhinos were scattered, with small populations located mainly in Sumatra. The Sumatran population was estimated to number between 425 and 800 in the early 1980s. There were lesser numbers in Peninsular

---

* In this aspect, it resembles the great Pleistocene wooly rhinoceros. But *Coelodonta* required a heavy coat for life in Ice Age Eurasia, while *Dicerorhinus* lives mostly in hilly Southeast Asian jungles. "As presently defined," wrote Groves and Kurt (1972), "*Dicerorhinus* is the genus that gave rise to all living Rhinocerotidae; in this sense, and that it closely resembles certain Miocene species, the Sumatran rhino may be regarded as a living fossil."

Sumatran Rhinoceros

Malaysia and in Sabah and Sarawak on Borneo. The largest and possibly most viable populations of Sumatran rhinos now exist in Sumatra, primarily in the Gunung Leuser, Kerinci-Seblat and Barisan Selatan National Parks.

The Sumatran rhino is classified as endangered by the World Conservation Union (IUCN), and is listed on Appendix I of CITES, which means that it is critically endangered. Its habitat has been replaced by agricultural and logging enterprises. With between 500 and 1,000 animals in Malaysia and Sumatra, and a dozen or so in zoos, its extinction appeared imminent as of 1984 (Nowak 1991). The population fell to about 600 in 1994, and has continued to fall. In less than a decade more than half the population has been lost; there are no more than 300 alive today. Although hunting is now illegal, poaching continues, fueled by high demand for rhino products in traditional medicine. More recently, there has been an increase in demand for horn products from more prosperous communities in Asia, where powdered horn is being prescribed for the treatment of a variety of ailments including epilepsy, fevers, strokes, and AIDS. "The result," wrote Edward O. Wilson in 2002, "has been a free-market death spiral for the Sumatran rhino." In *The Future of Life*, Wilson chose the Sumatran rhino as one of the paradigms for the current worldwide extinction crisis. He wrote:

Can the Sumatran rhino, like the California condor and the Mauritius kestrel, be pulled back from the grave? Of the two standard methods, captive breeding has been so far unproductive, while protection against poachers in existing reserves seems tenuous at best. The small circle of rhino experts working on the problem agree that *Dicerorhinus sumatrensis* has entered the endgame. Whatever the solution, they say, it must be found now or never.

The black rhino—which is not black, but gray—grows to a length of twelve feet, stands close to five feet high at the shoulder, and can weigh up to 3,900 pounds. It is recognizable by its long, pointed, prehensile upper lip and two prominent horns, the longest of which averages twenty inches, but in some animals can get to be several feet long. The horn is made up of fibrous keratin, the same stuff as your hair and fingernails. The longest black rhino horn on record—nearly five feet—belonged to a female named Gertie who lived in the Amboseli Reserve in Kenya in the late 1950s. Her horn, described by Guggisberg in 1966, "was certainly not under forty inches, probably quite a bit longer, inclined forward at an acute angle, and showing a slight, but graceful curve at the tip." With their calves, Gertie and another long-horned female named Gladys were the most photographed rhinos in East Africa. After giving birth to an earless calf called Pixie, Gertie lost her horn, perhaps in a fight, and was not seen after 1960 (Huxley 1963).

The black rhino inhabits the acacia scrub that grows at the edge of the plains of East Africa. It browses on a wide variety of ground plants, especially regenerating twigs, which it gathers into its mouth with the prehensile upper lip. Both black and white rhinos typically have two horns, formed of fused hair and growing continuously throughout the animal's life; horns that are broken off can regrow. Because of their notoriously poor eyesight, black rhinos sometimes charge a disturbing sound or smell, and they have been known to toss people into the air and up-end cars. Black rhinos were hunted by Africans for their horns and skin, and by Europeans and Americans for sport. At one time, black rhinos were considered premier big game trophy animals, along with the other four creatures that made up the "big five": lion, leopard, Cape buffalo,

Black Rhinoceros

and elephant.* In his 1952 book titled *Hunter*, the famed white hunter, whose name actually was J. A. Hunter, claimed to have killed more rhinos than any other man who ever lived. Hired by the Colonial Game Department of Kenya to clear the land for agricultural development, he killed 996 rhinos from August 1944 to October 1946. (The land proved to be too poor for agriculture, so the rhinos died for nothing.) Because their scrubland habitat was not much in demand for agriculture elsewhere, black rhinos were not seriously threatened until the depletion of Asian rhinos caused poachers to go after them too. As Guggisberg wrote, "Up to that time, the horns of the two African species had mainly served for the carving of drinking cups, snuff boxes, handles for knives, swords and war-axes, no superstitious significance whatever being attached to them by any of the indigenous people of Africa."

Living in Africa has provided no special protection for the black rhino from Asian pharmacological prescribers; its horns are collected by African poachers and sold to dealers to be converted to traditional medicines in

---

* In *Green Hills of Africa,* Ernest Hemingway described a black rhino he has just shot: "There he was, long-hulked, heavy-sided, prehistoric-looking, the hide like vulcanized rubber and faintly transparent looking, scarred with a badly healed horn wound that the birds had pecked at, his tail thick, round, and pointed, flat many-legged ticks crawling on him, his ears fringed with hair, tiny pig eyes, moss growing at the base of his horn that grew outward from his nose. . . . This was a hell of an animal."

Asia. Also, because young men in the Arab country of Yemen covet rhino horns for the handles of the elaborately carved daggers called *jambias*, another major threat to the already beleaguered black rhino has arisen. Until the 1970s, few men could afford these status symbols, but Yemen, like many other Middle Eastern countries, is rich in oil, and oil prices climbed dramatically because of a worldwide shortage. The result was a seven-fold increase in per capita income in Yemen, and a *twentyfold* increase in the price of rhino horn. The increase in price, of course, led to a corresponding increase in poaching. As the Martins (1982) have written,

> From 1971 to 1978 North Yemen was importing about three tons of rhino horn a year, which was three times more than was taken by Japan, the world's second largest rhino horn importer. From 1969/70 to 1976/77 the total amount of rhino horn legally entering North Yemen was 22.5 tons, which represents the death of about 7,800 rhinos. Nonetheless, with 50,000 young Yemeni men coming of age every year, this could satisfy only 17 percent of the potential market.

For the Yemenis, there is evidently nothing that makes a better *jambia* handle than the horn of a rhino; it takes a high polish and improves with age. When a Yemeni boy approaches manhood, he is given his own dagger, not as an ornament; as Martin and Martin have written, "they make use of them as offensive weapons and as deterrents to personal attack." All worldwide trade in rhino horn is now prohibited, because rhinos are protected under Appendix I of CITES, but the ban has not been very successful because of the thriving black market. In 1993, the United States threatened to ban legal imports of wildlife from China, which has a large wildlife trade with the United States, if that country did not start taking measures to stop illegal wildlife trade. In response, China made it illegal to sell, buy, trade, or transport rhino horns and tiger bones. Illegal stockpiles of rhino horns and tiger bones remain, however.

African black and white rhinos live in some of the same parks and reserves that provide habitat for elephants. Rhinos were often killed in protected areas because governments could not afford to patrol the parks to stop poachers. Now there are so few left that many are literally kept under armed guard. They forage during the day accompanied by guards with rifles, and they are locked up at night. Rhino horn is so valuable though, that poachers have killed guards to get at their prey. The rhino's plight has become so desperate that in some places conservation officials tranquilize rhinos and saw off their horns so poachers will have no cause to kill them. Martin and Martin (1982) described this plan as "a hare-brained scheme [that] would be fantastically expensive and would also have to be extremely thorough—and repeated at intervals since rhino horns do grow back." However, Malan Lindique and K. Peter Erb (1996) at least found no evidence that dehorning affected the mother's ability to protect her calf from lions and hyenas. "It is unlikely," wrote Hillman-Smith and Groves (1994), "that the black rhinoceros will ever go totally extinct, but their recent decline has been the most precipitous of any large mammal." According to the black rhino studbook maintained by the Berlin Zoo, as of 1999 there were 103 males and 127 females in captivity at seventy-four locations.

Originally, black rhinos occurred throughout eastern and southern Africa, in numbers that have been estimated at well over 400,000. Now there are fewer than 2,500 in all of Africa, living in small pockets in Zimbabwe, South Africa, Kenya, Namibia, and Tanzania. "The black rhinoceros," wrote Milliken, Nowell and Thomsen in 1993, "has declined at a faster rate than any other large land mammal in recent times, making a rapid transition from abundant to endangered. The estimated 65,000 black rhinos which roamed Africa's wildlands in 1970 have now been reduced by over 95 percent and the distribution of the species has been drastically altered." It is impossible to exaggerate the magnitude or the significance of the collapse of the black rhino population. Nowak (1991) called it "the greatest single mammalian conservation failure of the twentieth century," and wrote,

> This situation has been brought about entirely by an irrational demand for the horn and to a large extent through a strictly ornamental utilization by a single class of persons in one small country. If such a narrow and needless desire has led to the near extinction of one of the world's most spectacular and popular animals at a time when wildlife conservation is receiving immense international interest and support, how can we ever hope to save the multitude of other creatures and ecosystems that are

White Rhinoceros

jeopardized by much more trenchant problems of human population growth and development or by far broader and more substantive commercial pressures?

The white rhino (*Ceratotherium simum*) is no more white than the black rhino is black. Its name is actually a corruption of the Dutch word *wijdt*, which means "wide," and refers to its broad muzzle; the species is sometimes called the "square-lipped rhinoceros." Standing more than six feet high at the shoulder and weighing as much as three tons, the white rhino is the largest living land animal after the African and Indian elephants. Like the black rhino (the other African species), the white rhino has two horns, but the front one is usually much longer than the rear one. Also like the black rhino, the record length for the front horn is just over five feet. Unlike the black rhino, however, the white has a pronounced hump between the shoulders and a long, squarish head, which is so disproportionately large that the animal appears unable to hold it up, and it is often held with the muzzle close to the ground. An animal of open grasslands, the white rhino was once widely distributed throughout sub-Saharan Africa, wherever there was suitable savanna country. Like all other rhinos, the white was shoved off its land by people who wanted to raise cattle on the plains, and slaughtered for its valuable horns.

White rhino horns are just as desirable for *jambias* as black, and they are both powdered for traditional Asian medicines. Also, rhino horns (of all species) can be hollowed out, polished, and made into cups, or carved into small statues, buttons, hairpins, combs, and walking-stick handles. The feet have in the past been made into the (mercifully obsolete) ash trays and umbrella stands.

There are two recognized subspecies of white rhinoceros, a northern (*C. simum cottoni*), and a southern (*C. simum simum*); they have been shown to be genetically distinct. The northern subspecies, which was once found in southern Chad, the eastern Central African Republic, Sudan, the Democratic Republic of the Congo (DRC, formerly Zaire), and northwestern Uganda, is now found only in Garamba National Park in the DRC. The twenty-four animals left there—counted in an aerial survey in 1997 (Hillman-Smith 1998)—have earned it the IUCN designation of "critically endangered." In 1960, there were more than 1,000 northern rhinos in Garamba, but three years later rebel forces entered from Sudan, occupied virtually the entire park, and killed more than 900 of them (Fisher *et al.* 1969.) The southern version was formerly found throughout southeastern Angola, central Mozambique, Zimbabwe, Botswana, eastern Namibia, and northern South Africa, but a directed hunt in the nineteenth century came close to eliminating them altogether. Unlike the often belligerent black rhino, the white is a docile creature, and hunters were able to walk right up to them and shoot them where they stood. Others were trapped in spike-floored pits, and some were speared from horseback. Whatever the method, thousands upon thousands of white rhinos were killed, and their horns hacked off and shipped to Asia. Most of Africa's white rhinos are now confined to protected areas and game farms, but there are well over 10,000 of them in the south, and they are no longer considered endangered. Moreover, there are more than 700 southern white rhinos in zoological parks around the world. For some reason, northern white rhinos do not do well in captivity, and as of 1999, there were only nine individuals in two zoos, one in the Czech Republic and the other in San Diego.

When it was recognized that rhinos were being slaughtered out of existence, conservationist groups like WWF actively encouraged the hunting of the saiga, promoting its horn as an alternative to the horn of the endangered rhino. What is a saiga? It is *Saiga tatarica,* a funny-looking antelope of the Russian and Mongolian steppes, with a bulging proboscis like that of a tapir, a chunky body, spindly legs, and a yellowish gray coat that turns lighter in winter. Only the males bear horns, and these are eight to ten inches long, semitranslucent, amber-colored, and ringed on their lower two-thirds. The saiga's exaggerated nasal passages are an adaptation to the swirling dust of its arid habitat; the nostrils point downwards and the passages are lined with a complex arrangement of hair, glands, and mucous membranes that filter, warm and moisten the air as it is inhaled. Saigas wander for miles every day, marching with their heads low to the ground, their specialized noses filtering out the stirred-up dust. At the end of April, the males start their seasonal spring migration, forming herds of up to 2,000 animals and setting out northwards. In the meantime, females wander in huge congregations to a suitable birthing ground, where they all drop their calves within about a week of each other. Eight to ten days after the calves are dropped, the females and new babies set off after the males in groups that may exceed 100,000 animals. Once the migration is finished, the streams of animals break up and disperse into smaller herds. The large groups re-form in the autumn, and the saigas move en masse back south. A timid species, the saiga can be easily startled, even in the huge migrating herds, precipitating a panicked stampede. They flock together even when being shot at, so they are particularly easy to kill in large numbers.

By the time of the Soviet revolution, there were only a few thousand saigas left. To forestall their total eradication, the Soviets protected them in Europe in 1919 and in Soviet Central Asia in 1924. In the 1950s commercial harvests of saigas by local groups began. They were shot from vehicles, which spooked them and worked poorly for large-scale

Saiga

harvesting. In some cases, they were jacklighted at night—dazed by bright spotlights and easily hit with buckshot. This method, wrote Chan, Maksimuk, Zhirnov, and Nash (1995), "is the most effective on dark windy nights, and an experienced team of four or five hunters can harvest 100 to 150 animals during a five- or six-hour hunt." Despite this sort of thing, the antelopes made a remarkable comeback, and by 1957, a loose herd estimated to number between 150,000 and 200,000 animals was observed east of the Caspian Sea. At that time there were believed to be more than 2 million saigas in the world (Nowak 1991). However, more recently their fortunes have turned for the worse. "In 1993," wrote Fred Pearce (2003), "over a million saiga roamed the steppes of Russia and Kazakhstan. Today, fewer than 30,000 animals remain, most of them females." The males are being killed for their horns, which can fetch as much as $100 per kilogram (about three pairs), to fill the almost insatiable demand of Chinese who believe in their magic pharmacological properties.

The saiga is not mentioned in the *Chinese Materia Medica* of 1597, because, according to the 1995 *Traffic* report by Chan *et al.*, "as late as Li Shinzen's time (the late sixteenth century), the real Saiga Antelope and its horn was largely unknown to the Chinese, let alone utilized in any way . . . there is no historical record of any Saiga horn trade between China and Central Asia along the Silk Road . . . [which] would have been the most likely route for trade in Saiga horn." By 1989, however, in Zhang Enquin's *Rare Chinese Materia Medica*, an entire section is devoted to *Lingyangjiao*, or *Cornu Saigae Tataricae* (Latin for "horn of *Saiga tartarica*"). We are warned to be on the lookout for counterfeits, such as the horn of the Mongolian gazelle (*Procapra subgutturosa*) or the Tibetan antelope (*Pantholops hodgsoni*), which might be processed to simulate the horn of the saiga. "After being soaked, dried, and ground to a powder, the horn, which is 'salty in taste and cold in nature,' can be used to check hyperactivity of the liver and relieve convulsion, treat the up-stirred liver wind, infantile convulsion and epilepsy; calm the liver and suppress hyperactivity of the liver-*yang*; it is efficacious in the treatment of dazzle and vertigo due to hyperactivity of the liver-*yang*; it improves acuity of vision, cures headache and conjunctival congestion; clears away heat and toxic material; and can be used to treat unconciousness, delirium and mania in the course of epidemic febrile diseases." Like rhino horn, saiga horn is classified as a product "salty-cold in character and which can detoxify the body and reduce 'heat'" (Chan *et al.* 1995).

There are four populations of saiga in Kazakhstan and Russia: Kalmykia, Ural, Ustiurt, and Betpak-Dala, and a small one in western Mongolia. All are declining, but in Betpak-Dala the numbers plummeted from half a million in 1993 to 4,000 in 2000, a drop of 99 percent. Between 1993 and 1998, the overall million-plus population was essentially halved as the horn-bearing males were culled. Eleanor Milner-Gulland and her colleagues wrote in 2001, "The lack of males in Kalmykia is causing the dramatically reduced conception rates, which, in addition to the high hunting mortality, could lead to population collapse." Pearce's article, in the *New Scientist*, is entitled, "Going the way of the dodo?" and includes this phrase: "In a bid to save the rhino, conservationists suggested using saiga horn instead of rhino horn in traditional medicines. Their plan has backfired as hunters run amok." Rhino advocate Esmond Bradley Martin spearheaded the movement to substitute saiga horn for rhino horn, but when he realized that the saiga was close behind the rhino on the fast track to extinction, he publicly recanted. As Milner-Gulland *et al.* wrote in *Nature* in 2003:

> Reproductive collapse in the critically endangered saiga antelope is likely to have been caused by a catastrophic drop in the number of adult males in this harem-breeding ungulate, probably due to selective poaching for their horns. . . . Horns are borne by males and are highly favored in traditional Chinese medicine, which has led to heavy poaching since the demise of the Soviet Union, when the Chinese-Soviet border was opened.

The Mongolian population (*Saiga tatarica mongolica*), smaller than those in Russia and Kazakhstan, is now estimated to be at its lowest point ever. In a 1999 article in the conservation journal *Oryx*, Anna Luschekina and her colleagues from the Russian and Mongolian Academies of Science wrote:

> The subspecies has suffered a considerable decline in its range because of hunting and competition with domesticated stock. In 1997 a survey was made of almost all the known range, which

consists of two disjunct areas and covers a total of 2200 sq. km. A total of 609 animals was recorded and analysis of the census results suggests that *c.* 1300 saiga remain in total. The authors recommend strengthening the nature reserve established in 1993 in the Shargyn Gobi, and creating several sanctuaries outside this area, where Mongolian saiga from the main remaining population could be reintroduced.

It has now been estimated that less than five percent of the total saiga population survives, and the species may be beyond the point of no return. As of 2002 the saiga was classified as "critically endangered" by the IUCN, along with the wild Bactrian camel and the Iberian lynx.

Soon to make its appearance on the "endangered" list is the Tibetan antelope known locally as chiru and scientifically as *Pantholops hodgsoni*— yet another species threatened by human commerce, though not for its horns. The males have slender, forward-curving black horns that can be two feet long. The females, somewhat smaller, have no horns at all. Males stand about three feet high at the shoulder, and weigh about eighty pounds. Coloration ranges from white on the belly to fawn and reddish brown on the sides, with a pinkish suffusion. Males exhibit distinctive black markings on the face and legs in winter.

The slender, gazelle-like chiru has the finest, softest wool in the world, which is known as "shahtoosh." The term, Persian for "king of wools," is

Chiru

applied to the undercoat of the chiru as well as the wool that is spun from it. Shahtoosh fibers measure nine to twelve microns in diameter, or one-fifth that of a human hair. The shawls are so fine they can be threaded through a wedding ring—hence the nickname "ring shawls."

The chiru's principal habitat is the Tibet Autonomous Region, Qing-hai Province, and Xinjiang Autonomous Region of China, but a couple of hundred move seasonally into the Ladakh region of northwestern India. The geographical range of the chiru thus encompasses more than 200,000 square miles, extending for 1,000 miles across the Tibetan Plateau between the eastern limit in Qinghai to the western limit in Ladakh. Tibetan ante-lope prefer flat to rolling terrain, usually above 13,000 feet, although their habitat ranges from 11,000 feet in the north to as high as 18,000 feet in Ladakh. They favor alpine steppe, alpine meadow, and desert steppe habi-tats characterized by below-freezing average annual temperatures and a brief growing season. The largest Tibetan antelope populations survive in the Chang Tang region of northwest Tibet. This region is so inaccessible to humans that the expeditions of George Schaller, the man who first rec-ognized and publicized the perilous status of the chiru, are documented in the *Explorers Journal*. In 2002, George Miceler wrote:

> We are following a route that has not seen Westerners since the British surveyor Captain H. H. P. Deasy traversed the area in 1896. . . . The more than 400-square-mile area in which we hope to find the chiru, is a region of high desert averaging 16,000 feet in elevation, scattered with mountains reaching well over 20,000 feet. We plan to cross the desert from east to west, keeping the Kun Lin Range to our north and then recross the range via sev-eral passes before emerging on the southern rim of the Tala-makan Desert. Our only landmarks are the three huge basins that we must cross and the massive snow peaks of the Kun Lin.

As inhospitable as this region is to humans, it is home to the chiru. These antelopes are among the world's hardiest animals and can survive in −40° Fahrenheit temperatures. It is the dense layer of fur next to the skin—the shahtoosh—that insulates them, and at the same time has proved their undoing. More than a million chiru may have roamed the Tibetan Plateau at the beginning of the twentieth century, but the population has been reduced by more than 90 percent, principally due to poaching, and has now been estimated at less than 75,000. "In 1993," wrote George Schaller (2003), "one poaching team was arrested with more than 1,300 hides. In 1999, an antipoaching sweep netted 66 poachers, 18 vehicles, and 1,658 hides." As many as 20,000 Tibetan antelope are killed annually to supply the trade, with males, females, and young slaughtered indiscriminately.

Four protected areas in China have been set aside specifically to safeguard Tibetan Plateau wildlife species, including the chiru, but the remoteness of their habitat, an influx of settlers, and a lack of management capacity makes these reserves inadequate for effectively protecting the chiru or its habitat. Ironically, the deaths of thousands of Tibetan antelope, and indeed the entire shahtoosh trade are unnecessary. Attractive, fashionable alternatives are available, including products made from the finest Tibetan cashmere (made from the wool of goats, and sometimes known as "pashmina"). Yak wool is another excellent shahtoosh substitute. By purchasing cashmere and yak wool products, consumers can contribute to the protection of the Tibetan antelope, as well as assist with supporting the livelihoods of traditional communities in the Tibetan Plateau region.

Popular myth paints a picture of a bucolic world were the wool is collected by peaceful nomads who follow the herds collecting the hair that has caught on shrubs. The truth is different—merciless slaughter as part of a multimillion-dollar trade that threatens to wipe out the chiru throughout its range. The only way to obtain shahtoosh is to kill the original owner; it takes three to five dead antelopes to yield sufficient wool for one shawl. The chiru are skinned and the raw shahtoosh is collected and smuggled to India, where it is manufactured into shawls in the states of Jammu and Kashmir, the only locations in the world where shahtoosh possession and manufacture are legal. Shahtoosh products are then illegally transported to fashion capitals worldwide, where they sell for phenomenal prices. In a 2003 article on the chiru, George Schaller told of an ad in the fashion magazine *Harper's Bazaar* for a $2,850 shahtoosh shawl, but then he wrote that "the shahtoosh industry is bound to collapse within a few years as chiru populations are reduced to scattered remnants, and the wool spinners, weavers, dyers, and others involved in Kashmir return to processing mainly goat cashmere."

Tibetans formerly hunted chiru on a subsistence basis only, principally for their meat, using traps, dogs, and muzzle-loading rifles; traditional

Tibetan culture discourages hunting as a rule. Use of Tibetan antelope horn has been documented in traditional Tibetan and Chinese medicine, although only a few chirus are killed for these purposes. As of October 2003, the U.S. Fish and Wildlife Service determined that the chiru should be listed as "endangered" under the Endangered Species Act, which means that shahtoosh products are now prohibited from sale in interstate or foreign commerce. Shahtoosh shawls may not be sold legally in the United States.

Not as large as the rhino, but much more beautiful and in a similarly precipitous decline, the tiger represents the quintessence of raw energy and power. It was these qualities that made the tiger the object of centuries of "big game hunting"; today not only can these powerful qualities not protect it, they are the very attributes that make it—or almost any part of it—highly desirable to human beings, who believe they can acquire the tiger's wild energy by consuming the essence of the great cat.

Although they are all classified as *Panthera tigris*, two forms are recognized: the larger, heavier-coated Siberian tiger (known by the subspecies name of *P. tigris altaica*), and the somewhat smaller, slenderer Bengal tiger. There were once eight living tiger subspecies, but three have already been exterminated. The living five are the Amur (Siberian) tiger (*P. tigris altaica*), the Bengal tiger (*P. tigris tigris*), the Indochinese tiger (*P. tigris corbetti*), the South China tiger (*P. tigris amoyensis*) and the Sumatran tiger (*P. tigris sumatrae*). The Bali tiger (*P. tigris balica*) was the first to go—the last confirmed sighting was in 1939—followed by the Caspian tiger (*P. tigris virgata*), last seen alive in 1968. The Javan tiger (*P. tigris sondaica*) has not been spotted since 1979.

Most familiar is the so-called Bengal tiger (*P. tigris tigris*), found largely in India, although some range through the more isolated parts of Nepal, Bangladesh, Bhutan, and Myanmar. The most numerous of tigers, the Bengal ranges from the high-altitude, cold, coniferous forests of the Himalayas through the steaming mangroves of the Sunderbans to the arid forests of Rajasthan. The estimated wild population of Bengal tigers is approximately 4,000; there are about 300 in captivity, primarily in zoos in India. White tigers are a color variant of Bengal tigers and are rarely found in the wild. Indian zoos have bred tigers since 1880, when the first

Tiger

captive birth occurred at the Alipore Zoo in Calcutta. In the last two decades they have bred so successfully that there are now too many. Unfortunately other subspecies of tigers brought by dealers from outside India
over the years have been mixed with Indian tigers, so that many zoo tigers
are of questionable lineage and therefore not appropriate for conservation purposes.

In 1969, when it appeared that the world's tiger populations were becoming dangerously low, the IUCN held a conference in New Delhi to
discuss the problem. Three years later, together with the World Wildlife
Fund, IUCN initiated "Operation Tiger," which was supposed to raise
support for tiger conservation programs in Asia. Indian Prime Minister
Indira Gandhi set aside nine national parks—Manas, Palamau, Similipal,
Corbett, Ranthambhore, Kanha, Melghat, Bandipur, and Sundarbans—
for the protection of India's tiger population, then estimated to total
about 1,500. Project Tiger was set up with the help of the World Wildlife
Fund in 1973; its first director was Kailash Sankhala. In the early years of

Project Tiger, every reserve showed a decrease in hunting and an increase in tigers. In his 1977 book, Dr. Sankhala wrote, "I am greatly encouraged by the response of the habitat, the tigers, and their prey in the Tiger Reserves. It may be too early to predict the outcome of this effort, but it is surely not too much to hope that ultimately the tiger will be restored to a less precarious position than he is at present."

Alas, it was not to be. The tigers are now in a *more* precarious position than they were in 1977. Since the inception of Project Tiger in 1973, the human population of India has increased by 300 million, and livestock numbers have risen by 100 million. Indira Gandhi was assassinated in 1984; without her support, Project Tiger all but evaporated. Although the project still officially exists—and now oversees another eighteen reserves—almost all the reserves have been invaded by settlers in search of food and fodder. They regard the tigers as a nuisance, or in some cases, a threat. Tigers have been killed because they interfere with farming, or because they interfere with the very lives of the farmers. Starting in the late 1980s, they began to be killed to supply the traditional Chinese medicine market.

"Beginning about 1986," wrote Geoffrey Ward in *National Geographic* in 1997, "something began to happen, something mysterious and deadly. Tigers began to disappear. It was eventually discovered that they were being poisoned, snared, and shot so that their bones and other body parts could be smuggled out of India to supply the manufacturers of traditional Chinese medicines." Until recently, habitat loss was thought to be the largest single threat to the future of wild tigers in India, but it has now been established that the trade in tiger bones, destined for use in Oriental medicine outside India's borders, is posing an even larger threat. The Chinese tiger is extinct, so the manufacturers of traditional medicines began targeting India for their supply of tiger bones. Poaching of tigers for the traditional Chinese medicine industry started in northern India in the mid-1980s.

According to investigations carried out by the Wildlife Protection Society of India (WPSI), a total of thirty-six tiger skins and 667 kilos (1470 pounds) of tiger bones were seized in northern India in 1993-94. A tiger can be killed at a cost of just over a dollar for poison, or $9 for a steel trap. Much of the poaching is done by tribals who know the forests well. They are usually paid a meager amount; in one case near Kanha Tiger Reserve, in May 1994, a trader paid four poachers $15 each for killing a tiger. It is the traders and middlemen who make the substantial profits in the illegal tiger part trade. Tiger poaching occurs in all the areas where large numbers of tigers have been recorded, particularly in the states of Madhya Pradesh, Uttar Pradesh, West Bengal, Bihar, Maharashtra, Andhra Pradesh, and Karnataka. Firearms are used where hunting can be carried out with little interference, but where shooting is impractical, poison or traps are employed. Poison is usually placed in the carcasses of domestic buffaloes and cows, and sometimes dropped into forest pools. Major traders operate a sophisticated and well-organized supply route, to distribute poison and collect tiger bones in the remotest villages. Steel traps, made by nomadic blacksmiths, are immensely strong; in a tiger-poaching case near Raipur in 1994, it took six adult men to open a trap. In one area in central India, investigators found that so many steel traps had been set that the villagers were fearful of going into the forest.

According to its Web site (www.wpsi.india.org),
 The Wildlife Protection Society of India works with government enforcement agencies to apprehend tiger poachers and traders throughout India. WPSI also makes every effort to investigate and verify any seizure of tiger parts and unnatural tiger deaths that are brought to our notice. The following figures represent only a fraction of the actual poaching and trade in tiger parts in India. The central and state governments do not systematically compile information on tiger poaching cases and the details come from reports received by WPSI from enforcement authorities, work carried out by WPSI, and other sources. WPSI has documented the following cases:

    95 tigers killed in 1994
    121 tigers killed in 1995
    52 tigers killed in 1996
    88 tigers killed in 1997
    44 tigers killed in 1998
    81 tigers killed in 1999
    53 tigers killed in 2000

72 tigers killed in 2001

43 tigers killed in 2002

WPSI also has records of a large number of tigers that were "found dead." Without verification of poaching evidence these deaths have not been included in the above figures. To reach an estimate of the magnitude of the poaching of tigers in India, it may be interesting to note that the Customs authorities multiply known offenses by ten to estimate the size of an illegal trade.

Poaching constitutes the most direct threat to India's tigers, but India's immense population constitutes a much larger threat. People often want to live in regions inhabited by tigers, and while this association might occasionally end badly for a person, it almost always ends badly for the tigers.

The Panna Preserve, in the central state of Madhya Pradesh, is tropical dry forest, which characterizes some 45 percent of India's tiger habitat; the Ranthambhore habitat is another typical tropical dry forest. To date, however, most studies of tigers have taken place in tropical *moist* forests (such as Bandipur-Nagarahole) and alluvial grasslands (such as Chitwan, Corbett, and Kaziranga). Raghundan Chundawat, Neel Gogate, and A. J. T. Johnsingh (1999) undertook a study of Panna from 1996 to 1997. They collared and radio-tracked tigers to evaluate their activities and range, and approached feeding tigers on elephant-back to see what they were eating. They found that the dry forest habitats "support a relatively low large ungulate biomass and have high human disturbance." The scarcity of large wild ungulates—particularly deer and pigs—means that the Panna tigers have to eat smaller mammalian prey, such as monkeys, and are also likely to take cattle, which does not endear them to the local people. The authors wrote, "Tigers in fact take less than 1 percent of the available cattle each year, but taking cattle on any scale places tigers at the risk of poisoning and creates bad feeling towards them." In order to protect the tigers of Panna (the 1997 census found twenty-one), "disturbance-free zones must be created, and as many as 4,000 people must be moved out of the reserve . . . without this, the future of the tiger population in Panna is very bleak."

Rather than encouraging tiger protection, the government of India now seems prepared to discourage it. India is entering a new era in conservation politics—one that bodes ill for her tigers. Poachers killed eighty-one tigers in 1999, fifty-three in 2000, seventy-two in 2001, and forty-three in 2002. If the population estimates are wrong because the methodology of counting is inadequate, then the only available accurate figures are for the number of tigers killed every year—and these numbers might be wrong too. In an article in *Cat News* of Spring 2001, Valmik Thapar is quoted as responding to the statement that 200 tigers were killed in the year 2000. "The figures are terrible," he said, "but the ground realities are worse because a number of deaths still remain unrecorded." He added that the country had about 3,000 tigers at the start of the year 2000, so today their population could stand at between 2,500 and 3,000. At the same meeting, Project Tiger Director P. K. Sen said that tiger numbers were declining because of "ever-growing habitat vandalism, depletion of the tiger's prey base, illicit trade in tiger parts, [and] lack of infrastructure facilities, staff and money to effectively protect the tiger."

When Richard Perry published *The World of the Tiger* in 1965, tigers had been hunted for centuries throughout their range, but only the Balinese subspecies was known to have been driven to extinction. The Caspian tiger would be seen for the last time four years after Perry's book was published, and the Javan subspecies was barely hanging on; it would be declared extinct fifteen years later. In 1964, therefore, tigers were hardly numerous in India, China, and Russia, but people were not worried that they would become extinct. Perry wrote:

> The present world population of tigers is perhaps 15,000, perhaps rather more. Therefore, one might think, *Panthera tigris* is in no immediate danger of extinction; but only a super-optimist would expect more than a few hundred wild tigers to be alive by the end of this century. National Parks on their present scale—and this is especially true of Asia—have hardly begun to touch the problems of wildlife conservation, and are in any case entirely at the mercy of political, national and tribal pressures. We have very little time left in which to find the final solution to this world-wide twentieth-century conflict between human and animal populations. One suspects that in the end it will probably be solved in a manner disastrous for both men and animals; but we must continue trying to solve it humanely.

But humans cannot be defined by their humane behavior, especially to animals. In the forty years between the publication of *The World of the Tiger* and the publication of this book, the Javan tiger disappeared, and tiger populations seemed to be falling everywhere in India, Project Tiger was initiated to protect them, and, just when it looked as if tigers might recuperate, poachers began killing them with such celerity that the remaining Indian tigers careered toward extirpation. A hundred years ago European hunters and maharajahs tried to eliminate the tiger by their wanton hunting; in their wake, the poaching brigade is nearing success. Due to some bizarre and utterly fallacious ideas of Chinese traditional medicine, the last of India's tigers are being wiped off the face of the earth. In his 1996 book *Tigers*, biologist John Seidensticker wrote:

> The 500 or so remaining Sumatran tigers are restricted to some forest areas on Sumatra. The 3,000 to 5,300 remaining Bengal tigers are found in some fragmented forest tracts in India, Nepal, Bangladesh, Myanmar, and China. The 900 to 1,500 Indo-Chinese tigers live in scattered forest tracts in Myanmar, Thailand, Laos, Cambodia, Vietnam, China, and Malaysia. Fewer than 20 Chinese tigers live in southeastern China. The 250 to 400 Amur or Siberian tigers live in the Russian Far East, with perhaps a few remaining in northeastern China.

Almost always invoked as examples of convergent evolution (where unrelated species develop similar modifications that allow them to function more or less in the same fashion) are the Australian marsupials, which evolved on their isolated island continent to fill niches comparable to those filled by placental mammals elsewhere. Thus the kangaroos and wallabies are said to be the functional equivalents of herbivorous placentals like deer and antelopes, and the predatory thylacine is the equivalent of the dog or wolf.

Just as Australia has marsupial "dogs," so also does it have marsupial "cats" and marsupial "mice." The dasyures are the "native cats" (although not domesticatable), and there are several families that closely resemble mice and rats. They are all proper marsupials, carrying their young in a pouch. In *Walker's Mammals of the World* (1991), they are always referred to as marsupial "mice," with "mice" in quotes. There are even marsupial "moles."

Although they do sometimes call the quolls "native cats," and the thylacines "Tasmanian tigers" (or "wolves"), the Aussies rarely use this convergent terminology. They much prefer to give their native animals very Australian common names, such as koala, bilby, yallara, bandicoot, bettong, dunnart, quokka, and, of course, kangaroo, wallaroo, and wallaby, all derived from Aboriginal nomenclature.

Another element of convergence with placental species is endangerment. As Derrick Ovington wrote in his *Australian Endangered Species* (1978), "Although Australia is a large, sparsely populated and relatively recently settled country, the impact of people on the native fauna has been no less profound than in other countries."

The bilby (*Macrotis lagotis*) is a member of the bandicoot family; because of its large ears, it is also known as the rabbit-eared bandicoot. Bandicoots are long-snouted, rather delicate animals, with soft coats, furred tails, clawed feet, and hind legs that are longer than the forelegs. They are found in Australia, Tasmania, New Guinea, and certain nearby islands. There were once two species of bilby: *M. lagotis,* known as the greater bilby, and *M. leucura,* the lesser bilby. If one Aboriginal name wasn't enough, the lesser bilby was also known as the yallara. According to Flannery (2001), the yallara had a nasty temperament, and "repulsed the most tactful attempts to handle them by repeated savage snapping bites and harsh hissing sounds."

The bilbies are the largest of the bandicoots, with a body length of about twenty inches and a nine-inch tail; adult males weigh up to six pounds, the females about half that. The bilby is bluish gray above and white below, with a black, white-tipped tail. The large, almost hairless ears provide keen hearing, and are also believed to help keep the animal cool. Its excellent senses of hearing and smell compensate for the bilby's poor eyesight. Because the lesser bilby has not been seen since 1950, the greater has become sole heir to the name; apparently, there is only one species of bilby left.

Once a favorite food of Aborigines, bilbies are largely solitary, widely dispersed, and found in low numbers. They have strong claws and are very efficient burrowers. In sandy soil they can disappear from sight within three minutes. Their burrows go down in a steep spiral to a depth

Bilby

of around ten feet, which makes it very difficult for predators such as foxes and cats to unearth them. They dig burrows wherever they go and may use as many as two dozen at any one time. Their main food items are bulbs and insects such as termites, witchetty grubs, and honeypot ants. The bilby is almost completely nocturnal, emerging from the burrow about an hour after dusk and retreating at least an hour before dawn. A full moon, strong wind, or heavy rain can keep bilbies in their burrows all night. Like the koala, the bilby doesn't drink water, and gets all it needs from its food. Bilbies breed all year round. Their gestation period is only twelve to fourteen days, and the young, between one and three in a litter, remain in the pouch for seventy-five to eighty days, becoming independent about two weeks later. Bilbies could once be found in over 70 percent of mainland Australia, but they are now found only in the Tanami Desert (Northern Territories), the Greater Sandy Desert and Gibson Desert (Western Australia), and in southwestern Queensland. Reduction of the bilby's range has resulted from habitat destruction by people, cattle, and rabbits, as well as from predation by cats, dingoes, and foxes. Its range has contracted alarmingly; a total of around 600 to 700 animals live in far southwestern Queensland near Birdsville. The bilby was chosen by the Australia Endangered Species Program as a mascot representing all endangered Australian species. Rabbits have never been very popular in Australia, so a campaign was launched to replace the Easter Bunny, and Australians now celebrate with the Easter Bilby.

It appears that just being a bandicoot is dangerous to your health. The pig-footed bandicoot (*Chaeropus ecuadatus*), a delicate little creature that looks like a long-nosed, long-tailed rabbit on stilts, was discovered in 1836, but hasn't been seen much since. Nocturnal and secretive, these animals inhabited forests, grasslands, and saltbush plains in south-central Australia and a small area of Western Australia. They had only a single functional toe on their hind feet, like a tiny horse's hoof; their forefeet were equipped with two toes, like miniature cloven hooves, which obviously gave the animal its common name. The pig-footed bandicoot has not been seen in the wild since 1907, though there were a few unconfirmed sightings up until 1926. In Menkhorst's 2001 *Mammals of Australia,* it is listed as extinct.

With three toes on each foot, the barred bandicoot (*Perameles bougainville*) is built along more traditional lines, but it is no less threatened for that. Also known as the marl, the barred bandicoot is about a foot and a half long, counting the short, tapered tail. Grizzled brown in color, it has a couple of dark stripes on its rump, giving it its common name. Like most bandicoots, it is a consumer of small prey, mostly insects, worms, and spiders, but it is also known to eat roots and tubers. It comes out of its burrow in the early morning and the evening, the schedule biologists refer to as "crepuscular." The marl is said to leap high in the air when disturbed, and upon landing, burrows for safety. It was once widely distributed throughout Australia, but is now found only on Bernier and Dorre Islands in Shark Bay, Western Australia. And we know what happens to small populations of endangered species on islands.

What do the spot-tailed quoll, the brush-tailed phascogale, and the black-footed rock wallaby have in common? Other than the fact that you've probably never heard of any of them, they are all endangered Australian mammals.

The spot-tailed quoll (*Dasyurus maculatus*), also known as the tiger quoll, is the second largest surviving marsupial carnivore on mainland Australia. (The largest *surviving* marsupial carnivore is the Tasmanian devil; the title used to belong to the thylacine, now extinct.) There are three species of quolls, the eastern (*D. viverrinus*), also known as the native cat; the western (*D. geoffroii*), also known as the chuditch; and the spot-tailed. All of them are in trouble. Quolls used to be widespread throughout Australia. The eastern is now rare and found in significant numbers only in

Spot-tailed Quoll

Tasmania; the western is found only in the southwest corner of Western Australia. The spot-tailed quoll is more widely distributed. It is about the size of a large housecat, dark brown above and pale below, with white spots all over the body and tail. (The eastern and western quolls have no spots on their tails.) It is largely nocturnal and solitary, ranging over large territories in search of birds, eggs, gliders, possums, rabbits, and even small wallabies. Logging and fragmentation of habitat have greatly reduced the area in which quolls can live, and they are considered endangered in Victoria and vulnerable in South Australia and New South Wales.

In keeping with the tradition of giving native species native names, the brush-tailed phascogale (*Phascogale tapoatafa*) is also known as the tuan. A black, squirrel-sized marsupial, the tuan has a tail whose terminal half is covered with long black hairs, which, when erected, can produce a totally unexpected "bottle-brush" effect. Tuans are largely arboreal and nocturnal animals, preferring to hunt in the trees for even smaller mammals, birds, and lizards, and especially for large insects such as spiders. In addition to these appetizing items, tuans also spend a lot of time harvesting nectar from flowering trees. Females come into estrus in the Australian winter (June to August). The males die after mating, before they reach the age of one. According to Nowak (1991), "Captive males have lived for over 3 years, but are not reproductively viable after their first breeding season." The females produce a litter of eight young, which remain attached to the nipples for about forty days, after which they are left in the nest while the female forages for food. Logging and habitat destruction by grazing cattle and sheep has greatly reduced the places where tuans can live, and they are also preyed upon by nonnative cats, feral dogs, and foxes. Because they live in scattered, small populations (mostly in Victoria), tuans are extremely susceptible to predation. The males exercise their own population control, but if, say, a cat kills three lactating females and owls kill two more, the small population will immediately go extinct.

Even kangaroos, the everlasting symbol of Australia, have not been spared, and some species of wallabies are already extinct. Flannery (2001) identifies the Toolache wallaby (*Macropus grayi*), as "the most elegant, graceful and swift member of the kangaroo family." Their speed was legendary; Australian "sportsmen" ran them down on horseback or with greyhounds, but the wallabies often outran their pursuers. The Toolache (pronounced too-late-shee) was plentiful until around 1900, but hunting and farming took their toll, and by 1920 the species was considered rare. In 1924, an attempt was made to relocate some of the survivors to Kangaroo Island, but they all died. The last Toolache lived in captivity for twelve years, and died in 1939.

Another lost kangaroo is the crescent nailtail wallaby (*Onychogalaea lunata*). The "crescent" in its common name comes from the white band on its shoulders, and "nailtail" from the horny projection at the tip of its tail, whose function has never been understood. Only about fifteen inches

Toolache Wallaby

tall, with soft and silky fur, the crescent nailtail was once widely distributed throughout the scrublands and thickets of central Australia. When running, it carried its forelimbs at an awkward angle and moved them in a rotary motion, which caused it to be nicknamed "organ grinder." It often sought refuge in hollow trees, entering at the bottom and clambering up until it appeared at an opening high above (Flannery 2001). Farming and grazing greatly reduced the habitat of the nailtails (there are two other wallaby species, *O. unguifera* and *O. frenata*), and foxes, feral dogs, and dingoes finished the job. Although the exact cause and date of death of the last crescent nailtail is not known, Flannery gives 1956 as the date of the last sighting.

The bonobo straddles two of the categories in this section: (1) it was recently discovered, and (2) it is already endangered. Before 1933, it had never been seen in the wild by a zoologist or an explorer. The first scientific evidence for its existence turned up in 1928 in the Congo Museum in Tervueren, Belgium, when Ernest Schwarz examined some skins and skeletons that had been collected earlier along the banks of the Congo River. To differentiate this new animal from the common chimpanzee, *Pan troglodytes*, Schwarz named it *P. paniscus*.* The Tervueren specimens were smaller than common chimpanzees, so for a while these animals were known as "pygmy chimpanzees;" but since the living animals are not appreciably smaller than common chimps that appellation now seems off the mark. The bonobo is not a chimpanzee, but a separate species altogether. The name "bonobo," first used by German zoologists Eduard Tratz and Heinz Heck in a 1954 paper that differentiated these apes from chimpanzees, seems to have come from the town of Bolobo on the east bank of the Congo River in the Democratic Republic of the Congo (formerly Zaire; formerly the Belgian Congo), where some of the first specimens were collected.

---

* Schwarz's account was published in 1929, and his name is appended to the description. In 1933, Harold Coolidge, an American anatomist, published a much more detailed description; 50 years after that, Coolidge published still another account, in which he claimed that his description had priority. The rules of zoological nomenclature are clear: even though Schwarz's account was published in an obscure journal, his description has priority, and therefore the proper name of the bonobo is *"Pan paniscus Schwarz 1929."*

Bonobo

Bonobos differ from chimpanzees in body proportions: they are slenderer in build, with a smaller, rounder skull, and a flatter face with less-prominent brow ridges. Chimpanzees have a large head, a thick neck, and broad shoulders, while bonobos have a thinner neck and narrower shoulders. Bonobos have a black face and pink lips, compared to the chimpanzee's black lips. The young bonobos are born with a black face and hands, and small ears that are hidden behind distinctive side whiskers. With her colleagues, physical anthropologist Adriennne Zihlman (1978) suggested that the posture and proportions of the bonobo are closer to those of the Australopithecines, the protohumans that lived in Africa some 3.5 million years ago. She does not suggest that we are closely related to bonobos; only that we might look to bonobos for a prototype of the common ancestor of humans, gorillas, and chimps.

The bonobo is a highly social animal, with groups usually based upon a female and her male offspring. Like humans, bonobo females are sexually receptive throughout most of their estrus cycle. (Chimpanzees only mate during the few days when a female is fertile.) Generally, the ranking males in chimp society are sexually dominant; they make macho displays to impress females, and can be quite violent in their demands. Consequently, chimp females do not have much control over who they mate with. Bonobo males tend to be a bit more polite. They ask first, by displaying themselves in a persuasive but nonaggressive manner, offering food or making other propositions—and bonobo females have the right to refuse. Bonobos can peacefully exist in large groups, sometimes with as many as a hundred members; generally groups of bonobos forage together. Female chimpanzees will often retreat with their infants to forage alone, to avoid bullying by the larger males or other females. While infanticide and killing of males by males has been recorded among chimpanzees, such behavior has not been seen in bonobos.

These acrobatic apes spend a lot of time high in the rain forest canopy. They move through the trees swiftly and gracefully, maneuvering through the forest to forage on fruit and other foods. They also travel on the ground, often single file along their own sort of trail system. On the ground they usually walk on all fours ("knuckle-walking"), but on average, they spend more time walking upright than chimps. They tend to like swampy areas, where sometimes they dig for grubs or beetles. Bonobos have complex "mind maps" of the forest, and coordinate travel through vocalizations and other forms of communication that we do not yet understand. They live in groups of up to a hundred, breaking up into foraging groups by day and gathering to nest at night. When bonobos gather in the trees to make their night nests, they fill the twilight with a symphony of soprano squeals. Their high-pitched vocalizations sound like those of a flock of exotic birds, unlike the more guttural hoots of chimpanzees. "Their calls are so high-pitched and penetrating," wrote psychologist Frans de Waal (1998), "that they do not even remind one of the typical drawn-out 'huu . . . huu' hooting of the chimpanzee. The difference in timbre between the voices of the two species may well be of the same magnitude as that between a small child and a grown man."

Bonobos have been described as "pansexual" by de Waal. Sex permeates the fabric of bonobo society, weaving through all aspects of daily life. It serves an important function in keeping the society together, maintaining peaceful, cooperative relations. Besides heterosexual contact, both male and female bonobos engage in same-sex encounters; even group sex occurs. Female-female contact, or "genital-genital rubbing," is actually the most common sexual activity. Unlike other apes, bonobos frequently copulate face-to-face, looking into each other's eyes. When bonobo groups meet in the forest, instead of fighting they greet each other, bond sexually, and share food. Likewise, almost any conflict between bonobos is eased by sexual activity, grooming, or sharing food. The behavior of bonobos in the wild has been observed and described by the Japanese researcher Takayoshi Kano, who built a camp at Wamba in 1973 and has been studying the elusive apes since then. Like many endangered species not easily observed in the wild, the bonobo has been heavily studied in captivity, with results published in numerous scientific papers and books. With wildlife photographer Franz Lanting, Frans de Waal has written *Bonobo: The Forgotten Ape*, a beautiful compendium of incisive text and spectacular photographs that more than fills in the gaps of this summary.

Bonobos are found only in the Democratic Republic of the Congo, a resource-rich region that has been ravaged by war for decades. They live in scattered populations, dispersed over an extensive area. More specifically, the bonobo has a discontinuous distribution in the central Congo basin, south of the Congo River. Studies in the last decade have confirmed viable populations only near the towns of Befale, Djolu, Bokungu, and Ikela, and in a 1,800-square-mile area between the Yekokora and Lomako Rivers. No reliable estimates of total numbers are available, largely because of the problems of survey work in war-torn forest habitats.

Encompassing almost a million square miles, the Democratic Republic of the Congo (DRC) is one of the largest countries in Africa—and one of the poorest. Government corruption, neglected public services, tribal fighting, and depressed coffee and copper prices have thrown the DRC into a prolonged economic depression. Since 1994, the country has been rent by ethnic strife and civil war, touched off by a massive inflow of refugees from the fighting in neighboring Rwanda and Burundi. To feed themselves, soldiers and refugees frequently subsist on "bushmeat," which is anything in the forests they can kill, mostly monkeys and apes. Traditional slash-and-burn agriculture and commercial forestry operations are continuing to expand in the heart of the DRC, eliminating suitable bonobo habitat, especially near villages and roads. Few domestic animals are kept by resident people; many are dependent on wildlife for a large proportion of the protein in their diet. In addition, bonobos and chimpanzees are occasionally hunted for traditional medicinal or magical purposes; specific body parts are thought to enhance strength and sexual vigor. Such charms are widely available in some parts of Zaire, suggesting that large numbers of bonobos are being killed annually. Hunting may therefore be an important factor underlying the species' fragmented distribution. Bonobos are especially vulnerable to the increasing use of firearms, since they flee into trees which can be easily surrounded by hunters. Infant bonobos are captured for the local wildlife trade after the mother is killed, and are kept as pets. Small numbers are illegally traded with countries in Europe (particularly Belgium) and East Asia. Infants and juveniles are currently sold to zoological gardens, laboratories in Europe and Asia, and for the pet trade.

In March 2000, Gretchen Vogel wrote, "The war that has gripped the Democratic Republic of Congo for the past 18 months, killing thousands and displacing many more, is also taking a devastating toll on the great apes." The war cuts through the heart of bonobo country, but because all the researchers have fled for their lives, there is no one to count the remaining apes. There are, however, a large number of bonobo orphans brought in for sale in Kinshasa, the DRC capital, suggesting that adults are being killed for meat. Vogel quotes Belgian bonobo researcher Ellen van Krunkelsven: "The civil war might take several more years. We cannot just sit and wait, because bonobos might not have that long." Tucked away in an article (Bohannon 2002) about the plight of all the great apes is this sentence: "At its present rate of decline, the bonobo (*Pan paniscus*) is predicted to go extinct within a decade." (That sentence is followed by "The Sumatran orangutan (*Pongo abelii*) is thought to have five years left. Only 150 individuals remain of the Cross River gorilla (*Gorilla gorilla diehli*)."

When they were first discovered by European explorers in Central Africa, gorillas were believed to be savage beasts who would beat their chests before charging and tearing apart any unfortunate creature that got in their way. Think of the hunters' original attitude to King Kong, before he was shown to be a gentle, misunderstood giant. But now that the roaring, chest-thumping gorilla has been shown to be mostly apocryphal, gorillas are probably the best-loved and most popular of all primates. In zoos, gorilla exhibits always rank among the public's favorites. George Schaller wrote two books about them, *The Year of the Gorilla* and *The Mountain*

Gorilla, and Dian Fossey wrote *Gorillas in the Mist* about her battle to save the mountain gorillas. But despite this publicity—or perhaps even *because* of it—they remain critically endangered. In a 1995 *National Geographic* article, Schaller wrote:

> Gorilla numbers plummeted. In 1960 I estimated abut 450 in the Virunga region. Censuses during the 1970s showed around 275, and by 1981 there were only 350. During this critical time Dian Fossey, assisted for varying periods by Craig Sholley, David Watts, Kelly Stewart, Ian Redmond, Alexander Harcourt, and others, was at Karisoke. Dian harassed poachers with obsessive zeal. And she made the world aware of the gorilla's plight. However, her unyielding confrontational approach with local people, one that she termed "expedient action," cannot save wildlife.

There are three subspecies of gorillas: the western lowland gorilla (*Gorilla gorilla gorilla*); the eastern lowland gorilla (*G. gorilla graueri*), and the mountain gorilla (*G. gorilla beringei*). The three subspecies are very similar and show only minor differences in size, build, and coloring. All three gorilla subspecies are endangered. There are currently about 50,000 western lowland gorillas living in the wild in west central Africa. The eastern lowland gorilla population has declined significantly in recent decades, and today there are about 2,500 remaining in the wild; only a few dozen more live in the world's zoos. The mountain gorillas are the rarest of all gorillas and are on the verge of extinction. Like their lowland relatives, they are black, but adult males have silvery gray markings on their back and hips and are referred to as silverbacks. Only about 600 of these magnificent animals are left in the wild, about 300 in the Parc National des Volcans in the Virunga Mountains of Rwanda, and another 300 in the Bwindi Impenetrable Forest National Park in Uganda. There are none in captivity. Hunting and poaching reduced their numbers to about 250 by 1981, when the protection efforts of the late Dian Fossey and others brought the decline to a halt. The long-term survival of the Virunga population is threatened by natural changes and disasters, hunters and poachers, and the chronic political instability that swirls around the edge of their forest home.

In the *Condé Nast Traveler* of May 2003, Graham Boynton wrote of his visit to the mountain gorillas to see Annette Lanjouw, Dian Fossey's

Gorilla

successor as director of the International Gorilla Conservation Program. Fossey originally went to Africa in 1966, lived among the gorillas for years, and, as Boynton wrote, "became steadily closer to the animals and more isolated from the humans. She became autocratic with the Rwandans who worked on gorilla conservation, and suspected poachers were treated ever more harshly. . . . However, Fossey's behavior grew more erratic, her alienation from the locals ever deeper, and in the end nobody was surprised when she was murdered in her bed in December 1985." Political chaos in Rwanda and the massacre of hundreds of thousands of Tutsis by Hutu tribesmen has left the country—and the mountain gorillas—in a state bordering on anarchy. Even so, gorilla tourism in Uganda brings in more than 3 million dollars annually; ecotourism represents the best opportunity—perhaps the *only* opportunity—for the financial resuscitation of Rwanda. "Amazingly," writes Boynton, "not only have the mountain gorillas survived this roiling human unrest but over the past twenty years, they have actually begun to flourish. Both the Bwindi and the Virunga gorilla populations have increased almost ten percent." He quotes Lanjouw: "This is not about saving gorillas; it is about preserving the habitat of which the gorillas are a part. It is about saving ourselves."

If we cannot save our closest relatives, can we save anything at all? In May 2001, the United Nations Environment Programme (UNEP) launched the Great Ape Survival Project (GRASP) headed by three UN Special Envoys: Dr. Russell Mittermeier, President of Conservation International and one of the world's leading authorities on primates; Dr. Jane Goodall, the British conservationist and UNEP Global 500 laureate best known for her pioneering work with chimpanzees; and Professor Toshisada Nishida, a world-famous Japanese primate researcher. Also involved with GRASP are Ian Redmond of the British-based Ape Alliance, and Dr. Richard Leakey, the celebrated Kenyan authority on wildlife conservation, who was nominated as the project's special advisor. According to the GRASP Web site (http://www.unep.org/grasp/), "The purpose of the Great Apes Survival Project is to provide a framework into which all the individual conservation efforts of governments, wildlife departments, academics, non-governmental organizations (NGOs), UN agencies and others can be layered to ensure maximum efficiency, effective communication and successful targeting of resources." In a press release dated November 25, 2001, GRASP plaintively announced:

> Great ape populations are declining at an alarming rate worldwide. The continuing destruction of habitat, in combination with the growth in the commercial bushmeat trade in Africa and increased logging activities in Indonesia, have lead scientists to suggest that the majority of great ape populations will be extinct in the next ten to twenty years. Even if isolated populations were to survive, the long-term viability of great apes is in doubt due to their limited numbers and the fragmentation of their habitat.

"In west and central Africa," wrote John Whitfield in a 2003 article in *Nature*, "it's estimated that one million tonnes of forest animals are killed for meat each year. . . . In 2000, Miss Waldron's red colobus monkey, once resident in Ghana and the Ivory Coast, was hunted to extinction." In

Africa, where the forest is often referred to as "the bush," both wildlife and the meat derived from it are referred to as "bushmeat" (*viande de brousse* in French). The term is applied to all wildlife species used for meat, including elephant, gorilla, chimpanzee, all sorts of monkeys, forest antelope (duikers), crocodiles, porcupines, bush pigs, cane rats, pangolins, monitor lizards, and guinea fowl. Rural communities have always relied on food from the forests, but in recent years, due to the commercialization of the bushmeat trade and the opening up of forests by logging companies, the practice has spiraled out of control. The problem is particularly severe in West and central Africa, where the deadly ratio of rising populations to declining food supplies has led to an exponential increase in illegal hunting of endangered species. It is estimated that in ten to twenty years the great apes will be extinct. This is not only an environmental tragedy; it is also a humanitarian crisis—there will not be enough wildlife left in the forests to provide food for the people who live there.

As the wildlife population of West and central Africa diminishes, the human population increases. In an article entitled "The Bushmeat Boom and Bust in West and Central Africa" in the conservation journal *Oryx* (2002), Richard F. W. Barnes wrote:

> The human population of sub-Saharan Africa grew from about 84 million in 1900 to 168 million in 1950 and to 612 million in 2000. The mean rate of human population growth for West African countries is 2.6 percent per annum, but the rate is higher in the forest zone because of internal migration. . . . I have argued that human pressure on bushmeat populations are growing more rapidly than national population statistics suggest, that forests produce smaller harvests than people assume, that the growing hunting pressure . . . will produce a period of good bushmeat harvests followed by a collapse, and that collapse will be sudden. . . . By the time the collapse is noticed, it may well be too late to do anything about it.

African immigration into countries like the United Kingdom (UK) creates a market for bushmeat far from the forest. According to the Web site of the UK Bushmeat Campaign (http://www.ukbushmeatcampaign. org.uk/), between 3 and 5 million tons of bushmeat are collected in central and West Africa each year (Wilkie and Carpenter, 1999). On the Web site we read that

> Illegally imported bushmeat bypasses normal health and safety procedures and hence poses a severe threat to the health of both humans and animals in the UK. More than 1000 tonnes of meat are illegally imported into the UK each year. There are about 200 airport seizures a month but it is thought that airports detect, at most, one tenth of illegal meat. Raids at Heathrow on 4, 12 and 15 April 2001 on Ghana Airways, Flight 770 resulted in a haul of 698 kg of meat. In an interview with the CITES customs at Heathrow airport, the campaign was told that, on average, 90 percent of passengers on flights from West Africa are illegally importing meat into Britain.

The recent outbreak of foot-and-mouth disease in the UK has highlighted the dangers of illegally imported meat—it is thought that this is how the disease entered the country. Other diseases such as Ebola, anthrax, swine fever, salmonella, E. coli infections, sheep pox, pseudorabies, African horse sickness, and the Nipah virus (a disease which causes agonizingly painful deaths in pigs and humans) could all be introduced into the UK via meat. These diseases are present in countries which illegally export bushmeat to Europe and the US, and the threat from them is very real. In 2000 there was an Ebola scare at Heathrow Airport, when the carcasses of fifteen monkeys were found hidden in a cargo of fruit and vegetables, along with a pangolin and some tortoise legs. Fortunately, this was just a scare; another time we might not be so lucky. In December 2001, there was an outbreak of Ebola in Gabon which killed at least eleven people in one week. It is thought that these people caught the virus from primates, since an unusually high number of dead primates were found in the area at that time. The Ebola virus can only be passed on through contact with bodily fluids such as mucus, saliva, and blood. The hunting and eating of primates therefore seriously raises the risks of more Ebola outbreaks. There is no cure for Ebola. The virus causes death in 50 to 90 percent of all clinically diagnosed cases, due to severe internal bleeding.

In early 2003, a threat to gorillas of such frightening magnitude appeared that the mind reels. Here is the article from *Newsweek*, dated January 20, 2003:

Gorillas in the Midst of an Outbreak?

Apollo, the world's best-known gorilla, is missing, and the Ebola virus may be the culprit. The alpha male of a 24-member family hasn't been seen since early December, when two members of his family were found dead—along with three other endangered western lowland gorillas and several chimps—in the remote Odzala National Park of the Republic of the Congo. Less than a year ago, contact with a dead ape was blamed for an Ebola outbreak in the area that killed at least 53 people. Specialists have again found Ebola in the dead apes, *Newsweek* has learned. Last week government officials began warning locals to not eat monkeys or ritually wash any relatives who die of fever. Public-awareness efforts may have paid off. So far there have been no confirmed human deaths. But keeping the epidemic at bay is a daunting challenge. Some 3,000 Pygmies and others in the area live from hunting monkeys, a prized source of protein. The area is thick with apes—as many as nine per square kilometer. That adds up to 80 percent of the world's remaining lowland gorillas. Efforts to protect the apes until recently have centered on ending the traditional trade in "bush meat." But Ebola may prove far more devastating to man's closest relatives. The Wildlife Conservation Society, based at the Bronx Zoo, suggests that huge numbers of gorillas and chimps may have died in an Ebola epidemic in the area five years ago. And the new outbreak may not be over—another chimp was found dead in the park last week, according to Jean-Marc Froment of ECOFAC, a regional conservation group. "We may be heading into a catastrophe," he says (Masland 2003).

Less than a month after the *Newsweek* article appeared, the catastrophe materialized. In February 2003, Ebola was reported in the Cuvette-Ouest region of the Republic of Congo. Within three weeks, 89 out of a

total of 110 infected people in the districts of Kelle and Mbomo had died
of Ebola; 600 of the 800 gorillas living in the Odzala park had also died.
Ebola is a hemorrhagic fever transmitted through direct contact with the
body fluids of infected humans or other primates. Outbreaks of Ebola
have been associated with people eating primates infected with the virus;
when Ebola appears, it is not difficult to conclude that bushmeat is the
vector. According to a 2003 online report from the Environmental News
Service, "bushmeat vendors in Ouesso, the largest town in the Republic of
Congo's region of Sangha, have reported a sharp drop in sales due to consumers being frightened by the Ebola virus ravaging a nearby area."
Ebola is not only killing the gorillas, it is killing the people who eat gorilla
meat.

On April 6, 2003, the journal *Nature* devoted its lead article to the issue. Entitled "Catastrophic Ape Decline in Western Equatorial Africa,"
the article was signed by 23 biologists, politicians, and conservationists
from the United States, Africa, France, Spain, and the U.K. (Walsh, *et al.*).
It began like this:

> Because rapidly expanding human populations have devastated
> gorilla *(Gorilla gorilla)* and common chimpanzee *(Pan troglodytes)*
> habitats in East and West Africa, the relatively intact forests of
> western equatorial Africa have been viewed as the last stronghold
> of African apes. Gabon and the Republic of Congo alone are
> thought to hold roughly 80 percent of the world's gorillas and
> most of the common chimpanzees. Here we present survey re-
> sults conservatively indicating that ape populations in Gabon de-
> clined by more than half between 1983 and 2000. The primary
> cause of the decline in ape numbers during this period was
> commercial hunting, facilitated by the rapid expansion of mech-
> anized logging. Furthermore, Ebola haemorrhagic fever is cur-
> rently spreading through ape populations in Gabon and Congo
> and now rivals hunting as a threat to apes. Gorillas and common
> chimpanzees should be elevated immediately to "critically en-
> dangered" status. Without aggressive investments in law enforce-
> ment, protected area management and Ebola prevention, the
> next decade will see our closest relatives pushed to the brink of
> extinction.

This book contains many references to other books that have been written about extinct or endangered species. In addition to these compendia, however, there are many books that pay homage to a particular species, such as Errol Fuller's books about the great auk and the dodo; Tony Juniper's *Spix's Macaw;* Frans de Waal's moving tribute to the bonobo; Robert Paddle's *Last Tasmanian Tiger;* John Walker's book about the last of the giant sable antelopes; C. W. W. Guggisberg's *S.O.S. Rhino,* and an almost uncountable number of books about *Panthera tigris,* one of the most charismatic animals on earth and now one of the most endangered. Dian Fossey wrote about gorillas, and Jane Goodall wrote (and still writes) about chimpanzees. We mourn the passing of those species that are gone, and we agonize for those whose days we know to be numbered, but there is no more poignant impending extinction than that of our fellow great apes, and no more anguished cry than Dale Peterson's *Eating Apes* (2003), the horrifying story of the slaughter of gorillas and chimpanzees for human consumption. Written before Ebola began killing the gorillas of Gabon, Peterson's book focuses on the painful practice of humans eating creatures that are toolmakers, capable of laughter and grief, and are among the few "animals" that know it is themselves that they see in a mirror. We share at least 98 percent of our DNA with chimpanzees, making them our closest living relatives; this alone ought to argue against the idea of our using them for food.

But the indigenous peoples of central Africa have hunted and eaten the animals that share their habitat from time immemorial, making no distinction between gorillas and chimps and the other edible creatures of the forest, such as elephants, antelopes, porcupines, cane rats, and monkeys. (Domesticated cattle and pigs cannot easily be raised in the jungle.) So if they have always eaten gorilla and chimpanzee meat, why should it be a problem now? Simple: the great apes have now been so decimated by hunting and disease that they are heading toward the off-ramp to extinction. In May 1999, the *New York Times Magazine* published an article by Donald McNeil entitled "The Great Ape Massacre," in which he wrote, "Until recently, the great rain forest stretching from Nigeria to Rwanda belonged largely to the Pygmies hunting with poisoned arrows. Nibbling at the forest's edges, they posed no more of a threat to the species than the plains Indians did to the vast American bison herds." Now contributing to the demise of the great apes is the massive invasion of European logging companies into the region, who are cutting down the vast tracts of African hardwood to supply the huge market for exotic woods. These companies need to feed the thousands of loggers brought in to do the work, and because the number of mouths to be fed jumps exponentially, professional hunters are hired to provide food from the forest. A side effect of the timber business is the building of roads to allow the logs to be brought out, and these roads also provide easy access for the bushmeat hunters who would otherwise have to cut their way in.

Donald McNeil's 1999 *New York Times* article was one of the few discussions of this alarming situation to appear in the Western media. In the bushmeat market in Yaounde (the capital of Cameroon), McNeil met with Karl Ammann, a Swiss photographer, who told him that environmentalists once thought that the adult apes were being killed so their babies could be sold in the illegal pet trade, but this turned out to be wrong; the adults were being killed for food, and if they survived, the babies were often taken home by the hunters for their children to play with until the babies starved. McNeil also talked to Steve Gartlan, head of the Cameroon office of the World Wide Fund for Nature, Gartlan's view is that "vivid photographs of dead apes and reminders of their human qualities are 'emotionally understandable but biologically unsound.'" The entire ecosystem need protection, he believes. Gartlan also "expressed sympathy for the hunters, saying, 'We share an even larger percentage of our genes— 100 percent—with the rural poor.'"

The chapters in Peterson's book have titles like "Death," "Blood," "Flesh," and "Business." Perhaps the most disturbing—and believe me, they are *all* disturbing—is the chapter called "Denial," where the author explains why you've heard so little about this problem. It seems that Karl Ammann, the book's photographer and, in a sense, its protagonist, tried to place articles with such magazines as *National Geographic, International Wildlife,* and *Wildlife Conservation,* only to be told that the editors felt that the subject matter was "inappropriate" for their readers, who didn't need to be reminded of such terrible goings-on in the African forests. Peterson wrote, "We can then recognize what a scandal it is that the most significant North American natural history and conservation media were too timid and complacent to present this story as it emerged, too fearful that it would offend the sensibilities of some readers, disturb the progress of ongoing conservation projects in Africa, demonstrate a lack of cultural sensitivity, or fail to emphasize the positive."

Eating Apes is not a supermarket tabloid exposé. It is a serious study, published by the University of California Press, with maps, thirty-three pages of endnotes, and a fifteen-page bibliography for those who might be inclined to check on the veracity of Peterson's reporting. As unpleasant as it is, however, this is a story that must be told, and more important, a story that must be *known*. In *Eating Apes,* Dale Peterson presents an articulate, reasoned, outraged account of the destruction of endangered gorillas and chimpanzees for food. This is in no way a "feel-good" book. It is a "feel-bad" book, even though the author manages to make some suggestions about reversing the terrible downward spiral. You may not like this gory, unhappy tale, and you may shudder at the picture of a dead baby chimp in a freezer or a gorilla head in a bowl, but if you care at all about the status of wildlife conservation in Africa, or if you believe we have a responsibility not to wantonly eliminate other species from the face of the earth, you have to read this book.*

An article by Gretchen Vogel in the June 13, 2003, issue of *Science* begins with this statement: "Ongoing Ebola outbreaks in central Africa are taking a gruesome toll on both humans and great apes. Conservationists, primatologists, and disease experts agree on that much, but in an increasingly heated debate, they are arguing over whether they can or should do anything to limit the spread of the disease." So far, Ebola has killed 150 people and thousands of apes, and while some want to employ whatever measures are necessary to curb the outbreak, others feel that any attempt to stop the spread of the disease will be a logistical nightmare (and probably futile anyway), and that all we can do is stand by helplessly and watch. In their 2003 article in *Nature*, Peter Walsh and his colleagues suggest that the separate African outbreaks are all part of one large epizootic, spreading primarily from ape to ape, that can be controlled by erecting barriers between infected and uninfected populations. What sort of barriers? Removing fallen trees that form natural bridges over rivers, for example, would present a barrier, but how much of the forest has even been sur-

---

* If this section on *Eating Apes* reads like a book review, it is. I was so upset by Peterson's book that I wrote a review and sent it—unsolicited—to the *Times* of London. In a slightly modified form, it ran on May 21, 2003.

veyed, let alone cleared? Transporting uninfected animals to uninfected areas might also work, but "this unprecedented undertaking could end up killing more apes than it saves" (Vogel 2003). Probably the best way to curtail Ebola in apes and people is to stop poachers from killing the gorillas (and stop the people form eating the infected meat). To date, this has proven to be nearly impossible.

Peter Walsh, now at Princeton, felt that the great apes crisis was so severe that he put up a website (www.ApeEbolaCrisis.org) to explain it and to seek a resolution. On the Web site, which so far contains sections on "Ape Decline," "Ebola," and "Hunting," Walsh wrote:

> Less than two decades ago, healthy gorilla and chimpanzee populations swept almost continuously across western equatorial Africa, the region where most of the world's apes live. A wave of commercial hunting and an epidemic of Ebola hemorrhagic fever have now reduced ape populations by more than half. Some large concentrations of apes still remain, but if the current trend continues apes will soon be reduced to scattered pockets. While the threat from commercial hunting is severe and accelerating, the threat from Ebola is more urgent. The epidemic is now burning through the densest ape populations on earth and shows no sign of abating. If it is not checked, it will kill perhaps ten thousand gorillas and chimps in the next year or two, a substantial proportion of the world population. At the current rate of spread, it could reach all major ape populations in the region within a decade. We have created this website to alert you to the threat, provide information on apes and Ebola, and enlist your help in stopping the slide towards extinction.

Asia's only great ape, the orangutan (*Pongo pygmaeus*) is the largest arboreal (tree-dwelling) mammal in the world. The word *orangutan* translates as "person of the forest" in Malay and Indonesian. Except for occasional forays to the forest floor by adult males, these animals spend their entire lives in trees, where they subsist on a diet composed almost exclusively of fruit. While the other great ape species (gorillas, chimpanzees, bonobos, and gibbons) are social animals that live in family groups or larger

Orangutan

troops, orangs are largely solitary animals, coming together only for breeding. They do not swing from branch to branch like gibbons, but rather stretch out their long arms and hooklike hands to grasp the next branch.* The armspread of an orangutan is greater than its height. Once a day they construct sleeping platforms in the trees made up of branches and twigs, and during periods of heavy rain, they often add a rooflike covering. Male orangs can grow to twice the size of females, and in the wild can weigh upwards of 200 pounds. (In captivity, where they are usu-

---

* Gibbons are usually classified as "lesser apes." They vary in size and color, but all nine species belong to the genus *Hylobates*, which means "tree-dweller." All are found in Southeast Asian tropical forests. Like all apes, they have no tails. They are smaller than chimpanzees, and their arms are proportionally longer. They do most of their traveling through the trees by swinging from branch to branch with their hands held like hooks, a process known as "brachiation." When walking on the ground on their hind legs, they hold their arms high and bent at the elbow. Gibbons are incredibly noisy creatures, calling out in screams, hoots, howls, and trills. The howling serves a communicatory function, and is also employed in mating rituals. Many of the species in Malaysia, Java, Sumatra, and Borneo are endangered by the destruction of their habitat by logging, and the reclamation of the forests for agriculture.

ally overfed and underexercised, male orangs grow sedentary and grossly obese, and may balloon to 400 pounds.) The forehead is high and the muzzle protrudes. Adult males develop fatty cheek pads and a pronounced beard which enhances the sexual dimorphism of the species. Males and females wear a coat of long, sparse, brownish orange hair, which darkens with age.

At one time, orangs occupied much of what is now Indonesia, particularly the islands of Sumatra and Borneo (now Kalimantan). Zoologists once separated *Pongo* into two subspecies, the Bornean orang, *P. pygmaeus,* and the Sumatran, *P. abelii,* but they have now been lumped together as *P. pygmaeus.* Whatever their designation, their numbers have been steadily reduced, mostly because they have been hunted since humans first settled their islands 40,000 years ago, and, more recently, because their tropical rain forest habitat is being eliminated. Illegal logging, and forest fires deliberately set in order to convert virgin forest to timber and palm oil plantations, have been responsible for the loss of most of the available orangutan habitat over the last twenty years. In Sumatra they are disappearing at a rate of 1000 individuals a year, while in Borneo the rate of decline is likely to be even higher. Recent estimates suggest that wild orangutan numbers could be as low as 15,000. The World Bank has stated that if habitat destruction continues at its current rate the forest will be gone in Kalimantan by 2010, and as the forest goes, so goes the orangutan.

Sumatra (and therefore the Sumatran orangs) is in Indonesia, while the island of Kalimantan is partly in Indonesia and partly in Malaysia (Sabah and Sarawak). Both countries have established protected reserves for the orangutan, but they are poorly maintained and ineffectually patrolled, and illegal logging goes on under the noses of the wardens. The biggest threat is from permanent agriculture in the form of palm oil plantations and, more recently, illegal logging and gold mining inside protected areas. Orangutans breed more slowly than any other primate, the female producing a baby on average once every seven to eight years. This makes the population extra-vulnerable to loss, and accelerates the decline in numbers. At the conclusion of the entry about orangs in the 2001 *Encyclopedia of Mammals* we read:

Orangutan numbers are down by over 92 percent compared with a century ago, and fell by fully half in northern Sumatra between

1993 and 2000. Remaining populations are restricted to small islands that will remain separated, since orangutans are poor colonizers. Hence, serious protection and active management of the remaining forests is needed to avert the orangutan's extinction in the wild.

Such as it is, the world's largest remaining orangutan population is found in northern Sumatra, in the region known as the Leuser ecosystem. The precipitous decline of the orang in its last stronghold has been documented by van Schaik, Monk, and Yarrow-Robertson in a study published in *Oryx* in 2001. From a high of 12,000 in 1993, the orangs have been killed off at a rate of approximately 1,000 per year, and the total is now estimated at around 5,000. "It has long been known," wrote the authors, "that orang-utans do not respond well to logging." They are driven into unlogged areas by the disturbance, but as legal logging and tree-poaching spreads, the tree-dwelling apes are deprived not only of the fruit that grows in the trees, but of the trees themselves. "At this rate," they conclude, "further losses in the near future are expected to put the survival of Leuser's orang-utans in serious doubt." How serious? *They will be extinct within ten years.*

We should not need a "reason" to avert an animal's extinction in the wild, but the recent discovery that orangs, like chimpanzees, exhibit signs of what primatologists refer to as "culture," might foster an even deeper concern for the survival of the species. In a 1999 article entitled "Cultures in Chimpanzees," Andrew Whiten, Jane Goodall, and seven other primate experts identified "39 different behaviour patterns, including tool usage, grooming and courtship behaviours" that different groups of chimpanzees approached differently." If culture can be defined as the "non-genetic transmission of habits" (de Waal 1999), then the finding of some behaviors that are practiced by some groups and not by others strongly suggests that these apes have it. In other words, the behavior of one group differed substantially from the behavior of another in ways that could only be acquired by association or observation. In the Kinabtangan Forest of Borneo, orangs that are bedding down for the night utter a sound between a hoot and a sigh, but orangs elsewhere are completely silent. Carel van Schaik (one of the authors of the "orangutan" entry in the *Encyclopedia of Mammals*) and several others (including Birute Galdakis, the best-known of all orang researchers) declared that "great-ape cultures exist, and may have done so for at least 14 million years."

In November 2003, the United Nations Educational, Scientific, and Cultural Organization (UNESCO) declared that $25 million was urgently needed to save the great apes from extinction. Koïchiro Matsuura, UNESCO's director-general, said, "Great apes form a unique bridge to the natural world. The forests they inhabit are a vital resource for humans everywhere, and for local people, in particular, a key source for food, water, medicine, as well as a place of spiritual, cultural, and economic value." Matsuura was quoted in an Environmental News Network Web page, dated November 26, 2003, which went on to say, "Every one of the great ape species is at high risk of extinction, either in the immediate future or at best within 50 years." Chimpanzees are gone from Benin, the Gambia, and Togo, the site reported, and are down to the low hundreds in Ghana and Guinea-Bissau. At current infrastructure growth rates, by the year 2030 there will be only 10 percent of the gorilla's original habitat left in Nigeria, Gabon, Rwanda, and Uganda. And for bonobos, only 4 percent of their original habitat will be undisturbed and available by 2030. The Southeast Asian habitat of orangs already has almost no undisturbed areas.

The Anti-Extinctions

Hard as it is to believe, with Mother Earth groaning under the weight of billions of human beings, there are at least a few flickers of hope for the planet's beleaguered wildlife. Tucked away in an otherwise grim report on endangered species last week was news that two humble creatures long thought to have vanished have resurfaced.

According to the Switzerland-based World Conservation Union, the Lord Howe Island stick insect, a five-inch-long segmented bug native to a tiny island group off eastern Australia, turned up on a rocky outcrop last year after lying low for some eighty years. Known locally as the "land lobster," *Dryococelus australis* was thought to have been wiped out after black rats were accidentally introduced to the islands in 1918.

Meanwhile, the Bavarian pine vole (*Microtus bavaricus*), whose only known habitat—a single dewy meadow—was destroyed by the construction of a hospital, reappeared in northern Tyrol, just across the German-Austrian border, after an absence of nearly forty years.

Such comebacks are relatively rare, though. The conservation union, which issues a periodic "Red List" that tallies disappearing flora and fauna, said 11,167 plant and animal species are threatened with extinction by habitat destruction and other environmental pressures.

"We certainly find so-called 'Lazarus species' from time to time," said Craig Hilton-Taylor, the program officer for the Red List. "But we tend to lose far more than we discover."

—Scott Veale, *New York Times*, October 13, 2002

Most of the earth's animal species were here long before they had the good fortune to encounter *Homo sapiens,* but in our typically anthropocentric fashion, we have been inclined to view them in terms of their "discovery." Consider the Linnaean system of nomenclature: the generic and specific names identify the animal, but the name of the describer is also part of the name, along with the date that the first description was published. Thus the full name of the green turtle is *Chelonia mydas* Linnaeus, 1758. In the tenth edition of his *Systema Naturae*, Linnaeus named those animals known to him, but of course, thousands of species were named after 1758, and their names include the names and dates of their first describers. For example, the leatherback turtle is *Dermochelys coriacea* Vandelli, 1761. But the real definition of anthropocentrism is that many species are no longer viable, or will not last very much longer. We had nothing to do with their arrival, but we have more than a little to do with their departure.

If there is such a thing as an antithesis to extinction, it might be found in the recent discovery of animals whose existence was not previously suspected, although of course, they do not take the place of those species that have vanished; those that are gone far outnumber those that have so unexpectedly appeared. Early in the twentieth century, when most people assumed that all the large animals had already been discovered, the okapi, a hitherto unknown relative of the giraffe, was found in the dense rain forests of central Africa. During his 1890 search for Dr. David Livingston, Henry Stanley penetrated the Ituri Forest of the Congo, and learned of the existence of an enigmatic creature called "o'api." In his book *In Darkest Africa*, Stanley wrote, "The Wambutti knew a donkey and called it 'atti.' They say that they sometimes catch them in pits. What they can find to eat is a wonder. They eat leaves." Rumors of this strange, donkey-like animal reached Sir Harry Johnston, the governor of Uganda, and he planned an expedition to the Congo in 1899 to search it out. After winning the confidence of the Wambutti, Johnston was able to learn more about the mysterious atti—including its real name.

After hearing it described as a dark brown animal resembling a donkey with striped legs, Johnston was sure that the o'api was a species of forest zebra still awaiting a scientific description. Later that year, in the Belgian fort at Mbeni, he was able to obtain two headbands made from striped pieces of okapi skin, which he sent to the Zoological Society of London in 1900. Based on these pieces of skin, an announcement of a new species, *Equus? johnstoni* was made. Back in the Congo, Johnston was shown a set of tracks by the natives which they insisted were made by an okapi. However,

Okapi

as the tracks were cloven-hoofed, Johnston dismissed them; they did not fit his notion of the okapi as a member of the horse family. Meanwhile, Karl Eriksson, commandant at Fort Mbeni, was able to secure a complete skin and two skulls, which he sent to Johnston. Armed with these findings, Johnston wrote back to the Zoological Society of London, sending the priceless cargo along. The new animal was named *Okapia johnstoni* by E. Ray Lankester of the British Museum of Natural History. In *On the Track of Unknown Animals* (a 1995 book dedicated to the proposition that there are still lots of undiscovered animals), Bernard Heuvelmans wrote,

> It was an odd animal—reminiscent of those mythical monsters made up from bits of different creatures. It was as big as a medium-sized horse, a little like a giant antelope in shape, but without visible horns; on the other hand it had a long tongue like an anteater's and ears as big if not bigger than a donkey's; its thighs and hindquarters were covered with stripes. Because of this detail alone, it had been taken for a new species of zebra. Actually, as Sir Harry wrote, "Upon receiving this skin, I saw at once what the okapi was—namely, a close relation of the giraffe."

Named for the elongated rostrum of the skull, which to some resembles the beak of a bird, the beaked whales are the most poorly known of all cetaceans. Ranging in length from fifteen to forty feet, most species have been classified in the genus *Mesoplodon* ("middle tooth") for the two teeth in the middle of the lower jaw of the males (the females are mostly tooth-less). The mesoplodonts are spindle-shaped, tapering noticeably at both ends. The blowhole is crescent-shaped, with the horns facing forward, and there are two throat grooves. The dorsal fin is usually small and lo-cated far back, and the tail lacks the median notch present in most other cetaceans. Many of the beaked whales are poorly known, but *Indopacetus pacificus* is the least known of all.

In a 2002 article in *Natural History*, Robert Pitman wrote, "It had to be out there somewhere. It wasn't just a set of car keys that had gone missing—somehow an entire species of whale had been lost for a cen-tury." The Indo-Pacific whale was not a wraith or a distant sighting at sea; its existence was documented by two skulls that were obviously beaked whales, but were different enough from the other members of the group to warrant their own species designation. One of the skulls had been found on a beach in Queensland, Australia, in 1882, and the other was dis-covered in Somalia, on the east coast of Africa, in 1955. The Queensland skull is almost four feet in length, so it was assumed that this was one of the larger of the beaked whales. At first it was thought to be a rare mem-ber of the *Mesoplodon* genus, but in 1972, Joseph Moore assigned it to its own genus, *Indopacetus*, or "Indo-Pacific whale." Most of the other beaked whales are known from strandings around the world, but *In-dopacetus* was known only from these two skulls. Because we knew it ex-isted, it seemed only a matter of time before another *Indopacetus* specimen showed up on a beach somewhere.

A little cetological detective work has shown that *Indopacetus* has been visible for some time—we just didn't know what it was. For at least thirty years, there have been reports of an unidentified species of beaked whale in the tropical waters of the Indian and Pacific Oceans. In his 1971 *Field Guide of Whales and Dolphins,* W. F. J. Mörzer Bruyns wrote of "the possibility that the author observed these animals [*Indopacetus*] in the Gulf of Aden and the Sokotra area, being very large beaked whales and cer-tainly not *Ziphius* [another genus of beaked whales]." Mörzer Bruyns also wrote that "Mr. K. C. Balcomb of Pacific Beach, Washington USA took a photo of a school of 25 beaked whales on the equator at 165° West [in the vicinity of the Gilbert Islands], which were almost certainly this species."

In Balcomb's and other photographs the animals looked more like bottlenose whales (*Hyperoodon*) than beaked whales, but the two known
species of bottlenose whales (*H. ampullatus* and *H. planifrons*) are from
high northern and southern latitudes respectively, and had no business being in tropical waters. Both bottlenose species have bulging foreheads,
and although they are born dark gray or brown, they become lighter with
age, especially around the head. Males differ so markedly from females
(they are considerably larger and have a much more pronounced forehead
bulge, commonly known as the "melon"), that early cetologists classified
them as two different species. Maximum length for a male is thirty feet,
and for a female, twenty-three feet.

In 1999, Pitman *et al.* published an article in *Marine Mammal Science*
with the intriguing title, "Sightings and possible identity of a bottlenose
whale in the tropical Indopacific: *Indopacetus pacificus?*" The article incorporated a collection of photographs (one of which was Balcomb's 1966
picture), and when the photographs and eyewitness descriptions were
compared, it was clear that the whale was a bottlenose whale, and it was
neither the northern nor southern version. The photographs clearly show
a bottlenose whale, whose body color "has been variously described as
tan, light brown, acorn brown, gray brown or just gray" (Pitman *et al.*
1999). When an adult female beaked whale stranded in late 1999 in the
Maldives in the northern Indian Ocean, it was identified as *Indopacetus*,
and comparison with the museum specimens showed once and for all
that *Indopacetus* was actually the tropical bottlenose whale. In his 2002 article, Pitman wrote,

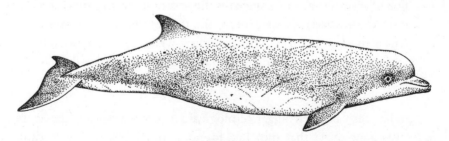

Indopacetus

For a hundred years, cetologists had nothing to work with but two skulls on the shelf. We now have specimen material from six individuals (including five skulls and one complete skeleton) records of more than two dozen sightings, numerous photographs of large animals in the field, recordings of their vocalizations, and (welcome to the twenty-first century) eight minutes of digital video footage.

And then, in July 2003, a complete description of *Indopacetus pacificus* appeared in *Marine Mammal Science* (Dalebout *et al.* 2003). It was written by nine authors who came from New Zealand, Australia, the Maldives, South Africa, Kenya, and California (Robert Pitman was the California contributor). They examined the two original skulls from Queensland and Somalia, and added four new specimens to the list: a skull that was found in the National Museum of Kenya; several ribs and vertebrae that had been found on a Natal beach in 1976; a skull, mandible, teeth, ribs and earbones of a specimen (mistakenly identified as *Hyperoodon planifrons*) in the Port Elizabeth (South Africa) Museum; and the adult female that was collected in the Maldives in January 2000. DNA sequencing showed that all these specimens belonged to the same species, *Indopacetus pacificus*, now known as Longman's beaked whale. The range of this species is now known to incorporate "the western reaches of the tropical Pacific Ocean . . . and the western, northern, and southern latitudes of the tropical Indian Ocean." The authors concluded:

> The discovery of these four new specimens has extended the known range of Longman's beaked whale and led to the description of its external appearance for the first time. As this species is now known from six specimens, the title of the world's rarest whale must pass to the spade-toothed whale [another beaked whale] *Mesoplodon traversii* (= *M. bahamondi*) which to date is known from only three specimens.

In late 2003, an even bigger cetological surprise was sprung: Japanese scientists announced that they had found a completely new species of baleen whale. In the journal *Nature* (November 20, 2003), Wada, Oishi, and Yamada described a "newly discovered species of living baleen whale," which they named *Balaenoptera omurai,** after the late Hideo Omura, one of Japan's foremost cetologists. In the 1970s, according to the report, eight specimens of unknown identity were collected in the South China Sea as part of Japan's research whaling program, and in 1998, another specimen, also of unknown identity, was found washed ashore on an island in the Sea of Japan. In coloration and proportions, these nine specimens resembled fin whales (*B. physalus*), but they were much smaller—thirty feet maximum length to the finner's seventy-five. After analyzing the skull, baleen plates, enzymes in the liver and muscle, and DNA, the Japanese cetologists could affiliate their specimens with no known baleen whale species, and announced the new species.

Around 1960, during the heyday of modern Japanese pelagic whaling, scientists had announced the discovery of a new subspecies of blue whale, which they called *B. musculus brevicauda*, and because it was considerably smaller than the "ordinary" blue whale—which is, in fact, the largest animal that has ever lived on earth—it was nicknamed the "pygmy blue whale" (Ichihara 1966, Omura *et al.* 1970). Finding a "new" species at that time meant, at least to the Japanese, that the quotas for blue whales did not apply to the pygmies; they slaughtered them in large numbers. This earlier episode does not mean that the latest is merely a subterfuge to allow Japanese whalers to hunt a species not covered by International Whaling Commission (IWC) regulations. However, some researchers have suggested that the new discovery conveniently seems to justify Japanese "research whaling," which has been under attack for years, largely because tons of meat end up in the supermarket after the "research" on the whale carcasses has been done. If *B. omurai* is a new species of baleen whale, it is a cause for celebration, but the fact that it was unknown until 2003, in waters heavily fished by the Japanese for centuries, suggests that there are not very many of them, and if they are being hunted, their discovery may coincide with their incipient extinction.

Many species of whales and dolphins have been hunted so mercilessly that they have been driven to the brink of extinction. That we have failed

---

*At the same time, and using the same techniques, they concluded that Bryde's whale (*B. edeni*) was actually composed of two different species, *B. edeni* and *B. brydei*, bringing the total number of balaenopterids to eight. The previous six were the blue whale (*B. musculus*), the fin whale (*B. physalus*), the sei whale (*B. borealis*), Bryde's whale (*B. edeni*), and two types of minke whale, the northern species (*B. acutorostrata*), and the Antarctic (*B. bonaerensis*).

to eliminate an entire species is certainly not for want of trying. The entry of Longman's beaked whale and "Omura's whale" into the catalog of living species does not offset the massive depredations of the past, but it is somehow fitting that among the unexpected "anti-extinctions" is a large whale, and a baleen whale at that. Just as the sperm whale was the predominant species in nineteenth century whaling, the baleen whales virtually defined commercial whaling in the twentieth century.

Despite these fancy new whales, the paradigm for recently discovered, unexpected species is a fish. In 1938, fishermen from the South African city of East London hauled in a five-foot-long specimen that was steely blue in color, with large bony scales and fins that appeared to be on leg-like stalks. It was first examined by Marjorie Courtney-Latimer, and when she could not identify it, she contacted J. L. B. Smith, an amateur ichthyologist and professor of chemistry at Rhodes University at Grahamstown. Smith correctly identified it as a relative of a lobe-finned fish of the genus *Macropoma*, extinct for about 70 million years, and known since 1836 as "coelacanth" (pronounced *see*-la-canth). The name means "hollow spines," and refers to the first dorsal fin. Smith named the living species *Latimeria chalumnae*, after Miss Latimer and the Chalumna River near which it was found. No other coelacanth was seen until 1952, but since then many more have been caught, usually in the vicinity of the Comoro Islands between Mozambique and the island of Madagascar. Local fishermen catch them unintentionally, usually while fishing for the oilfish (*Ruvettus pretiosus*). What was once believed to be a stable

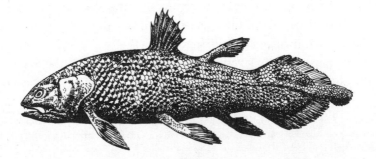

Coelacanth

population of about 650 animals is now thought to number no more than 300, and this rare and zoologically significant creature is probably on the brink of extinction. Female coelacanths give birth to live young, and did so long before the arrival of mammals. The fish spend the day in lava caves, and descend to around 2,000 feet to forage at night. In 1997, a coelacanth was spotted in a fish market on the Indonesian island of Celebes (now known as Sulawesi), and another specimen was hauled up in July 1998. Some 6,000 miles from East Africa, it appears that there is another, heretofore unexpected, population of coelacanths.

The 1976 discovery of the first megamouth shark (*Megachasma pelagios*) off the Hawaiian island of Oahu was another of the more remarkable zoological finds of the twentieth century. Completely unexpected because nothing like it had ever been seen, this shark, tangled in the parachute sea anchor of a U.S. Navy research vessel, was fourteen and a half feet long and weighed 1,653 pounds. A male, it had big rubbery lips lined with thousands of tiny teeth; the examination of its stomach contents showed it to be a plankton feeder (Taylor, Compagno and Struhsaker 1983). It was originally classified as its own family, genus, and species, but it has now been shown to be related to the Lamnidae or mackerel sharks. Since the original find, eighteen more specimens have been collected, usually dead, but in 1990, a healthy specimen was caught in a gill net by a California fisherman, filmed, tagged, and released. Other specimens have come from Japan, Australia, Brazil, and the Philippines, indicating a wide if still unknown distribution. In January 2003, the eighteenth megamouth specimen was caught in a gillnet off Cagayon de Oro on the island of Mindanao in the Philippines.

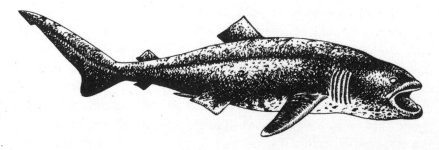

Megamouth Shark

*   *   *

In 2002, a camera trap in Tanzania's Udzungwa Mountains caught an image of a little carnivore known as Lowe's servaline genet, the first photograph ever taken of this mongoose-like creature. It was previously known from a single skin collected in 1932 by British explorer and naturalist Willoughby Lowe. (One of Lowe's other contributions was the description of Miss Waldron's red colobus, a species of African monkey that was declared extinct in 2000.)

In 1993, the saola or Vu Quang ox, a previously unknown deer-sized animal with horns like an antelope, was discovered in a mostly unexplored rain forest on the Vietnam-Laos border. Its generic name *Pseudoryx* is derived from the Greek *pseudes*, which means "false" and *oryx*, an antelope with straight horns that are ringed at the base. Because the creature was found in the Vietnamese provinces of Nghe An and Ha Tinh, it was given the almost unpronounceable name of *nghetinhensis*. Known to Vietnamese hunters for centuries, it was "discovered" by Do Tuoc and John MacKinnon in 1992 and described in the journal *Nature* by Dung, Giao, Chinh, Tuoc, Arctander, and MacKinnon (1993). Because of the proximity of the original discovery to Laos, George Schaller and Alan Rabinowitz (1995) sought evidence of this rare creature in that country too, and learned from hunters that they occasionally found saola in the steep forests of the Annamite Mountains. Later, Kemp, Dilger, Burgess, and Dung would extend the known range of *Pseudoryx* to other regions of Laos and Vietnam, all "mountainous with steep river valleys, covered by evergreen or semideciduous forests between 300 and 1800 m, with low human disturbance."

The saola is a chestnut brown animal, with darker legs, white markings on its face, and a short white "sock" above each hoof. It is antelope-like in appearance, with a long, heavy neck and slightly curved black horns in both sexes. The word *saola* (also spelled *sao la*) means "spindlehorn" or "spinning-wheel post." The largest "new" herbivore since the discovery of the okapi in 1915, the saola stands about three feet high at the shoulder, and weighs about 200 pounds. It is so different from any currently known species that a separate genus has had to be constructed for it, and so far, it has not been determined if it ought to be placed with the goats, antelope, or cattle.

Saola

In January 1995, Boris Weintraub, writing in the "Geographica" section of the *National Geographic,* wrote:

> She lived barely four months after farmers captured her last year in central Vietnam's Vu Quang Nature Reserve, but this female *sao la* was a revelation—the first live example of her species seen by scientists. She was nearly eight months old when she died of respiratory and digestive complications in a botanical garden in Hanoi.
>
> The sao la came to the world's attention when John MacKinnon, a biologist at the World Wildlife Fund, and Vietnamese researchers saw three partial skulls during a survey of the Vu Quang. DNA analysis confirmed it was a previously unknown species—one of only a handful of large mammals discovered this century. A second sao la, a young male, brought to Hanoi last September, died a week before the female.

The saola lives in the highlands of Vietnam during the wetter summer months, moving down to the lowlands during winter when water at high elevations becomes scarce. Most specimens have been caught during the winter, when these shy animals live in closer proximity to people. Although it is not known how many saola survive in the dense jungles of Vietnam, they have already been declared endangered, because they are hunted for food by native people, and subject to loss of their habitat as it is converted to farmland. They are also threatened by hunting pressure as zoologists are eager to capture live specimens for exhibition. "This desire to shoot or capture saola because of their value cannot be good for the future survival of the species in Vietnam," wrote Kemp *et al.* (1997).

Alan Rabinowitz of the Wildlife Conservation Society participated in the early revelation of the saola. In 1997, working in the jungles of northern Myanmar (previously closed to western scientists), he found the world's smallest deer. Previously unknown to science, *Muntiacus putaoensis* is about the size of a large beagle and half the size of the smallest deer previously known. Examining a tiny carcass, Rabinowitz originally thought it was a juvenile of some known species, but, as he wrote in his account of his adventures in Myanmar, "I pried open the animal's mouth to examine her teeth. My heartbeat quickened. The teeth were worn and stained. I rolled her onto her back. The nipples were brown and wrinkled. She was an old adult!" Local hunters knew the animal well, and called it *phet-gyi*, or "leaf deer," because it was small enough to wrap its body in a single large leaf. Rabinowitz wrote,

> Except for the male's diminutive, unbranched antlers, which averaged more than an inch long, adult male and female leaf deer are identical in size and appearance. They both weighed about 25 pounds. . . . In addition, while all the other deer species I knew about were born with spots and retained them for the first two to six months of life, there was no spotting on the coats of either the newborn or the two juvenile leaf deer I examined.

Within a couple of years, Indian scientists working in Arunachal Pradesh, the part of northeastern India that borders on northwestern Myanmar, found specimens of what the locals called *lugi che,* which means the same thing as *phet-gyi,* a deer small enough to be wrapped in a leaf. The leaves come from a plant that the Burmese call *taungzin phet* and botanists call *Phrynum capitum*, but Rabinowitz writes that "it often took several leaves, not one, to wrap the entire deer's body for transport." The Indian scientists (Datta, Pansa, Madhusudan, and Mishra 2003) also noted

Leaf Deer

that both males and females had pronounced canine tusks, "which has not been reported in any other muntjac species."

Duikers (pronounced *dye*-kers) derive their name from the Dutch word for "diver," a reference to their ducking into the underbrush when threatened. They are small, forest-dwelling antelopes found in central Africa. There are some eighteen species, ranging from the hare-sized blue duiker (*Cephalophus monticola*); to the Great Dane–sized *C. sylivicultor*. Both sexes have horns, but they are short, spikelike projections that would interest no reputable big game hunter. Duikers come in a wide range of colors and patterns, from solid to boldly striped, but none is lovelier than Jentink's duiker (*C. jentinki*). It has a mouse gray body and a jet black head and neck, separated by a white band that encircles the shoulders. Like many duikers, Jentink's is nocturnal, spending the day holed up in a hollow tree, a fallen tree trunk or the buttress bay of a kapok tree. The species was first recognized by F. X. Staempfli in 1884, when a specimen was collected by F. A. Jentink near the coast of Liberia. The type specimen was described by Oldfield Thomas in 1892, and is now in the British Museum of Natural History. Nothing was heard about the species for another half century, and zoologists began to fear that it was extinct, but a fresh skull obtained in 1948 in Liberia restored the little antelope to the realm of the living. Since the 1960s it has been spotted in the southwestern forests of the Ivory Coast, and in Liberia and Sierra Leone, in the few remaining areas of undisturbed forest. Nowak (1991) wrote that it "is jeopardized by habitat destruction and probably numbers only a few hundred individuals; it is classified as endangered by the IUCN." Like almost any mammal larger than a squirrel, duikers are targeted by commercial bushmeat hunters and local residents. In 1971, at the Gladys Porter Zoo in Brownsville, Texas, Jentink's duiker was successfully bred; the second generation of the offspring of "Alpha" and "Beta" is now thriving.

Camels seem odd candidates indeed for the category of rediscovered animals, since they have been known and used as beasts of burden since pre-biblical times. Because of their early domestication, however, there are very few wild camels—except in Australia, of all places, where large numbers of one-humped camels now run wild. The early explorers imported them to transport people into the dry interior of the island continent. In 1866, the first camel stud farm was established in South Australia. An additional 10,000 to 12,000 camels were imported from India and Pakistan between 1860 and 1907, to be used as draft and riding animals by people pioneering the interior. Camels were also used in the construction of the overland telegraph line; they carried pipe sections for the goldfields water supply, and supplied goods to inland towns, mining camps, sheep and cattle stations, and Aborigine communities. Wagons hauled by camels moved wool from sheep stations to railheads, and pulled plows and other farm implements. With the introduction of motorized transport in the 1920s, the days of "working camels" were numbered, and they were simply released to fend for themselves. The population of wild camels in Australia is now estimated at between 150,000 and 200,000, distributed ·

Jentink's Duiker

throughout the semiarid desert areas away from human settlements and the coasts.

The camel family first evolved in North America; the llamas, vicuñas, guanacos, and alpacas of South America are descendants of the early North American camelids. There is very little fossil evidence to indicate how and when camels crossed into Asia and North Africa. Like those used by Indians, Arabs, and other Middle Eastern peoples, the Australian camels are of the one-humped variety, known as dromedaries (*Camelus dromedarius*). At one time they roamed wild, but except in Australia, there are no wild dromedaries left. According to Nowak (1991), there are "about 14 million domestic camels in Africa and Eurasia, the highest number being in Sudan, Somalia, and India." Most of the million and a half domesticated camels in China and Mongolia are of the two-humped variety, a species commonly known as the Bactrian camel (*C. bactrianus*) after the Central Asian region once known as Bactria. Shorter of leg than the dromedary, the Bactrian camel is a sturdier, more powerful animal not normally known for speed, although it is said that an unburdened Bactrian camel can outrun a horse. Like its one-humped counterpart, Bactrian camels are almost completely domesticated, although there are a few remote areas in Central Asia where they still run wild.

Almost as if they were designed as high desert pack animals, *C. bactrianus* can endure the searing hot days and the freezing nights of the Gobi Desert, while carrying 400-pound loads for days at a time; they can, if necessary, go for nine days without water, and a month without food. The humps store fat, not water, and provide nourishment when food is scarce. Mongolian herders have used domesticated camels for more than 3,000 years. In the almost impassible Gobi, they are still being utilized for transport, though they also serve as an important source of meat, milk, and wool.

The foothills of the Arjin Shan Range in Xinjiang province, China's nuclear test site, had been closed to outsiders for forty years; but in 1995 and 1996, John Hare, an Englishman with a passionate interest in double-humped camels and the founder of the Wild Camel Protection Foundation (http://www.wildcamels.com/), led two expeditions to the Chinese Gobi to look for wild camels. On the first expedition, which lasted for almost two months (March 31–May 28), they saw no camels, but the second time out, Hare spotted eight. He managed to photograph a mother and

Wild Bactrian Camel

young calf, but by and large, this expedition was not considered a great success (Hare 1997).

Hare's expeditions were conducted in the old-fashioned way; his team traveled by land through the designated areas, scanning the horizons for wild camels. He wrote: "Over 5500 km were covered in 1995, with some surveys being conducted on foot. Temperatures ranged from −7° to 30°C in the day, and on two occasions sand-storms occurred in excess of gale force 7. In 1996 the team covered over 3000 km by jeep and 65 km on foot." There had to be a better way to look for camels. Under the auspices of the Nature Conservancy, the Chicago Zoological Society, the Denver Zoological Society, and the Mongolian Academy of Sciences, Rich Reading led an aerial reconnaissance survey of southwestern Mongolia in March 1997. With Henry Mix, Badamjaviin Lhagvasuren, and Evan Blumer, Reading's team "observed 227 camels in 27 groups," and estimated that a total of 1,985 wild Bactrian camels lived in the study area. Reading *et al.* (1999) wrote: "Because virtually no camels exist outside the surveyed area, these results roughly estimate the total population of camels in Mongolia."

In 1999 the authorities permitted another land expedition to this barren corner of China; this time Hare found a total of 169 camels in the Kum Tagh ("sand mountains") over a period of six weeks. Probably cut off for centuries from others of the same species, the Kum Tagh and Mongolian camels were found to be genetically separate from the domesticated Bactrian camels; because there was no fresh water available, they had developed the ability to drink salt water. The camels of Kum Tagh managed to survive hunting and at least forty-five atmospheric nuclear tests, some reputed to have been larger than the bomb that devastated Hiroshima. The cessation of nuclear testing opened the area to hunters and to prospectors for gold and oil, who were said to be shooting the camels and blowing them up with land mines as they came to the bubbling saltwater springs. In May 1999, the Chinese government established the Arjin Shan Lop Nur Nature Reserve, incorporating 100,000 square miles of arid wasteland (an area roughly the size of Ecuador), to protect the camels primarily, and also the wild sheep, wild asses, and gazelles that live there. The Mongolian herd of wild Bactrian camels lives in the Great Gobi Strictly Protected Area, 5.3 million hectares (13 million acres) of protected land in the Southern Altai region, created in 1975. There are perhaps 300 wild camels in the Mongolian Gobi, though there have been no recent surveys and the number is necessarily a rough estimate. The total number of wild Bactrian camels is less than 1,000, fewer than the number of giant pandas. At the September 2002 conference of the UN-sponsored Convention on Migratory Species (CMS), held in Bonn, Germany, the Kum Tagh camels, along with the population in the Mongolian Gobi, were entered into Appendix I of the Convention (its highest preservation priority), along with the great white shark and three species of whales: the fin, Sei, and sperm.*

The 1991 edition of *Walker's Mammals of the World* lists three species of titi monkeys (genus *Callicebus.*) They are *Callicebus personatus* from the coastal forests of southeastern Brazil; *C. torquatus,* from southern

---

* When the news of the CMS listing was released, these rare and unusual camels were referred to in the press as "hairy-kneed camels," and this appellation stuck, even though, as John Hare informed me, "The wild Bactrian camel has no hair on its knees, unlike the domestic Bactrian. . . . The hairy-kneed species is a figment of some overenthusiastic journalist's imagination."

Colombia, Venezuela, and northwestern Brazil; and *C. moloch,* from central Colombia, eastern Ecuador, eastern Peru, and central and western Brazil. Titis, about the size of small cats, live in the dense understory of South American tropical forests in small family groups of a mated pair and their offspring. There are now twenty-eight recognized species, each with its unique and colorful fur pattern, occurring over a large part of the Amazon basin and the Atlantic forest of eastern Brazil (van Roosmalen *et al.* 2002).

Brazil has ninety-five known species of primates, far more than any other country. There are 134 species and subspecies, close to one-quarter of the global total for primates. Two of the recently discovered titis belong to the genus *Callicebus:* Prince Bernhard's monkey (*Callicebus bernhardi*), and Stephen Nash's monkey (*C. stephennashi*), both of which were discovered by Marc van Roosmalen, a Dutch scientist working at the National Institute for Amazon Research. Van Roosmalen found the *bernhardi* monkey by accident while searching the region for another kind of monkey. Named after Holland's Prince Bernhard, this titi has dark orange sideburns and chest, a reddish brown back and a white-tipped black tail. Its head and body are about fifteen inches in length and its tail measures about twenty-two inches. The *stephennashi* monkey, found in 2001 near the Purus River, is silver with a black forehead and red sideburns and chest. It was named for Stephen Nash, a biological illustrator with Conservation International, the organization that sponsored van Roosmalen's research. The head and body of *stephennashi* are about eleven inches long and the tail adds another seventeen inches. The two new monkeys bring Van Roosmalen's total of finds to five new species. He claims to have found another twenty species that are as yet unnamed in the same region of the Amazon. Also collected recently were three new species of marmosets: the manicore marmoset (*Callithrix manicorensis*), the Acari marmoset (*C. acariensis*), and the black-headed marmoset (*C. nigriceps*), as well as another monkey known as the Ka'apor capuchin (*Cebus kaapori*).

Lemurs are primitive primates that live in trees; they are found only on the island of Madagascar and the neighboring Comoro Islands. (Most of the known coelacanth specimens were caught off the Comoros.) The

Titi Monkey

forty-odd known species come in a variety of sizes and colors, and can be as big as a good-sized dog or as small as a chipmunk. Because of the massive destruction of Madagascar's forests, the natural habitat of all the species is under threat. The smallest of the lemurs are the mouse lemurs, less than six inches in length. Until recently, there were only two known species of mouse lemurs, the gray (*Microcebus murinus*), found in the dry forests along the western coast, and the russet (*M. rufus*), seen in the more humid eastern forests (Nowak 1991).

When a group of researchers including Steven Goodman, Jörg Ganzhorn, and Rodin Rasoloarison completed a survey of mouse lemur populations in Madagascar's western forests, they found that there were seven different species, including three that were completely new, and completely unexpected. In the forests, they found the four species they expected to find: the pygmy mouse lemur (*M. myoxinus*), the gray brown (*M. griseorufus*), the golden-brown (*M. ravelobensis*), and the gray. But they also collected and named three new species: *M. berthae* (Berthe's mouse lemur), *M. sambiranensis* (Sambirano's lemur); and *M. tavaratra* (the northern rufous mouse lemur). Madagascar's forests are home to an amazing variety of unique plant and animal life, including probably more than 12,000 species of flowering plants, half the world's chameleon varieties, 300 species of butterflies, and nearly a hundred species of mammals. Nearly one hundred percent of the mammals on the island are endemic, existing only there and nowhere else on earth. Despite Madagascar's biological riches, it is one of the world's poorest nations, with a per capita income of approximately $240 per year. About 80 percent of the population are subsistence farmers, many of whom practice traditional "slash and burn" agriculture. As a result, only 10 percent of the island's forests remain, and recent estimates suggest that one to two percent of that is being destroyed each year.

Rescuing Animals from Oblivion

> The beauty and genius of a work of art may be reconceived, though its first material expression be destroyed; a vanished harmony may yet again inspire the composer; but when the last individual of a race of living things breathes no more, another heaven and another earth must pass before such a one can be again.
>
> —William Beebe, 1906

In response to our collective guilt about eliminating so many species, efforts are now being made to rescue animals from the brink of extinction. In those cases where domestic husbandry is not an option, we must consider the environment in which the threatened species lives (or lived), and try to rescue that.

James Greenway's 1958 *Extinct and Vanishing Birds of the World* contains brief accounts of 136 species, some of which, like the passenger pigeon and the great auk, are well known, while others, like the Puerto Rican blue pigeon, the Ryukyu Island kingfisher, and Mata the fernbird, are probably known only to ornithological specialists. In the foreword, Jean Delacour, president of the International Committee for Bird Preservation, wrote,

> We are now witnessing the most tremendous changes in the world, and one of the saddest consequences is the awful threat to the existence of many forms of wildlife. Human populations increase; weapons are improved; new poisons are found and used; and remote areas, so far inaccessible, are penetrated more and more easily. As a result, plants and animals are fast decreasing, and may eventually disappear altogether. Birds, conspicuous and easily killed as many of them are, become particularly affected. Numerous species have been dwindling in numbers in the course of the last twenty years, and have reached a dangerously low level. Furthermore, a large proportion of them are narrowly adapted to certain types of habitats, the destruction of which they cannot survive.

Greenway lists forty-four species and forty-three subspecies that are gone, writing that "the vast majority [of these] disappeared during the last 270 years since the last dodo is thought to have died." I searched the book for macaws, and while I found the Cuban Macaw (extinct), the blue and yellow macaw (status unknown), and the green and yellow macaw (status unknown), I found no mention of *any* of the blue macaws, the hyacinth, glaucous, Lear's, and Spix's, which differ from other macaws in that they do not have the characteristic bare patch of facial skin, and that they are largely *blue*. The hyacinth macaw (*Anodorhynchus hyacinthinus*), the largest of all parrots, exists in the wild in low numbers only in the Pantanal, a huge tract that covers the southern Brazilian states of Mato Grosso and Mato Grosso do Sul and parts of eastern Bolivia and northern Paraguay, for a total area of some 92,000 square miles, half of that in Brazil. The glaucous macaw (*A. glaucus*) is probably extinct; Lear's (*A. leari*) exists in the wild only in the Raso da Catarina district in Bahia, where 246 wild birds were counted in 2000, though there are another twenty birds in the Rio and Sao Paulo zoos. Spix's macaw, a blue parrot with a grayish head, is always listed as one of the rarest birds in the world.

Macaws, all of which are found in South and Central American woodlands and jungles, come in a rainbow of colors; some in feathered finery of bright red, blue, and yellow; some in emerald green; and others in a startling shade of purple. The rarest of them all—and therefore the most desirable to "collectors"—are the blue or purple ones, including the hyacinth, Lear's, glaucous, and the one that lent its name to the title of Tony Juniper's book, published in 2002.* Named for the German naturalist who discovered it in Brazil in 1817, Spix's macaw (*Cyanopsitta spixii*) is certainly one of the rarest birds in the world. When Juniper began to write *Spix's Macaw*, he traveled to the woodlands of northeastern Brazil in hopes of finding some of the birds there. After an extensive search, he and his colleagues spotted only one. "The awful conclusion," he wrote, "was that the world population of wild Spix's macaws contained just a single example. The last of the little group of macaws found at Curaçá was the sole representative for an entire genus of unique birds." Called "the world's loneliest and rarest bird," the

---

* Because of their worldwide popularity as cage birds, there are any number of magazines and websites about parrots, including a Web site devoted exclusively to the blue macaws: http://www.bluemacaws.org/

Spix's Macaw

last wild Spix's macaw has disappeared from its home territory in Bahia, and is feared dead. It has not been seen since October 5, 2001, and extensive searches have failed to find any trace.

There are about sixty Spix's macaws in zoos or private bird collections, but they have not been successfully bred in captivity. For the most part, their owners regard them as valuable commodities—a pair was sold to a Qatar sheik for £50,000—and not as valuable elements of the world's biological inheritance. "This bird," wrote Juniper, "was like a Rembrandt or a Picasso, one of the finer things that only the really wealthy could afford." An effort was made in 1990 to establish a worldwide recovery protocol for Spix's macaws, but it fell victim to conflicting attitudes and policies among committee members, particularly some of the private owners outside Brazil, who decided they wanted to trade in the species. In July 2002, The Brazilian government's environmental protection agency (IBAMA) dissolved the Committee for the Recovery of the Spix's macaw. The remaining Spix's macaws will probably live out their lives in cages, waiting unknowingly for the inevitable extinction of their species.

In *Extinct and Vanishing Birds of the world*, Greenway wrote of whooping cranes: "The population was just holding its own in 1953. . . . It might have been expected that a wild population, with what is usually a mortality rate of 80–90 percent in the first year of life and 50 to 75 percent in later

years, might long ago have disappeared. . . . We may hope, but perhaps
we cannot expect, that the cranes will survive." When Peter Matthiessen
wrote *Wild America* in 1959, he said this about the whooping crane:

> The whooper, understandably, was one of the first birds re-
> marked upon by the explorers of this continent, and the wild
> horn note of its voice, of a volume suited to its stature, con-
> tributed to its early legend. Its dislike of civilization, as evidenced
> by its swift disappearance from the East Coast, gave rise to such
> misconceptions as the following, authored in the early nine-
> teenth century by Thomas Nuttall: "Ever wary, and stealing from
> the view of all observers, it is surprising that furtive and inhar-
> monious as owls, they have not excited the presence of the su-
> perstitious." Nuttall claimed to have observed these cranes in the
> Mississippi Valley, where "the passage of their nightly armies fills
> the mind with wonder."

Nuttall had probably confused the "nightly armies" with sandhill
cranes, smaller and much more numerous than whoopers have ever been;
but at one time in America, it was indeed possible to see flocks of the
great whooping crane (*Grus americana*) flying overhead, migrating north
or south depending on the time of year. It has been estimated that between 500 and 1,400 whooping cranes inhabited North America in 1870.
With numbers like these, the whooping crane population was extraordinarily vulnerable to accidental depredations, but as usual, the causes for
the crash of this population can be laid directly at the feet of man. These
include habitat loss from draining and clearing wetlands, disturbance in
breeding areas and along the migration routes, and the one thing absolutely guaranteed to reduce a population, shooting the birds for their
feathers and as meat for the table. Conversion of wetlands and prairie to
hay and grain production made much of the original habitat unsuitable
for whooping cranes, and indeed, mere human presence interfered with
the continued use of prairies and wetlands by breeding and migrating
whoopers. Reduced in 1926 to no more than a dozen, the population of
whooping cranes has now been brought back to a mighty 424 birds, 292 in
the wild and 132 in captive breeding programs (Teuke 2003).
    Their rescue—and it is by no means assured—was accomplished by

people who realized that this magnificent bird could not survive without help, and therefore worked to stave off what others regarded as the grim inevitability of extinction. They were never very numerous, and several factors contributed to their decline. Many cranes died from collisions with power lines. Substantial numbers were lost to illegal shooting for meat and sport. Some died of avian tuberculosis, others of avian cholera and lead poisoning. Along their long migration route, whooping cranes also are vulnerable to natural disasters such as tornados, hailstorms, or drought. By 1942, only sixteen birds remained in the population that migrated from Arkansas on the Gulf Coast of Texas to Canada. The remnant Louisiana nonmigratory population was reduced from thirteen to six following a hurricane in 1940, and the last individual was taken into captivity in 1950.

In 1945 the U.S. Fish and Wildlife Service and the National Audubon Society established the Cooperative Whooping Crane Project, which, among other things, brought the plight of America's tallest bird before the public's eye. At this time, whoopers were known to winter along the coast of the Gulf of Mexico from Florida to Central Mexico, and migrate northward using two routes, one between Louisiana and Manitoba and the other from Texas and the Rio Grande Delta region to the Canadian provinces. Often, as the great birds flew north to a then unknown breeding destination in Canada, they were shot by Saskatchewan farmers, who believed that the birds were feeding on their grain. In 1954, a bush pilot investigating a fire in Wood Buffalo Park in Canada's Northwest Territories spotted three cranes in a region of spruce bog and tamarack, and when his reports were filed it was realized that the northern breeding grounds of the whooping crane had finally been identified.

The whooping crane is snowy white with black wingtips that are visible only when the wings are extended; these wings can span eight feet. The bird's bill is long, dark, and pointed, and the legs are long, thin, and black. When it flies, the neck is stretched straight out in front, and the long legs trail behind; its silhouette is that of a broad-winged missile. There is a patch of reddish black bristly feathers on the top and back of the head, and black feathers on the side of the head below the bright yellow eye that look like long, dark moustaches. The most conspicuous spot of color is a red patch on the head, which becomes more prominent when the bird is agitated. Juveniles in their first fall migration usually have a brown head and neck, and a mixture of cinnamon brown and white on the body. The plumage becomes predominantly white by the following spring—if the bird makes it. "Hunters" are still not able to resist taking a shot at such large targets; the birds sometimes collide with power lines; and bobcats raid the nests.

In 1967, efforts were initiated to develop a captive flock of whooping cranes, in the event that the species was extirpated from the wild. The goal of the U.S. Whooping Crane Recovery Plan was to establish two wild populations of at least twenty-five breeding pairs in addition to the existing population, so that the species can be downlisted from endangered to threatened status. There are now two breeding populations in captivity, one at the Patuxent Wildlife Research Center in Maryland and one at the International Crane Foundation (ICF) in Baraboo, Wisconsin.*

Efforts to establish an additional wild population began in 1975 when whooping crane eggs from Wood Buffalo National Park were placed in the nests of sandhill cranes at Grays Lake National Wildlife Refuge in Idaho. After hatching, the chicks were adopted and raised by the foster parents. The young whoopers then migrated with their adoptive parents and wintered in New Mexico. Initial results were promising, but the whoopers failed to form pair bonds, and breeding never occurred. The next attempt to establish an additional population was made in January 1993, when the first group of fourteen whooping cranes hatched in captivity was released in Kissimmee Prairie, Florida, where the objective was to establish a nonmigratory, self-sustaining population.

From 1992 to the present, eighteen whooping cranes have been transferred to a facility in Calgary, Canada, to establish a third captive flock. Presently, biologists are evaluating sites in Canada for the reintroduction of a migratory flock of whooping cranes later this decade. Thanks to these efforts, the whooping crane population has survived and

* Under the inspired leadership of George Archibald, the ICF, in the words of its Mission Statement, "works worldwide to conserve cranes and the wetland and grasslands communities on which they depend. ICF is dedicated to providing experience, knowledge, and inspiration to involve people in resolving threats to these ecosystems." There are fifteen species of cranes; they are found on every continent except South America and Antarctica. The ICF pursues two techniques for crane preservation: captive breeding and reintroduction into the wild. In 2001, Peter Matthiessen published *The Birds of Heaven*, in which he chronicled his travels—often in the company of George Archibald—to see and learn about the plight of the world's cranes.

Whooping Crane

continues to increase. The U.S. Fish and Wildlife Service's whooping crane recovery program has been so successful that other countries have adopted similar methods to protect other species of crane that are also threatened.

In *Wildlife in America,* described as "The first history of man's effect on the fishes, amphibians, reptiles, birds, and mammals of the North American continent," Matthiessen could devote only a few pages to the whooper, but in the 2001 *Birds of Heaven: Travels with Cranes,* he gives a whole chapter to *Grus americana*. Matthiessen may be the best nature writer in America (many regard him as the best *writer,* regardless of subject), and he does the whooping crane proud. He begins: "A magnificent bird was vanishing from the Earth, and any hope that the last flock might prevail seems small indeed." In exquisite and often painful detail, Matthiessen chronicles the recent vacillating fortunes of the whooping crane, a bird that could not possibly survive as a species without the intervention of man, the very agent who brought the bird to its spindly knees:

> In 1999 the project celebrated a 70 percent survival rate among the first-year cranes, higher than had ever been expected, and the first year of the new millennium may be even better. Of the thirty new birds released in 1999–2000, twenty-five are still alive and flying around Florida—in effect, about as many *Grus americana* as existed on earth four decades ago, when I first wrote about them. In the brief annals of crane conservation, this is an exciting success, yet *americana* at the turn of the new century remains much the rarest of the cranes, effectively limited to that single flock that leads its young on a perilous migration some twenty-six hundred miles from the border of the Northwest Territories to a small eroding corner of the Gulf Coast very vulnerable to oil spills and winter storms.

In 2000, a mated pair flew from Florida to northeastern Michigan, where they were seen foraging in wheat and soybean fields. They built two nests, but no eggs were laid. As of this writing, the whooper's prospects are looking better than ever. According to the ICF Web site (http://www.savingcranes.org/).

(Spring, 2002). Whoopers Return to Wisconsin! The flock of whooping cranes raised at the Necedah National Wildlife Refuge last year, and led south to Florida by ultralight airplane are back in Wisconsin! The "Florida five" initiated spring migration on their own on April 9. They returned to Wisconsin after 10 days on migration. By April 19th, four of the five cranes returned to the Necedah National Wildlife Refuge, where they fledged last summer. The whooping cranes are part of a reintroduction effort that hopes to bring whooping cranes back to the eastern flyway after an absence of more than 100 years. The project is led by the Whooping Crane Eastern Partnership (WCEP), a cooperative team of governmental and non-profit agencies including the International Crane Foundation. WCEP led the small flock of whooping cranes from the Necedah National Wildlife Refuge in Wisconsin to Chassahowitzka National Wildlife Refuge, Florida last fall.

Where endangered species are concerned, it is not wise to be too optimistic. Things happen that could not have been predicted, such as the color photograph of two whooping cranes that appeared on the cover of *Science News* for March 29, 2003. The caption read: "Next? West Nile Threatens Wildlife"; the story, written by Susan Milius, was about the spread of the West Nile virus, which appeared in North American bird populations around 1999, and eventually made the transition from birds to humans. West Nile virus, discovered in the West Nile area of Uganda in 1937, has spread to South Africa, the Mediterranean, temperate parts of Europe, and now to North America. It can cause fatal inflammation of the brain (encephalitis), or of the brain lining and the spinal cord (meningitis), in birds, horses, and humans. It is carried and spread by the common household mosquito, *Culex pipiens,* when it feeds on a blood meal from infected birds. Scientists believe that the most likely "reservoir" for the virus in North America is the common sparrow, which can tolerate the infection, but the virus has had the greatest impact among crows. In 1999, in the New York area, the crow population crashed by about 90 percent in a few months.

Milius wrote that "by the end of 2002, the virus had struck a remarkable range of animals, including both dirt-common ones and nearly extinct species. It infected at least 186 wild and captive species, including pigeons, house sparrows, chickens, cardinals, chickadees, mockingbirds, mallards, parakeets, peacocks, macaws, flamingoes, bald eagles and whooping cranes." House sparrows and pigeons exist in large enough numbers so their populations can probably withstand a debilitating epidemic, but species whose populations are already low would be threatened by such a disease. Now we know why the whoopers were used for the cover.

In his 1958 book, Greenway gives only a passing mention to the Mauritius kestrel. The entire entry (in a book that devotes six pages to the Eskimo curlew and eleven to the great auk) consists of the following:

> *Falco punctatus*
> Range: Known only from Mauritius
> Status: Still to be found in small numbers. M. Georges Antelme believes that the species is gradually becoming extinct. The implication of its local name, "mangeur-des-poules" [eater of chickens] is the probable reason. It is a close relative of *Falco newtoni*, the kestrel of Madagascar.

Twenty years later, in *The Sinking Ark*, Norman Myers selected this bird as one that might have to be abandoned to its inevitable extinction, so that we could devote our resources to saving species that are more likely to survive. Neither M. Georges Antelme nor Norman Meyers had reckoned with Carl Jones. Or for that matter, with David Quammen. Because Quammen's book is entitled *The Song of the Dodo*, we expect to find a certain emphasis on Mauritius's most famous extinct avian inhabitant, but there is also a discussion of other troubled species on that island in the Indian Ocean 500 miles east of Madagascar. Quammen quotes Carl Jones, a Welsh-born autodidact biologist who lives on Mauritius:

> "If you look at the early illustrations of this island, and you read the Dutch accounts, there's this overwhelming feeling that Mauritius was a paradise. There were dugongs in the lagoons, there were multicolored shells and fishes everywhere, there were turtles, there were giant land tortoises. Herds of land tortoises. There were vast colonies of seabirds, frigates and boobies, fairy

terns and sooty terns and noddies. There were many species of parrot. . . . There was a very large species of parrot, a ground parrot, that probably couldn't fly. The largest parrot ever known. Called *Lophopsittacus mauritianus.* Herons and egrets, flamingoes, cormorants. All types of birds. Many of these, regrettably, have become extinct."

But not the Mauritius kestrel.

Before Mauritius was settled in 1598, the kestrel probably occurred throughout the island, but as the island's forests were cut down for lumber, the range of the little falcon was restricted to the cliffs, ravines, and remote forests of the southwestern plateau. The Mauritius kestrel is the island's only bird of prey, but the nickname, "mangeur des poules," is inaccurate— it is not big enough to take anything but a chick. While other falcons have long pointed wings that enable them to reach great speeds in aerial pursuit of their prey, the Mauritius kestrel has short, rounded wings for dashes through the trees as it picks off geckos, small birds, dragonflies, cicadas, cockroaches, and crickets.

At one point in the 1970s, the Mauritius kestrel was the most endangered bird of prey in the world, with only four surviving in the wild. (It is still the rarest falcon in the world.) On the island of Mauritius it suffered from habitat loss, the introduction of monkeys and mongooses which ate its eggs, being hunted as a pest, and widespread spraying of DDT. By 1973, the Mauritius kestrel was officially declared an endangered species. The International Council for Bird Preservation (ICBP) launched a campaign to save the species, and sent a biologist to Mauritius to census the kestrels. He counted eight, but when Cyclone Gervaise stuck the island in 1974, it destroyed much of the birds' habitat, and the kestrels were more endangered than ever. In 1978 (just about the time that Norman Myers was suggesting that the bird be a victim of triage), the ICBP told Carl Jones to shut down the effort to save the bird. Lucky for the kestrel, Jones refused to follow orders. Instead, he developed a captive breeding program and was able to raise enough birds to release some back into the wild. Virtually singlehandedly, Jones brought the species from the verge of extinction to more than 200 wild kestrels in the early 1990s. Since then, the Wildlife Preservation Trust of Jersey (UK), in collaboration with the Mauritius government, has maintained an intense conservation program, which consists of captive breeding, supplemental feeding of wild birds, provision of nest boxes, improvement of natural cavities, forced double-clutching (in which eggs are harvested for captive rearing, causing the females to lay replacement eggs), captive-bred bird releases, and control of introduced predators. In 1994, on the Ile Aux Aigrettes off the coast of Mauritius, in a frustrating moment of irony for conservationists, a Mauritius kestrel ate a newly hatched offspring of a recently reintroduced pair of Mauritius pink pigeons, the world's rarest pigeon. The kestrel was captured and removed to a forest site on Mauritius. With the release of the 300th kestrel in 1994, the captive breeding and release program was stopped (Jones *et al.* 1995).

From a low of two breeding pairs in the 1970s, the Mauritius kestrel wild population now numbers more than a hundred breeding pairs and some 350 individuals. Releases of captive birds have ended, thanks to the program's success and the dedicated efforts of many organizations and individuals. The population is monitored carefully during the summer breeding season, and monkeyproof, cycloneproof artificial nesting boxes are placed in selected areas. The young are tagged and measured. The Mauritius kestrel is still a highly vulnerable species; to ensure its survival, its population will have to be managed carefully for the foreseeable future. In 1998 the bird was removed from the endangered species list, and the effort is regarded as one of the most successful species reintroduction and recovery programs ever undertaken.

Although our record of saving endangered Hawaiian bird species is quite terrible, we can point to one success. The nene (*Branta sandvicensis*) is a small goose with only partially webbed feet. Somewhat smaller than the Canada goose but similar in appearance, the nene (pronounced nay-nay) has a black face and hindneck and buff-colored cheeks. The front and sides of the neck are diagonally striped in black and white. When Captain Cook arrived in 1778 there were probably 25,000 nenes, but, like many island species, the nene could not stand up to the introduced animals that colonists brought with them, and the environmental changes that were made in the pursuit of agriculture. By 1950, the population, which had also been decimated by hunting, was teetering on the brink of extinction, with an estimated thirty to fifty birds left in the wild.

By 1957, conservationists began breeding the birds in captivity in hopes of preserving a remnant of the declining population and, someday, re-establishing them in the wild. Early programs for returning captive-bred birds to the wild proved difficult, but recent efforts have been more successful. There are now small but stable populations of nenes on the islands of Hawaii, Maui, and Kauai. They became wild on Kauai in 1982 after Hurricane Iwa destroyed the cages of captive birds on the southeast side of the island. These birds rapidly adapted to the mongoose-free, lowland grassy habitat. Because the birds were so successful, state biologists recently have introduced the nene on the north and northwest coasts of Kauai. There are now about 800 wild nenes in Hawaii, with the numbers climbing each breeding season, and another 1,000 in zoos and private collections outside Hawaii. The largest of these is at Wildfowl and Wetlands Trust in Slimbridge, England. A small group of nenes (probably escapees from Slimbridge) also seem to be enjoying life in the heart of London at St. James Park, a tranquil waterfowl haven situated in front of Buckingham Palace. Hunting is no longer permitted in Hawaii, but the nene is still threatened today by introduced mongooses and feral dogs and cats, which relentlessly prey upon the goose's eggs and young. Preservation efforts are continuing and the future of the nene in Hawaii, although not certain, is promising. It seems fitting that the nene, still perched on the endangered species list, is the State Bird of Hawaii.

Nene

If you were to look up "California Condor" in Alexander Wetmore's 1934 *National Geographic* article on "Eagles, Hawks and Vultures," you would read: "Formerly quite abundant, according to recent estimate by Mr. Harry Harris, possibly ten individuals still exist in California. Little is known of them in Baja California, save that Indians hunt them for ceremonial purposes. But it is certain that few remain, and the species is one that may easily become extinct." Then in Greenway's 1958 *Extinct and Vanishing Birds of the world,* this is (part of) what you would find:

> Rare, local, and in great danger of extinction. . . . No enemies of the California condor are recorded save man alone. The vast increase in human population and the increased range and efficiency of the rifle are the reasons (at least apparently) for the diminution in the numbers of birds. During the same period great cattle ranches have been transformed into fruit farms, and it seems probable that the consequent reduction of food has been a powerful contributing factor.

Never mind that the condors made a perfectly good living even before cattle conveniently died to feed them; in 1958, people who worried about this sort of thing were concerned that this great bird was in danger of disappearing forever.

There are two species of living condors, the Andean (*Vultur gryphus*), and the California (*Gymnogyps californianus*), both of which are candidates for the title of "largest flying bird." At eleven feet, the wandering albatross (*Diomedea exulans*) may equal the condors in wing length, but the broad wings of the condors gives them a far greater wing surface. Soaring high above the mountains that give it its name, the Andean condor is relatively safe from human influences, but in California, America's most populous state, there was little sanctuary for this broad-winged scavenger. On the wing, they are impossible to confuse with any other bird; David Sibley (2000) says, "at a distance it can be mistaken for a small airplane." Far from pretty, these huge black birds develop a ruff of back feathers out of which sprouts a featherless, reddish orange head. Unlike other vultures that rely on their exceptional sense of smell to locate carrion, condors

California Condor

depend on their eyesight, often investigating the activity of ravens, coyotes, eagles, and other scavengers on the ground.

Condors attain adult plumage and coloration by five or six years of age, and reach sexual maturity soon thereafter, first breeding between six and eight years old. The nest is usually in a cave on a cliff or a crevice among boulders on a steep slope. If the nesting cycle is successful, one egg is laid every other year. Instead of having many young and gambling that a few will survive, the condor produces very few young and provides extensive parental care. The average incubation period for a condor egg is about fifty-six days; after hatching, the nestlings remain in the nest for six months. In some cases, juvenile condors may be dependant on their parents for more than a year. In historical times, the California condor occurred from British Columbia south to northern Baja California and other parts of the southwestern United States. The main reason for their decline was an unsustainable mortality rate of free-flying birds combined with a naturally low reproductive rate. Most deaths in recent years have been directly or indirectly related to human activity. Shootings, poisoning, lead poisoning, and collisions with power lines have been the major threats.

In 1953, the first legal protection was given to the California condor, when the State of California wrote into its Fish and Game Code, "It is unlawful to take any condor at any time or in any manner." A statewide survey was conducted in 1965, and it was estimated that sixty birds remained at large. Some fifteen years later, with the number of birds observed in the wild dropping lower and lower, The California Department of Fish and Game issued permits to the San Diego Zoo and the Los Angeles Zoo to breed captive California condors, and allowed the U.S. Fish and Wildlife Service to conduct a three-year research study, including trapping of three condors for captive breeding. In 1982, a chick named "Xolxol" (pronounced *hol*-hol) was taken from a nest and brought to San Diego's Wild Animal Park. At this time, there were three condors in captivity, and an estimated twenty-one to twenty-four in the wild. The following year, four condor eggs were taken from the wild and hatched at San Diego. The first of these was "Sisquoc," the first condor chick to be artificially incubated. It took a variety of techniques developed by scientists and bird keepers to do this, including "double clutching." To make the hand-raised condor chicks feel like they were being raised by their parents, they were fed and cared for by people using adult condor hand puppets, while taped adult condors calls were played. In 1987, when the final wild condors were brought into captivity, the total population—captive and wild—was believed to number twenty-seven birds. One year later, a captive pair bred at San Diego, and produced a chick. In 1990, the San Diego and Los Angeles zoos produced eight chicks in the most successful breeding year to date; two pairs of adult birds double clutched; two pairs triple clutched. A total of forty birds existed in captivity.

The species seemed at least temporarily secure, so plans were made to reintroduce the California condor into the wild. Two birds were released from the Arundell Cliff facility in Los Angeles County on January 14, 1992. When it was seen that they did not know how to avoid power lines, the birds were captured and relocated. More and more birds were released, and by 1999, the total population of California condors was 172—118 in captivity and 54 free-flying. In 2002, a wild pair successfully hatched a chick from an egg laid in the wild. AC9 (adult condor nine), the last free-flying condor to be taken into captivity in 1987, was returned to the wild after fifteen years in the captive breeding program at the San Diego Wild Animal Park. California condors are currently being reintroduced to the central coast of California by the Ventana Wilderness Society (VWS), an organization dedicated to the preservation of native plants and animals. Thanks to the efforts of the VWS, condors are now seen throughout the mountains, coastal canyons and valleys near Big Sur. Biologists from the Peregrine Fund have released condors in the vicinity of the Grand Canyon in Arizona, the first California condors to fly outside of California in seventy years. There are now thirty-two California condors in the wild in Arizona, more than there were in the world in 1982. As of May 1, 2002, the total population of California condors was around 200, and the free-flying population was sixty-eight. In April 2002, three chicks hatched in the wild; they were the offsping of adults raised in captivity and then released.

The Condor Release Program, often hailed as one of conservation's great success stories, has not been without its setbacks. On September 4, 2002, the U.S. Fish and Wildlife Service and the Peregrine Fund each offered a reward of up to $10,000, and the Arizona Game and Fish Department offered another $1,000, for information regarding the recent death of California Condor #186. He died sometime between August 28 and 30, 2002, in the Kaibab National Forest in northwestern Arizona. Hatched at the Los Angeles Zoo on April 15, 1998, he was transported to the Peregrine Fund's Hurricane Cliffs release site on October 8, 1998, and released with eight other condors on November 18. He was expected to begin breeding at six or seven, the normal age for California condors. Furthermore, the three chicks hatched in the wild have all died. On October 23, 2002, AP and other news organizations reported that the last of the wild-born chicks had been found dead in a cave in the Los Padres National Forest near Fillmore, California. The other two chicks had died on October 4 and October 13. A necropsy on the third fatality proved inconclusive, but the first chick's cause of death was the ingestion of a dozen bottle caps and shards of plastic and glass.

Sadly, returning condors to the wild does not necessarily mean that they will live happily ever after and multiply. Of the first eight hand-raised birds that were released, one died from drinking anti-freeze, three died from collisions with power lines, and the rest had to be recaptured (Kaplan 2002). To solve the problem of the power lines, researchers installed wooden cross-beams in the enclosure, and then rigged them so that they would give birds landing on them a mild shock. The strategy worked; none of the birds released after 1994 died from accidental electrocution. Captive-raised specimens had no idea how to function in their natural habitat and instead of soaring aloft looking for carrion, many birds sought out human habitation, which they associated with feeding. When raised by hand and then released into the wild, some condors are so habituated to people that they land right in visitors' centers—such as those at Grand Canyon—and beg for food. Others have entered houses, landed on cars, and untied people's shoelaces. "Naughty condors," wrote Sandra Blakselee in the *New York Times* of June 3, 2003, "have pulled tents apart, ridden on backhoes, bounced on top of canvas-covered water tanks, sneaked up on people napping in remote places, started rock slides, chased helicopters, and played around with food thrown to them by tourists."

Because there were no wild birds to teach them how to behave outside the enclosure, the researchers reintroduced wild-caught adults to serve as "mentors." Concluding his 2002 article about the plight of the condors, Mark Kaplan wrote, "The California condor project shows that even animals on the brink of extinction can be returned to the wild. It also highlights the fact that human activities change the world irrevocably." In April 2003, a man was arrested and charged with shooting a California condor. According to a story in the *New York Times* of April 30, 2003, Brian Lewis, 29, shot and killed "Adult Condor 8," one of the few remaining condors hatched in the wild. She was captured in 1986, the last female of the species caught in the attempt to save the birds from extinction. She hatched a dozen eggs in captivity before she was freed in 2000, one of the first of the original wild birds to be released.

When the first settlers arrived in Bermuda in 1609, they found a plentiful little petrel that they named "cahow" for its call. And call it did; the first explorers called Bermuda the "Isle of Devils" because of the screaming of the bird multitudes. Although the actual number will never be known, it has been estimated that before the settlers arrived there were half a million of the white-bellied, black-backed birds with a three-foot wingspan. Shortly after the settlements, a grave food shortage caused the settlers to turn to the cahows, which were easily caught as they returned to their nests, and consumed in enormous numbers. Other man-introduced animals, such as rats, cats, and dogs, appeared with the early settlements; the impact of these new predators, combined with extensive burning, deforestation, and human capture of birds and eggs for food, greatly reduced the seabird population, bringing the cahows to the verge of extinction. Three centuries would pass before anyone heard the call of the cahow again.

In 1935, only a few years after William Beebe had made the historic dives in the bathysphere with Otis Barton, a boy on a bicycle brought a dead bird to Beebe at his laboratory on Nonsuch Island, Bermuda. He sent it to Robert Cushman Murphy at the American Museum of Natural History in New York, and immediately published an article on his "rediscovery" of the cahow in the *Bulletin of the New York Zoological Society*. (In 1932, in *Nonsuch, Land of Water*, he had written about hearing the song of the cahow, but he had probably heard Audubon's shearwater, which also breeds on Bermuda.) Another specimen was found freshly killed in 1941, and it appeared that *Pterodroma cahow* was not extinct after all, but was breeding on some of Bermuda's uninhabited offshore islands. Cahows, like most petrels, spend most of their lives at sea and come ashore only to breed and nest, so it is not that surprising that they were not noticed before. With Louis Mowbray, curator of the Bermuda Aquarium, R. C. Murphy, one of the world's foremost authorities on seabirds, organized an expedition in 1951 to search for cahow nests. He found seven, which led to a program of protection and support sponsored by the New York Zoological Society (Murphy and Mowbray 1951).

In 1961, eighteen breeding pairs—the entire breeding population at that time—were located on the Castle Harbour Islands. The Bermuda government declared Nonsuch Island a sanctuary for cahows, and removed all potential predators from the island. David Wingate, who had accompanied Mowbray and Murphy as a schoolboy in 1951, took up residence as warden of Nonsuch in 1962. Tropicbirds, larger than the cahows, were using the active cahow nests to lay their own eggs, and thereby displacing the newly hatched cahow chicks, so Wingate and Mowbray devised a series of wooden baffles that prevented the tropicbirds from entering the nests. By 1966, the total had risen to twenty-four pairs, and by the 1990s, there were more than forty breeding pairs. In the 1960s, high levels of pesticides such as DDT were identified in unhatched cahow eggs and dead chicks, presumably the cause of the observed decline in numbers during the first decades after the species' rediscovery. As the use of DDT declined in North America in the late 1960s and early 1970s, the reproductive success of the Cahows climbed back toward its earlier rate. In early 1987, a vagrant snowy owl (*Nyctea scandiaca*) found itself among the cahows on Nonsuch, and killed at least five prebreeding birds. Then an even larger natural disaster struck: in 1989, Hurricane Hugo was implicated in an

Cahow

unusually high mortality rate in the population, and by 1990, there was a substantial reduction in the numbers of breeding pairs. Cahows are well adapted to survive normal hurricanes, provided the storms remain over the ocean. But when very powerful hurricanes move ashore onto the continent, the cahows can be trapped against the shore and then blown far inland, where they crash into the forest or even end up lost on inland lakes. Not only were the cahows hard hit by hurricanes on their Gulf Stream foraging grounds, but their breeding islets in Bermuda were clobbered by Hurricane Gert on September 21, 1999.

Today, thanks to active conservation, approximately fifty breeding pairs of cahows survive to rear their single chicks, who leave to spend their first eight years of life on the open ocean before returning as adults to breed and complete the cycle. Since his first visit to Nonsuch in 1951, Bermuda naturalist David Wingate has almost single-handedly saved the precariously balanced cahow from extinction. Through his efforts, the Cahow Conservation Program has become one of the most successful endangered species recovery programs on the planet. Wingate, now President of Bermuda Audubon as well as Conservation Officer of Parks at Bermuda, envisions a thousand pairs of cahows crowded onto Nonsuch and nearby islets by the year 2020.

The kakapo is one weird parrot. It is nocturnal, it mates on display grounds ("leks") like the prairie chicken, and it breeds only in infrequent years when certain trees undergo mass seeding. As befits a ground-dwelling bird, it is the heaviest of all parrots, weighing in at up to eight pounds. Because New Zealand originally had no land predators, the kakapo (*Strigops habroptilus*) lost the power of flight; it is the world's only flightless parrot. Its inability to fly would render it terribly vulnerable to introduced predators, and would lead to another unfortunate superlative: it is also one of the world's rarest birds.

The Maori-named kakapo lives mainly on the ground, but it can climb trees, from which it descends by "parachuting," using its short wings for balance. In other words, it can fly down, but not up. Its soft moss green feathers are barred with black on the back, and it has pale yellow underparts. Females are smaller and less brightly colored than the males. The kakapo has an owl-like face with "whiskers" (another name is "owl parrot") and a large ivory and pale blue beak. The kakapo's unique bill structure is adapted for grinding food finely; the gizzard, the organ in which food is ground in most parrots, is small and degenerate. While walking on their sturdy legs, they keep their body held low and horizontal so that the whiskers touch the ground. They usually stand in this position as well, unless they are alert or defensive, at which time they hold themselves upright. Kakapos move more deliberately than most birds, and appear to live life in the slow lane. They may live up to sixty years, which would make them among the longest-living of all birds. (None of the known kakapos has died of old age, and some have been studied for twenty-five years.)

When the first Polynesians arrived in New Zealand, perhaps a thousand years ago, they found the pudgy, ground-dwelling parrots ridiculously easy to kill, and even before the first Europeans arrived (the Dutch explorer Abel Tasman was first, in 1642), the kakapo was in trouble. Captain Cook landed in New Zealand in 1769, and while the arrival date of the first rats cannot be documented, they probably followed hard on the heels of the European explorers. The settlers also chopped down the forest habitat of the birds, and introduced ferrets, cats, dogs, and stoats, who easily killed the hapless kakapos. When confronted by an enemy, the kakapo freezes and depends upon its coloring to blend into the background. This technique might have worked against aerial sight hunters such as an endemic eagle that is now extinct, but for mammals that hunted by smell, the kakapo's strong, musky scent was a dead giveaway.

Kakapo skins were highly prized by Maoris, who made cloaks from the feathers, and also used the soft feathers in their mattresses and pillows. European settlers found the kakapo particularly tasty, and around 1845 there began a slaughter of unprecedented proportions. During the New Zealand gold rush of the 1860s and 1870s, the settlers (known as "diggers") lived almost exclusively on a diet of kakapos. Exploring parties also made kakapo the principal item of their diet; in later times, tourists shot and ate kakapos as well. The slaughter of the kakapos by people and other predators continued well into the twentieth century. By the 1950s, the birds were extinct on North Island, and only eighteen remained in the remote mountain forests of the Fiordland region of South Island. Attempts were made to introduce the kakapo to small islands, such as Resolution in South Island's Dusky Sound, but rats and stoats quickly terminated such projects.

From 1958 to the 1970s, the New Zealand Wildlife Service made regular expeditions to the Fiordland area and northwest Nelson regions in search of the now elusive parrots. This work was very difficult, consisting of hiking in treacherous terrain along rugged trails in steep areas in all types of weather. Searchers found only eight birds until 1974, when helicopters were introduced into the search—not to spot the birds from the air, but to transport the workers. Kakapos were found only in two strongholds, the Milford catchment of Fiordland, and Stewart Island. In Fiordland, deer and possums had not yet encroached upon the rugged terrain, thus much of the vegetation survived intact. Predators had not yet arrived in the remote areas to any extent, but the situation was rapidly changing, and signs of stoats and cats were appearing in the birds' last safe havens.

Kakapos breed only in years when food is abundant, so to ensure their chicks' survival and encourage breeding, the Department of Conservation (DOC) decided to give the remaining birds a supplemental diet of nuts and sweet potatoes. (Their normal diet consists of foliage and other poor-quality foods that are insufficient to encourage breeding.) It is only when the podocarp tree undergoes mass fruiting ("masting") that female kakapos lay eggs. In 2002, the podocarps on Codfish Island fruited, so the DOC moved twenty-one adult females there, in hopes that they would lay more eggs. The strategy seems to have worked, and twenty-four young fledged that year. There are now eighty-six kakapos left. "Intensive manipulation of the kakapo has had its successes," wrote William Sutherland (2002), "but the long-term solution will surely lie in finding ways of restoring predator-free areas on the main islands of New Zealand." Curiously, the flightlessness of the kakapo might protect it, as it cannot leave a protected island for one where predators might await.

The National Kakapo Team of the DOC consists of Paul Jansen, Daryl Eason, Graeme Elliot, and Don Merton, who has devoted thirty years to the kakapo. On the Kakapo Recovery Web site,* Merton is quoted as saying, "I've been enthusiastic about the kakapo since childhood. It's a most remarkable animal and we just have to save it. To me, averting its extinction and facilitating its recovery was—and still is—the ultimate challenge in endangered species conservation." In 1974, the first kakapos were transferred to Maud Island in the Marlborough Sounds (off the northern end of South Island); some were moved to Little Barrier Island (known as *Hauturu* in the Maori language) north of Auckland; and others to Codfish Island (*Whenua Hou*) off the wild west coast of Stewart Island. (Stewart is New Zealand's third largest island, across the Foveaux Strait from the southern tip of South Island.) Because of the rat problem, the kakapos were moved off Little Barrier Island and relocated to Codfish, the center of kakapo recovery with seventy-two resident birds, but the kakapos are running out of room. On July 7, 2002, the DOC staff transferred seven males and seven females from Maud Island (*Te Hoiere*) and Codfish Island to Chalky Island (*Te Kakahu*) in Fiordland. The fourteen birds were flown by helicopter and plane to their new home, which is only two kilometers from the mainland. DOC staff have cleared the island of stoats and are confident their trapping programs on Chalky and nearby stepping-stone islands will keep the area stoat-free. The ultimate goal is to have a self-supporting kakapo population back on mainland New Zealand, living in a

---

* Much of the information in this account has been gathered from the Kakapo Recovery Web site (www.kakaporecovery.org.nz/), which is sponsored by the Kakapo Recovery Programme, consisting of the New Zealand. Department of Conservation, the Royal Forest and Bird Protection Society, and Comalco New Zealand, a company that operates an aluminum smelter at Tiwai Point on South Island. The smelter is near Codfish Island. The comprehensive Web site covers much more than could possibly be included in this brief account, including biographies of the people involved in the program, and even biographies of some of the better-known birds, including Richard Henry, Sinbad, Felix, Zephyr, Hoki, and Gumboots. The site presents the entire chronicle of the kakapo's history ("a story of drama, despair and hope") and its dramatic rescue from oblivion.

protected mainland 'island' environment, or on a remote island where the birds would not be threatened.

The final step in the kakapo recovery program is to create a self-sustaining population on Campbell Island, located in the southern reaches of the Tasman Sea, about halfway between South Island and Antarctica. Once the scene of fur sealing and elephant sealing, the forty-four-square-mile island is now uninhabited by humans, and will provide a safe haven for kakapos. In a 2002 article about the kakapo, Derek Grzelewski estimates that the Kakapo Recovery Programme costs the New Zealand government about $500,000 per year. Is the fuzzy parrot worth it? "If rescue and breeding efforts on the islands stopped for financial or political reasons," notes Grzelewski, "the bird would likely go the way of the dodo." He quotes Don Merton, "If we can't save the kakapo—our flagship species and number one conservation priority—what hope is there for all the other, less glamorous critters?"

Mammals Back from the Brink

Vanishing birds are not the only creatures that people feel guilty about. Mammals have also been the subjects of massive rescue attempts. Back in the Pleistocene, there were bison that stood seven feet high at the shoulder, ranging throughout Europe and North America. Some extinct species, such as *Bison latifrons*, had horns with a six-foot spread. For reasons not clearly understood, they vanished roughly 15,000 years ago, but their somewhat smaller descendants are still found on both continents. The American bison (whose scientific name is *B. bison*, but is still commonly called "buffalo") was the most numerous large mammal in North America until hunters dedicated themselves to its eradication. Estimates of the total numbers ran into the tens of millions. They were rescued from extinction by a few men from New York (one of whom was President Teddy Roosevelt); once again they range over parts of America's Great Plains. The European version, known as the wisent (*B. bonasus*), is rangier and taller than its American counterpart, and lived in forests rather than on open grasslands. Originally, the wisent ranged all over western and southern Europe, and as far north as the Lena River in Siberia, but hunting and the destruction of its woodland habitat brought the wisent's numbers to dangerously low levels.

In the sixteenth and seventeenth centuries, the remaining herds of lowland bison could be found only in protected hunting preserves in Poland's

Wisent

Bialowieza Forest, where 230 families of royal guards protected the forest and its inhabitants from any lowborn interlopers, allowing only Polish royalty to hunt the bison. By 1795, after Russia gained control of the region by way of Poland's partition, Czar Alexander I preserved the rights of the nobility to hunt the wisent. The Czar even removed farmers from the surrounding territory so they would not be tempted to hunt the royal quarry.

With wars raging throughout Europe in the early nineteenth century, the wisents were hunted for food, and by 1813, there were only about 300 to 500 remaining. After the fighting the animals were once more carefully protected—except from humans whose noble birth permitted them to shoot anything they wanted—and by 1857, the herd had increased to about 1,800. In 1862, a rebellion in the Bialowieza region decreased the herd to 875 in one year (Ricciuti, 1973). Wars are usually bad for animals that are large enough to feed several people. In 1915, when German troops occupied the Bialowieza area, they killed almost every wisent for meat, hides, and horns. A German scientist implored the soldiers to spare the remaining wisents, but by the end of World War I, retreating soldiers had shot all but nine of the remaining beasts. In 1923, with fifty-four wisents in zoos and private collections, Polish scientist Jan Sztolcman, along with Swedish, German, and British colleagues, establish the Society for the Protection of European Bison. The society, with headquarters in Frankfurt, established a studbook containing particulars on every specimen. Bialowieza, now

Europe's only remaining primeval forest, was established as a preserve in 1925, and the first calf was born there in 1930. Two years later Poland declared 11,000 acres of Bialowieza forest a national park.

Once again, the specter of war raised its ugly head, but in 1939, Polish foresters convinced Russian officials to protect the wood bison. The Russians posted signs in the forest prohibiting killing of wisents, and made it a crime punishable by death. When a bison was killed, the three soldiers who were responsible were executed. When the Germans took over the area they continued the prohibition on killing wisents. At the end of the war, twenty-four bison remained in Poland and twelve in Germany. Today there are thirty-one wisents in the Bialowieza forest that surrounds the preserve, and approximately 300 in the national park itself, the size of which was doubled in 1996. Others are at liberty in Belarus, Russia, Lithuania, Romania, and Ukraine, and while others survive in zoos, private preserves, and breeding stations.

The story of the North American bison is regrettably similar, a testimony to man's almost uncontrollable desire to shoot at large animals. One institution that tried to make amends was the famed New York Zoological Society (NYZS, commonly known as the Bronx Zoo). In 1993 the zoo officially changed its name to the Wildlife Conservation Society, partly because the traditional concept of a zoo suggested a jail, with animals pacing back and forth in barred cages or enclosures that were often too small and unclean. The Bronx Zoo had long since abandoned that kind of incarceration for its charges (they introduced open-air habitats such as the African Plains, where lions and zebras appeared to be in something approximating their usual haunts, but with cleverly hidden moats separating the predators from the prey), but they still wanted to distance themselves from the old zoological park concept, and move toward the newly defined educational mission of the WCS, to sustain and breed species that were threatened or endangered in the wild. Of course, there are still unthreatened species on exhibit, such as the giraffes, hippos, and giant snakes that fascinate children, but these creatures are often exhibited in an "ecological" context, to demonstrate the important connections between living things. (Among the recent additions to the Bronx Zoo are "Wild Asia," the "Congo Gorilla Forest," and "Tiger Mountain.") The Bronx Zoo was not the only institution to recognize a new "mission," but it was one of the first to change its name to conform to revised ecological and conservationist principles.

As early as 1899, in fact, the NYZS had entered the field of conservation of endangered species. Everyone knows the plains bison (commonly called "buffalo") that ranged over the American prairies in countless millions. The settlers, heading west in the nineteenth century, systematically destroyed the great buffalo herds, but nothing matched the "buffalo hunters," whose assignment it was to shoot as many buffalo as possible, to clear the way for farms and ranches and also to deprive the Plains Indians, with whom America was at war, of their major food source. Philip Sheridan, the Civil War general who led successful campaigns against the Cheyennes, Kiowas, and the Comanches, wrote:

> It is a sentimental error to legislate in favor of the bison. You should, on the contrary, congratulate the skin hunters and give each of them a bronze medal with on one side the image of a dead bison and on the other that of a distressed Indian. The hide hunters have done more to solve the Indian problem than the

Bison

whole of the American Army in thirty years. The extermination of the bison is the only way of founding a lasting peace and favoring the progress of civilization (Verney 1979).

Because they were so numerous and so easy to kill, bison were often shot just for their tongues, while the rest of the carcass was left to rot. In 1872, a million buffalo were killed for their hides alone. These buffalo "robes" were used as bedclothes, in open air carriages, and for the manufacture of clothing. Buffalo were killed to prevent them from knocking over telegraph poles, for sport, and often just for the hell of it. They were shot from moving trains, from boats as they swam across rivers, from horseback, and from "stands," where a man would shoot as many as he could without disturbing the herd (the record is 107, but fifty or sixty was not uncommon). "Buffalo Bill" Cody is said to have shot 4,128 in eighteen months to feed railroad construction gangs.

The number of buffalo on the plains will never be known, but it has been estimated at upwards of 50 million. The building of the transcontinental railroad was the immediate cause for the tremendous slaughter of buffalo in the 1860s and 1870s. When the Union Pacific Railroad was completed in 1869, it became possible to ship hides from the Great Plains to eastern markets for a profit. A second result was the division of the buffalo into two great herds, the northern and the southern. The southern herd was the larger and was exterminated first. The slaughter in the south began in earnest in 1874 and was over by 1878. In the north the great hunts began in 1880 and were over by 1884. In 1882, 200,000 hides were shipped out of the Dakota Territory; in 1883, 40,000; and the following year, only one carload. Even so, it was estimated that for every two hides shipped, three were left untouched where they lay.

From 1870 to 1884, millions of buffalo were slaughtered. In his 1938 book on the American bison, Martin Garretson asked, "What became of this vast number of animals in so comparatively short a time?" He answered, "Shamefully slaughtered with inexcusable prodigality." By 1905, there were fewer than 1,000 plains buffalo left. It was then that William T. Hornaday, the Bronx Zoo's director, founded the American Bison Society. The group managed to secure national protection for the bison, and established no-hunting reserves, first in Oklahoma, then in Montana, South Dakota, and Nebraska. There is now a sizable bison herd in Wyoming's

Yellowstone National Park (the first American national park) and the species is thriving, thanks to the actions of the Wildlife Conservation Society. Many zoos and aquariums are now engaged in the preservation and breeding of endangered species, because they are the best-equipped institutions to do so, and because many of them have begun to recognize their responsibility to wildlife, without which they would have no reason to exist.

But even bison that live in fully protected areas are vulnerable. Bison can get brucellosis, a disease of cattle caused by the bacterium *Brucella*. Ranchers whose land borders on Yellowstone fear that the bison will wander out of the park and infect cattle with the disease, which can cause spontaneous abortion and stillborn calves. Brucellosis also affects sheep, goats, deer, elk, pigs, dogs, and, on rare occasions, humans. (Humans become infected by coming into contact with animals or animal products that are contaminated with the bacterium; symptoms are similar to the flu, and may include fever, sweats, headaches, back pains, and physical weakness.) Yellowstone harbors the nation's largest herd of free-ranging wild bison, estimated at around 3,800 animals. Bison pay little attention to boundaries; when they leave the park, they are rounded up and shot by agents of the National Park Service. In 2002, more than 200 bison were shot, and in March 2003, close to 300 were killed. The purported reason is to protect the cattle from disease, but because there is no evidence of brucellosis transmission from bison to cattle, critics suggest that the culling is being done to protect the grazing rights of the ranchers.

The black-footed ferret (*Mustela nigripes*) is a two-foot-long, golden brown, weasel-shaped carnivore that is known for its black mask, black-tipped tail, black feet, and its status as the most endangered mammal in North America. The ferrets, which were once found throughout the American plains, depended on prairie dogs for food and shelter, using their burrows for dens during the day, since they were nocturnal hunters. But farmers, ranchers, and developers believed that prairie dogs (which are really rodents) were vermin, and therefore a nuisance that ought to be eliminated. A campaign to cleanse the Great Plains of prairie dog "towns" and their occupants has been going on for well over a century. They have been shot for "sport," poisoned, trapped, and gassed for so long and in such numbers, that in some

Black-footed Ferret

areas they too are now considered endangered.* There are no records of black-footed ferrets occurring anywhere but in prairie dog colonies, so the massive decline of the prey animal spelled doom for the predator. A small population of ferrets still hanging on southwestern South Dakota in 1964 was gone from the wild in ten years; the last survivor died in a zoo in 1979. With no further sightings until 1981, the black-footed ferret was believed to be extinct.

In that year, a rancher in Meeteetse, Wyoming, found a ferret that had been killed by his dog, and this led to the discovery of another 130 survivors. Almost as soon as the conservationists began rejoicing over this serendipitous discovery, the animals began to die. Prairie dogs have always been susceptible to sylvatic plague, which is spread by the same bacterium (*Yersinia pestis*) that causes bubonic plague in humans, and ferrets that feed on diseased prairie dogs die too. Dogs, coyotes, badgers, and skunks feed on prairie dogs too and somehow, canine distemper virus was introduced into the last known surviving population of black-footed ferrets. Canine distemper virus (CDV) quickly killed most of the Meeteetse survivors, though eighteen of them were successfully transferred to zoos,

---

* Prairie dogs are still being shot in large numbers. In a 1998 *National Geographic* article, Michael Long interviewed a member of the Colorado "Varmint Militia," who had already shot 20,000 and was planning on shooting more. But in the same article, biologist Tim Clark asks, "Does it make sense to try and eliminate prairie dogs and then declare ferrets endangered and spend a lot of money attempting to restore them? We'd never think of burning a museum, yet when it comes to our biological heritage, we don't hesitate to bulldoze it, burn it, or plow it under. What we're doing is creating threatened or endangered species. To avoid that, we should be protecting prairie dog habitat."

where they are being raised successfully. Attempts to reintroduce them to the wild have been unsuccessful—they are too easily caught by predators, and the prairie dog colonies that they require have been planted over or the inhabitants eliminated. So while *Mustela nigripes* still survives in captivity, it may never again take its rightful place among North America's wild native fauna. Looming also in the background is the threat of CDV, which is always fatal in ferrets, and may be easily transmitted to the survivors, now hanging on to existence by the sharp nails of their little black feet. As Peter Matthiessen wrote in the revised version of his 1959 classic, *Wildlife in America,*

> This very evening, one of the last ones will sit upright in its burrow, eyes glinting a strange green in the prairie starlight. At night, the stark landscape appears eternal, but it is cattle-pocked and fenced, and the buffalo and wolves are long since gone. The ferret is millions of years old. When this elegant slip of life returns to the earth for the last time, and a few grains of sand roll down behind it, the secrets of the long journey that brought it from Siberia to North America many thousands of years ago will disappear. There is starlight on the pale earth at the hole, but the hole has been taken over by spiders, and the land is empty.

African wild dogs (*Lycaon pictus*) live in tightly knit social groups and hunt cooperatively, preying primarily on grazing animals such as gazelles, springboks, wildebeest, and zebras. They are swift, tireless runners, and have been known to chase prey for an hour, for as much as three and a half miles. These bat-eared, brown, yellow, and white dogs were once common in virtually every environment in sub-Saharan Africa except rain forests and deserts, but human encroachment has drastically reduced their range and numbers. Because of land clearance, reorganization, and other factors, Africa's once-great herds of grazing animals are now restricted to scattered populations in parks and reserves. As of a 1997 IUCN study (Woodroffe, Ginsberg and MacDonald), it was estimated that there were between 3,000 and 5,500 wild dogs, in perhaps 600 to 1,000 packs, remaining in Africa.

More than half of these are in southern Africa, where the largest

African Wild Dog

population occupies northern Botswana, northeast Namibia, and western Zimbabwe. There are other populations in the Kruger National Park, South Africa, and Kafue National Park and the Luangwa valley of Zambia, all of which are probably viable. The only substantial wild dog population in East Africa is in southern Tanzania. Kenya and Ethiopia have small populations, but it is not clear whether these are viable in the long term. Wild dogs have been extirpated across most of West and central Africa, although there are populations in Senegal and Cameroon which might be viable. Because the dogs were widely regarded as pests, they were poisoned, shot, and trapped in many areas. Their most serious threat is now introduced diseases. Burgeoning human populations have brought the wild dogs into frequent contact with domestic dogs, many of which carry canine distemper virus (CDV) and rabies. These diseases are ravaging the wild packs. A century ago, African wild dog packs numbering a hundred or more animals could be seen roaming the Serengeti Plains. Today, pack size averages about ten, and the total dog population on the Serengeti is probably less than sixty dogs. In addition to the wild dogs, Roelke-Parker and her colleagues (1996) report that "uncounted hyenas, bat-eared foxes, and leopards were also affected by CDV."

The 1997 IUCN report reviews the factors that might cause the few remaining populations to decline or disappear altogether:

> Habitat fragmentation, persecution and loss of prey were the major causes of wild dogs' historic decline, and these factors still represent the principal threats today. Competition with larger carnivores keeps wild dogs' numbers low, so that even the largest habitat fragments may contain populations too small to be viable. Contact with human activity is directly responsible for over 60 percent of recorded adult mortality through road casualties, persecution and snaring. Even wild dogs living in large protected areas may stray over reserve borders where they are threatened by human activities. Disease represents another serious threat to wild dogs, which has already caused the extinction of one population. . . . As a result of these pressures, all of the wild dog populations remaining in Africa are under threat. In the long term, wild dogs living outside protected areas are unlikely to coexist with growing human populations without innovative management. Even in large protected areas, wild dogs' long-term survival will depend on reducing potentially fatal contact with people and domestic dogs on reserve borders.

A recent study of wild dog populations suggests that their imperiled status has been exacerbated by their reduced numbers. Wild dogs kill their prey by tearing it apart, the dogs grabbing it from all sides and pulling against one another. Obviously, this technique is better accomplished by many dogs, and hunting will be therefore be hampered by reduced numbers. Their unusual breeding strategy, where same-sex adolescents leave the pack until they meet up with a group of the opposite sex, means that there must be a minimum pack size to spin off the new breeding stock. If their numbers are reduced, say, by disease, the reproductive rate, and therefore the population, plummets. An idea introduced by ecologist Warder Allee in 1931 (and now known as the "Allee effect") holds that reduced populations, particularly of animals that tend to work together in groups ("obligate cooperators"), will have an adverse effect on their breeding ability. As Courchamp, Clutton-Brock, and Grenfell wrote in a 1999 study, "The Allee effect describes a scenario in which populations at low numbers are affected by a positive relationship between population growth rate and density, which increases their likelihood of extinction." Birds such as penguins are stimulated to breed by the presence of many other mating pairs; if the breeding population falls below a critical level, the birds simply do not lay eggs. Of course, animals that normally congregate together to protect themselves against predation, such as schooling fishes, are also affected by "undercrowding." For example, when populations of Peruvian anchovies were robust, predators would simply nibble around the edges of the huge schools, but when the population was decimated in the 1960s by overfishing, predators had a noticeable effect on the total numbers, and the once-plentiful anchovies have not recovered. (In the 1960s, the Peruvian anchovy fishery was the largest fishery in the world, with an annual haul that totaled more than 11 million tons; now fishermen rarely pull in more than 100,000 tons.)

If you're familiar with horses in cave paintings, you know what Przewalski's horse looks like. That's right; it's a stocky, heavily built horse with a black mane that does not flop over, but stands straight up. The mane ends between the ears, so this horse has no forelock. It is beige or light chestnut in color, with a black stripe that runs down the middle of its back, and faint zebra stripes on its dark lower legs. The neck is remarkably thick, and the head looks a bit too big for the body. The cave paintings were done about 30,000 years ago, however, so what has Przewalski's horse been up to lately? In the last couple of centuries, it has been trying not to become extinct. (Another race of this species, the tarpan, *Equus przewalski gmelini*, which was found from Spain and southern France eastward to central Russia, has already passed over into extinction.)*

---

* A somewhat questionable method of bringing animals back from extinction is known as "back-breeding," and consists of re-creating the extinct creature by crossing various animals that bear a resemblance to the desired target until something that looks like the extinct animal is born. The technique was pioneered by Heinz and Lutz Heck, brothers who did their work at the Tierpark Hellabrunn, the Munich zoo. (With Eduard Tratz, Heinz Heck also introduced the name "bonobo.") The Heck brothers used several European pony breeds, such as the Polish koniks, Icelandic ponies, Swedish Gotlands and "Polish primitive horses" from the preserve in Bialowieza. Even if the end result looks like the animal they were trying to re-create, it is not. Because its genetic makeup is a jumble, it is not actually a tarpan; it is just a tarpan look-alike.

That's where Colonel Nikolai Mikhailovich Przewalski (1839–1888) comes in. He was a Russian explorer who traveled in Central Asia in the service of Czar Alexander II. In 1878, he reported to St. Petersburg that the Mongolian wild horse known as the *takh* (plural: *takhi*), long believed to be extinct, had been seen alive in the Gobi Desert. Przewalski (pronounced she-VAL-ski) brought back a skull and a skin that had been given to him by a hunter, and took them to the Zoological Museum, where the new species was examined and named by conservator A. I. Poliakov. During his third journey to Mongolia, Przewalski saw two herds of these animals in the Tachin Schara Nuru Mountains near the edge of the Gobi Desert. In 1881, he filed an official report, describing their remote habitat, their appearance, and their characteristics. He noted that they lived in herds from five to fifteen animals led by an older stallion, and that they were alert and very shy, with acute hearing, very good eyesight, and a highly developed sense of smell. By the end of the century, news of Przewalski's discovery had spread throughout Europe and America, and "collectors" of rare and exotic animals all wanted *Equus przewalski,* which by this time could only be found in the semidesert plains of the Altai Mountains of southwestern Mongolia.

Frederic von Falz-Fein, German by birth and living in the Crimea in southern Russia, kept many rare species on his estates at Askania Nova, and when he heard of the discovery of these Mongolian wild horses he commissioned a dealer to organize an expedition to capture some for his private game park. The animals were too shy and too fast, and none were caught for von Falz-Fein until 1899, when seven foals were taken, and brought to Askania Nova. A foal caught in 1899 and a filly caught in 1900 were both presented to Czar Alexander III. The filly soon died and subsequently the Czar gave the stallion to von Falz-Fein so that he could start breeding his wild-caught fillies. In 1900 two more colts were caught and transferred to Moscow. The last captures of wild Przewalski's horses took place in 1938, when a single filly was caught in the Tachin Schara Nuru Mountains, and sent to Ulan Bator, where she was cross-bred with Mongolian domestic horses (Bouman and Bouman 1994).

Like von Falz-Fein, the first Duke of Bedford was a collector of exotic wild animals, and kept many rare birds and land animals at his Woburn estate. He was also Chairman of the Zoological Society of London. He commissioned Carl Hagenbeck, the great animal dealer in

Przewalski's Horse

Hamburg, to obtain Przewalski's horses for him. Fifty-one wild foals were caught in 1901, but only twenty-eight survived the seven month journey to Europe. Twelve of them went to Woburn Abbey, and two each went to London, Manchester, Halle, Berlin, and the Netherlands. Two were shipped to New York and later transferred to Cincinnati. Paris got one, and the last three remained in Hagenbeck's zoo in Hamburg. More foals were caught in the following two years, but only fourteen survived to reach their final destination. All in all, 53 Przewalski horses eventually survived the rough voyage from Mongolia to Europe. All the captive Przewalski's horses of today are descended from thirteen of these ancestors.

Until the Second World War, the number of Przewalski's horses living in the wild appears to have remained stable but not very high. After 1967, though, no herds of wild Przewalski's horses were seen in southwest Mongolia, despite several expeditions that were sent out specifically to look for them. The last wild Przewalski's horse was sighted in 1969 near a spring called Gun Tamga. After World War II, there were only thirty-one horses in captivity. Inbreeding raised the incidences of congenital diseases and defects, and led to a diminished life span and increased mortality among foals. The number of pregnancies also decreased. The situation of the Przewalski's horse was critical. Something had to be done to save the last wild horse from extinction.

In the 1950s, the largest herd of Przewalski's horses was in Prague. Dr. Jiri Volf, the director of the Prague Zoo, began to prepare a studbook of all the wild horses that had been kept in captivity, in order to better understand their genealogy. In 1977, when only 300 Przewalski's horses were left in various zoos and private collections, Inge and Jan Bouman founded the Foundation for the Preservation and Protection of the Przewalski Horse (FPPPH), which began advising zoos on breeding Przewalski's horses. Their program was especially geared to eliminate inbreeding by exchanging unrelated horses between zoos. The FPPPH maintained five reserves in the Netherlands and Germany, with the long-range goal of reintroducing Przewalski's horses into the wild. After having lived in zoos for many years, the horses needed to learn how to find their own food and to live in natural groups. In 1981 the foundation began purchasing horses from various zoos, selecting animals with as little common ancestry as possible. In 1986 the foundation began to collaborate with the Institute for Evolutionary Animal Morphology and Ecology of the Moscow Academy of Sciences. This collaboration resulted in a search for suitable steppe reserves in what was then still the USSR, and also in Mongolia. As of 1997, the foundation had sixty-one healthy Przewalski's horses living in five reserves, each measuring twelve acres or more. Every year 92 percent of all mares give birth, and the mortality among foals is exceptionally low. Many of the second- and third-generation offspring born in the foundation's semireserves have already been released into the wild, and more will be introduced in coming years.

Unfortunately, many steppe areas have already been lost to cultivation and to overgrazing by large herds of livestock, and the steppe itself is considered an endangered biotope today. In Mongolia, a number of relatively undisturbed steppe areas have been preserved, but these, too, are threatened by overgrazing. One such area, a 24,000-acre steppe known as Hustain Nuru, was turned into a national park, and immediately closed to Mongolian herdsmen, who must now graze their livestock elsewhere. The reintroduction project started in Mongolia in 1990 in cooperation with the Mongolian Association for Conservation of Nature and Environment. In June 1992, sixteen Przewalski's horses were shipped to Mongolia from private collections. (Earlier, in 1985, eleven Przewalski's horses had been reintroduced to their historic range in China, the Dzungarian Desert in the Xinjiang-Ugyar Autonomous Region; three years later, an additional five were brought in [Bouman, Bouman and Boyd 1994].) In 1994 and 1996, two groups of sixteen horses were brought into the the Hustain Nuru steppe reserve in Mongolia.

As of January 1, 1998, some sixty takhi live on the steppes of Hustain Nuru with 1,400 more spread throughout 135 zoos and private parks all over the world. In the conclusion of their 1994 book on Przewalski's horse, Lee Boyd and Katherine Houpt wrote,

> Today we deplore the death of so many wild Przewalski's horses at the turn of the century during attempts to catch and transport foals, but we must consider that this was the technology available at the time. Had those captures not taken place, the species almost certainly would be extinct. No one seems to have predicted the decline of the wild populations in time to prevent their demise. It is difficult to argue with those who today believe that preserving habitat is the best means of saving endangered species. But the example of Przewalski's horse conservation shows us that extinction events may be difficult to predict, and how important it is to have a captive population to draw upon should reintroductions become necessary. Maintaining such healthy captive populations has become a major role of modern zoos, receiving at least equal priority with their traditional mission of public education.

The July-August 2002 issue of *Natural History* magazine had a pair of Przewalski's horses on the cover, and a major story called "Reborn Free," written by Lee Boyd, who has been studying wild horses on two continents for twenty-five years. Boyd visited Hustain National Park, which now supports about a hundred takhi. "Each time I visit Hustain National Park," she wrote, "I must familiarize myself anew with the ever-changing bands. It has been exciting, though bittersweet, to see captive born stallions—after so successfully readapting to the land of their ancestors—be dethroned by a generation of rivals that they sired."

Père David's deer was first described to western science by Henri Milne-Edwards in 1866, from specimens that had been sent to France from

China. The Chinese name for this deer is *su bu xiang*, which means something like "none of the four," and refers to the animal's combination of the neck of a camel, the hooves of a cow, the tail of a donkey, and the antlers of a deer; it does not actually resemble a camel, a cow, or a donkey. These anomalous characteristics led the Chinese to revere the su bu xiang, but they also hunted them; eating the meat supposedly increased one's life expectancy fourfold.* "No wonder," wrote Robert Twigger (2001), "only the emperor was allowed to hunt them."

Found only in China, these deer were easily hunted in their wild habitat of open plains and marshes. The wild herds kept diminishing until the last known individual was shot in 1939 near the Yellow Sea. However, their extinction was avoided by an ancient Emperor of China, who installed a large herd in his Imperial Hunting Park (Nan Haizi Park) near Peking. While almost extinct in the wild, the deer thrived in the park, which was surrounded by a forty-five-mile-long wall and guarded by a Tartar patrol. The French missionary Père Armand David (1826–1900) had wandered around and wondered about the contents of this mysterious park, but strangers were forbidden to look inside. On May 17, 1865, he convinced the guards to allow him to look over the wall, and as luck would have it, a herd of these strange deer happened to walk by at that very moment. After many vain efforts, Père David was able to obtain two complete skins of the animal (which he believed to be a new species of reindeer). He sent them to Europe, enabling Milne-Edwards to provide the first scientific de-

---

* The Chinese practice of using animal parts—often from endangered species—to enrich or enhance human life continues today. In traditional Chinese medicine (TCM), the gallbladder and bile of endangered bears are used to treat a variety of inflammations, infections, and pain. There are "bear farms" that supply bile to Chinese practitioners, where steel catheters are surgically implanted into caged bears' gallbladders, enabling handlers to regularly "milk" the bears for their bile. North and South American bears are also being killed for their gallbladders, which are then smuggled into China. Other endangered species, such as various rhinos, are killed for their horns, which, when ground into a powder, are said to cure various ailments. Tigers, endangered throughout their range, are collected for their bones, which are made into "tiger bone wine," said to give the drinker the strength of the tiger. Tiger penis soup is thought to be an aphrodisiac; eyeballs rolled into pills are thought to cure convulsions; whiskers protect one against bullets; the tail mixed with soup cures skin disease, and the hair when burnt drives away centipedes. Musk deer are killed for their testicles. Russian hunters are decimating herds of saiga antelope for their horns. Tons of dried seahorses enter the TCM trade each year.

Père David's Deer

scription of Père David's deer, *Elaphurus davidianus*. (*Milu* is the Chinese name for the Japanese sika deer, but Milne-Edwards believed that the Père David's deer was called that too, so "milu" is sometimes used for the Père David.)

It is a large, heavy deer, reddish above and lighter below. It has a light muzzle and a white ring around each eye. The head is long and rather horselike, and the males have a kind of ruff around the throat. The antlers, found only in males, are unique among deer in that the hind prongs are forked. Another strange feature of the antlers is that they can grow in twice a year. The summer antlers are the larger set; they are dropped in November, after the June–August rut. The second set, if they appear, are fully grown by January, and are dropped a few weeks later. The legs are long, and the hooves relatively long and slender—an adaptation to walking on soft, marshy ground. The donkeylike tail ends in a black tuft. In *Deer of the World* (1998), Valerius Geist wrote that "Père David's deer is a very odd species":

It is associated with the extensive lowlands and river valleys. This plus its large hooves that are sensitive to hard ground; its slow, deliberate walk and low running speed (30 kph); its grazing-type incisors; large muzzle, and elongated snout; and its need for water and wallows in captivity indicate that it is adapted to moist, riparian grasslands.

The species had actually ceased to exist in the wild at the time of the Shang Dynasty, some 3,000 years ago, when the swamps in which it lived, the Chihli plains, were brought under cultivation. From 1122 B.C. onward, the only surviving herd was maintained in the Nan Haizi (Imperial Hunting) Park near Peking. After Père David had sent the skins to Paris, Milne-Edwards's report created such a desire for these deer in Europe, that the Chinese Emperor, who had given some to the French, could hardly deny a similar gift to the English and the Germans. Several pairs were subsequently sent to Europe, where they multiplied readily. Approximately two dozen deer in Europe, as well as the large herd remaining in China, seemed to ensure the survival of the species, but disaster struck in the form of massive flooding of the Hun Ho River in 1894. The rising waters of the river breached the brick wall surrounding the park, and while many of the deer escaped, they were killed and eaten by starving peasants. The survivors of that massacre were killed during the Boxer Rebellion of 1900, when international troops shot and ate every single one left in the Imperial Park. There were no more Père David's deer left in China, but the Duke of Bedford—certainly one of the seminal figures in the rescue of highly endangered animals—had sixteen at Woburn Abbey; by 1922, his herd had increased to sixty-four. In *The Extinction Club*, a book that is mostly about Père David's deer, Robert Twigger visits Woburn and sees the herd:

> This was my first in-the-flesh viewing of Père David's, and given that I had been building up to this moment for six months, I was glad not to be disappointed. I was happy that each part of the deer really did look mismatched: the neck is thick and shaggy as a camel neck is, the tail does swish like a horse's tail, and the feet have a hoofiness that is definitely more bovine than ungulate. The deer really had the four characteristics that did not match—and

when you see them alongside other deer, the backward-pointing antlers shout at you to be twisted forward.*

After World War II, Bedford made breeding stock available to collectors and wildlife parks in other countries, and by 1963, the world total—all of which had been born and raised in captivity—was up to 400. In 1956, the London Zoo sent four specimens back to China, where they were installed at the Beijing Zoo. (The name *Peking* was changed to *Beijing* after the Communist takeover of China in 1949.) To complete the rescue mission, in 1986, twenty-two more were flown from Woburn Abbey to Beijing, where, after a lengthy quarantine, they were released in the area of the old Imperial Park, where they had been discovered by Père David more than 130 years before. The last step—reintroduction to the wild—has yet to be taken, although a forest preserve has been selected for this purpose not far from where the last wild animal was shot.

Other deer species have not fared so well. Schomburgk's deer (*Cervus schomburgki*) was described in 1863 and named after Sir Robert Schomburgk, the British consul in Bangkok from 1857 to 1864. Known as *sa mun* or *nuar sa mun*, the animal lived in the swampy plains of central Thailand (then called Siam), especially in the Chao Phya River valley around Bangkok and surrounding areas. It was a graceful deer, in general appearance not unlike the swamp deer of Nepal and India (also known as the barasingha). Male Schomburgks had spectacular antlers, which were often described as basketlike, with all the main tines branching. This resulted in a large number of points, up to thirty-three per pair, which made the stag look as if it was wearing a gigantic crown or candelabra. The length along the outer curve of the antlers could be thirty-five inches. The females had no antlers. The deer's coarse coat was a dark, chocolate brown color, with lighter underparts. The legs and forehead of this deer

---

* I saw my first Père David's deer at the Denver Zoo. There were three of them—two hinds and a large stag with a full rack of antlers. The first impression is that of shagginess, more so than most other deer, but then I noticed the large feet and hairy fetlocks. Even the does, which should have been more delicate, had big feet, rather like reindeer. The deer were in an enclosure adjacent to that of the Przewalski's horses, so by positioning myself just so, I was able to see two of the rarest animals in the world, both rescued from extinction by human intervention.

were lighter, and the short tail had a white undersurface, like the North American whitetail. Schomburgks lived in small herds consisting of a single adult male, a few females, and their young, feeding in the early evening to the morning. Densely vegetated areas were avoided and most activity occurred on the open swampy plains. When flooding occurred during the rainy season, Schomburgk's deer were forced to crowd onto higher pieces of land, which often turned into islands, making it pitifully easy for hunters to kill them.

Once abundant in Thailand, they frequented the Chao Phya River plains, and roamed the area from Samut Prakarn to Sukhothai. In the east they were found from Nakhon Nayok to Chachengsao; in the west, from Suphan Buri to Kanchanaburi. Commercial production of rice for foreign export began in the late nineteenth century in Thailand, leading to the loss of nearly all the grassland and swamp areas this deer depended on. Intensive hunting pressure at the turn of the century restricted the species further. Efforts by foreigners to capture these animals for captive breeding all failed, because the Thais evidently could not understand the importance of saving an endangered species. The only known mounted specimen is in the Musée National d'Histoire Naturelle in Paris. This specimen was brought back from Thailand in 1862 and lived in the museum's menagerie until it died in 1868. The last wild individuals are thought to have died around 1932. The last known survivor was an adult male kept as a pet at a temple in Samut Sakhon province; it was killed by a drunk in 1938. No confirmed reports of this species have been heard since, and it was formally declared extinct in the 1996 IUCN Red List, although rumors continue to suggest a remnant population may still survive.

No other mammal on earth depends more on international politics for its survival than the giant panda. The state of the relationship between the People's Republic of China and the United States is critical to the survival of this animal. Everybody knows and loves *Ailuropoda melanoleuca*, the furry, black-and-white bear whose image appears as the World Wildlife Fund's symbol, and whose plush replicas adorn the bedrooms of countless children around the world. But despite all this adoration, the giant panda is in very, very serious trouble. These animals once lived in lowland areas of central China, but farming, forest clearing, and other development now restrict wild giant pandas to a few remote mountain ranges in Sichuan, Shaanxi, and Gansu provinces. There they live in broadleaf and coniferous forests with a dense understory of bamboo, at elevations between 5,000 and 10,000 feet. Torrential rains or dense mist throughout the year characterize these forests, often shrouded in heavy clouds. Although most of their diet consists of bamboo shoots, pandas in the wild have also been known to eat other grasses and occasional small rodents or musk deer fawns. (In zoos, giant pandas are fed bamboo, sugarcane, rice gruel, high-fiber biscuit, carrots, apples, and sweet potatoes.)

Like Père David's deer, the giant panda was brought to the attention of Western zoologists by Père Armand David in 1869; also like the Chinese deer, it was first described by Henri Milne-Edwards, who gave it the perfectly descriptive name of *Ursus melanoleucus*—black and white bear. The panda was later reassigned to the genus *Ailuropoda* when it was thought that the giant panda was a close relative of *Ailurus*, the lesser or red panda. Recent DNA analysis indicates that giant pandas are more closely related to bears, and red pandas are more closely related to raccoons. Accordingly, giant pandas have been placed in the bear family (Ursidae), while red pandas are now placed in the raccoon family (Procyonidae).

The first time a Chinese hunter shot a panda is not recorded, but the first American kill came in 1929, during an expedition to China funded by the Field Museum of Chicago, and led by Kermit and Theodore Roosevelt, Jr., the sons of former President Theodore Roosevelt; the Roosevelts bagged a giant panda and brought back the skin to be mounted in a habitat group. As the first specimen exhibited in the United States, the Chicago panda inspired other expeditions. In 1934, adventurer William Harkness set off to China, determined to capture giant pandas. Although he died within a year, his wife Ruth (a New York fashion designer) and her party found a three-pound giant panda cub in the wild and brought her to the United States. The first live giant panda ever seen outside of China, Su-Lin was installed at Chicago's Brookfield Zoo, where she won the hearts of an adoring American public. In 1938 Harkness brought another panda, Mei-Mei, from China to the Brookfield Zoo, where it survived until 1942. Also in 1938, The New York Zoological Society brought Pandora to the Bronx Zoo. Cute pandas seemed to require cute names, and while most of them were in Chinese, Pan Dee and Pan Dah were the names

Giant Panda

given to a pair donated to the Bronx Zoo by Madame Chiang Kai-shek in 1941 in gratitude for relief aid. They died in 1945 and 1951.

By the middle of the twentieth century, only a few giant pandas were surviving in U.S. zoos, and the world began to realize their numbers in the wild were dwindling. After Richard Nixon's historic visit to China in 1972, the People's Republic presented him with a gift of friendship to mark his visit: the giant pandas Ling-Ling and Hsing-Hsing, which were housed at the National Zoo in Washington. Ling-Ling lived at the National Zoo until 1992, when she died at the age of twenty-three. At the time of her death she was the oldest giant panda living in a zoo outside China. Hsing-Hsing was twenty-eight years old when he died in 1999. During their lives at the National Zoo, Hsing-Hsing and Ling-Ling thrilled millions of visitors a year, and National Zoo researchers greatly added to the world's understanding of giant panda biology. "Of the 304 pandas in zoos in 1985," wrote Lü Zhi (2002), "only 76 had been born in captivity, and all but 19 of those had died within a month." The arrival of Tian Tian and Mei Xiang at Washington's National Zoo in December 2000 marked the beginning of a ten-year research plan to improve the survival rate of giant pandas in zoos and maintain their numbers in the wild. One goal of this plan, to be carried out by National Zoo biologists and their colleagues in China and the United States, is to create a thriving, self-sustaining population of giant pandas in zoos. In turn, this zoo population may ultimately provide a reservoir for the reintroduction of giant pandas to the wild.

About the size of an American black bear, giant pandas look even heavier because of their larger heads and thick, fluffy coats. All other bears have round pupils, but the pupils of the giant panda are vertical, like those of a cat. These unusual eyes inspired the Chinese to call the panda *daxiong-mao*, which means "giant bear cat." (Its scientific name *Ailuropoda* means "cat-footed.") Pandas stand between two and three feet tall at the shoulder, and are four to six feet long. Males are larger than females, weighing up to 250 pounds in the wild; females rarely reach 220 pounds. Both have black ears, circular black eye patches, and a black muzzle, legs, and shoulders. The rest of the animal's coat is white. Giant pandas have large molar teeth and strong jaw muscles for crushing tough bamboo. Each forepaw has an auxiliary "thumb" used in grasping bamboo stalks. Pandas spend much of the day on the ground, but they are good tree-climbers. They cannot walk on their hind legs. They do not hibernate, but may den up during exceptionally cold periods.

Within recent Chinese history, humans encroaching on the pandas' lowland habitat have driven them higher up into the mountains. The giant panda is listed as endangered in the World Conservation Union's (IUCN) Red List of Threatened Animals. In fact, it is one of the most critically endangered species in the world, with only about 1,000 left in the wild, and about 140 more living in zoos and breeding centers around the world, mostly in China. The cover of *National Geographic* for February 1993 showed a mother panda cuddling a newborn pink thing that was heralded as a "Newborn Panda in the Wild." Born in the Qin Ling mountain forest, Xi Wang (Hope) was the first wild cub whose progress could be monitored. It was a triumph for Chinese scientists. Two years later, another *Geographic* article appeared, this time featuring the mother Jiao Jiao and two-year-old Xi Wang. The panda is not saved yet," wrote Pan Wenshi. "Poaching and loss of habitat are still serious threats. Pelts still sell for more than $10,000 in Hong Kong, Taiwan, and Japan. Although the government has gone to the extreme of executing several poachers recently, illegal hunting remains a problem." On August 21, 1999, a female panda named Hua Mei, who had been artificially inseminated, gave birth to a four-ounce baby at the San Diego Zoo. As of August 2002, the baby, named Bai Yun ("white cloud"), was doing fine.

★   ★   ★

The rescue from extinction of the Arabian oryx is a story that begins badly, gets better, and then, just when it appears to be heading for a happy ending, takes a nasty downturn. This large antelope is almost pure white, with dark brown markings on its face and legs. Its ridged horns are almost perfectly straight, and can be thirty inches long. Seen in profile, it looks as if the animal has only a single horn, and this might have contributed to the myth of the unicorn. (A 1965 book by Anthony Shepherd about the plight of the oryx is called *The Flight of the Unicorns*.) These antelopes once lived in nomadic herds on the gravel plains and dunes edging the Arabian deserts, feeding mostly on grasses, but also on herbs, seedpods, fruit, tubers, and roots. They could go for weeks without drinking water. As late as 1800 the Arabian oryx occurred over most of the Arabian Peninsula and the fertile crescent (which together include modern Israel, Jordan, Syria, Iraq, the Sinai Peninsula of Egypt, Saudi Arabia, Oman, Yemen, United Arab Emirates, Bahrain, Kuwait, and Qatar). By 1970, they were found only in the southeastern regions of the Rub' al Khali Desert of the Arabian Peninsula.

The main cause of the extermination of *Oryx leucoryx* in the wild was overhunting, by the Bedouin for meat and hides, and by others for sport. The antelopes used to be tracked on foot or by camel. The decline of the species was exacerbated by the introduction of jeeps and other all-terrain vehicles, which made it possible for hunters to follow these elusive creatures over the previously uncrossable deserts and trackless wastes. When oil became an important commodity in the Middle East, it not only fueled the vehicles that could cross the deserts, but it also brought into the region many people whose idea of sport was to chase down antelopes in a car. Multivehicle safaris were organized too; there is a story of one shooting expedition that used 300 vehicles to carry the hunters, their supplies, and their servants. "As a result of the incredible blood lust of the past twenty years," wrote Fisher, Simon, and Vincent (1969), "virtually all of the abundant wildlife of Arabia has been extirpated from areas accessible to automobiles."

In 1962, under the direction of Ian Grimwood, the chief game warden of Kenya, "Operation Oryx" was begun. Two males and one female were captured and brought to the Phoenix Zoo. They were joined by additional specimens that had been collected by Sheik Jabir Abdullah al Sabah of Kuwait and King Saud of Saudi Arabia; by 1966, there were

Arabian Oryx

eleven males and five females in the not unfamiliar desert atmosphere of Arizona. There were also seven in California and twenty-three in Slamy in northwestern Qatar. The goal was to raise enough oryx to allow for an eventual return to the wild. In 1982, several animals raised in captive populations were reintroduced into the deserts of Oman. Additional reintroductions have taken place in Jordan, Israel, and Saudi Arabia.

By 1990, a wild herd of Arabian oryx had taken hold in Oman; but soon the hunters were on the prowl again. In a 1999 article in the aptly named journal *Oryx*, Spalton, Brend, and Lawrence wrote:

> In February 1996 poaching resumed and oryx were captured for sale as live animals outside the country. Despite the poaching the population continued to increase and by October 1996 was estimated to be just over 400. However, poaching intensified and continued through late 1996 and 1997. By September 1998 it had reduced the wild population to an estimated 138 animals, of which just 28 were females. The wild population was no longer considered viable and action was taken to rescue some of the remaining animals from the wild to form a captive herd.

Between October 1996 and March 1999, the numbers fell from 400 to 100. The poachers, who had been capturing the animals and selling them to private zoos outside Oman, mainly targeted female oryx and their calves, many of whom died from stress or exhaustion during their capture. Others died during transportation or shortly afterward, their bodies were abandoned. Of the hundred or so oryx left in the wild, only eleven are female, reducing even further the species' chances of recovery. In 1999, a conservation journal headlined its story: "Wild Arabian oryx faces extinction—Again!"

The European lynx (*Felis lynx lynx*) occupied much of late Pleistocene Europe, including Britain and Ireland, but it is now very rare except in some remote parts of Scandinavia and northern Russia. The Iberian species, *Felis pardinus,* was once distributed throughout the Iberian Peninsula, perhaps as far north as the Pyrenees. There is some doubt, however, whether the Pyrenean population belongs to the Spanish or to the European species, which may possibly still survive in the Massif Central of France, but is otherwise now found no nearer to Iberia than Scandinavia, Poland, and Czechoslovakia. Widespread deforestation has resulted in the elimination of much of the animal's habitat, and like so many of the larger European carnivores, it has been relentlessly persecuted for the damage it inflicted on domestic livestock.

The Iberian lynx (*F. pardinus*) is a little smaller than the Canada lynx (*F. lynx*) but built along the same lines, with long legs, a short, black-tipped tail, a ruff around its face, and black tufts on the tips of its ears. Its thick coat is a grizzled gray brown, peppered with large, dark spots. At a maximum

Lynx

weight of eighty-three pounds and a shoulder height of about twenty-eight inches, the Iberian lynx is somewhat smaller than a mountain lion. It is the most endangered cat in the world, and is rapidly approaching another, more dubious title: it is close to becoming the first wild cat species to become extinct in modern times. On the IUCN Red List for 2000, its status was changed from "endangered" to "critically endangered." Found only in woodland thickets in Spain and (maybe) Portugal, *F. pardinus* has been reduced drastically from a 1980s estimate of 1,000 animals. Most if not all the surviving lynxes live in unprotected regions of southern Spain. Their precarious situation is the result of habitat deterioration (livestock raising), and a dramatic decline in the population of rabbit, its main prey species, because of myxomatosis and rabbit hemmorhagic disease (RHD). Road building and the erection of vacation houses have fragmented the lynx population to the point where there are now nine separate populations divided into forty-eight breeding areas. Hunters and trappers taking Iberian lynxes do not contribute much to the survival of the species. The Iberian lynx has now been exterminated in most of its range except certain controlled hunting areas, known as *cotos,* in the delta of the Guadalquivir in southern Spain. The total population is unknown, but may possibly amount to a few hundred, including an estimated 150 to 200 in the Coto Doñana, which is believed to hold the largest remaining population.

But all is not well in the Doñana. In response to a petition from Marbella-based businessman Alfonso Hohenlohe, Prince of Liechtenstein, the Spanish government has reclassified the land from greenbelt to urban, paving the way for the construction of a country club (complete with golf course), two hotels, a shopping center, two polo fields, a botanic garden, and two "typical" Andalusian villages of 3,000 houses apiece. At a meeting held in Andujar in November 2002, the Spanish Environmental Ministry, the Junta de Andalucía, the Council of Europe, Doñana Biological Station, WWF/Adena and the IUCN Cat Specialist Group, heard that the Iberian lynx had declined by nearly 90 percent in less than fifteen years. From a population estimated at 1,000 to 1,200, to 150 in two breeding populations in Coto Doñana and Andujar-Cardena. According to IUCN Cat Specialist Group cochair Dr. Urs Breitenmoser, there was strong evidence that the small isolated patches distributed along the Sierra Morena and all the way to Portugal had collapsed in recent years. Breitenmoser warned that spontaneous recovery may no longer be possible and that any further stress, such as a rabbit or lynx disease, would seal the fate of the species.

The IUCN Red List is a comprehensive inventory of the global conservation status of plant and animal species. According to a 2003 report by Kristin Nowell, the Red List authority on cats, the 2002 revision of the list classifies the Andean mountain cat, the Borneo bay cat, the snow leopard, and the tiger as "endangered," but of the felids only the Iberian lynx is "critically endangered"; that means that the two remaining but separate populations at Coto Doñana and Andujar, with only about 150 cats between them, "are insufficient for the survival of the species in the long term." The news from Portugal is worse. In the 1980s there was an estimated Portuguese population of forty to fifty animals, but a survey conducted by the Instituto da Conservacão da Natureza in 2001 failed to detect a single lynx.

Pitted against the most effective and ruthless predator that has ever lived, hunting cats like the lynx come off a poor second. Indeed, with the possible exception of the African lion, big cats around the world are all in serious trouble. Southern Asian tigers, along with their Siberian counterparts, have been reduced in numbers by hunters protecting livestock and people, and by those who require tiger parts—whiskers, claws, gall bladder, penis—for traditional Chinese "cures." Leopards are scarce in Africa and scarcer still in Southeast Asia; the snow leopard is hunted throughout its mountainous range; jaguars are being driven from their South and Central American habitats by deforestation and human activities; and the status of the cheetah is precarious wherever it can still be found. While we do not ordinarily think of Australian fauna in terms of large predators, there was one there in the past, but it was completely eradicated.

Its scientific name, *Thylacinus cynocephalus*, means "pouched dog with a wolf head," and it is an apt description. Commonly known as the Tasmanian tiger, Tasmanian wolf, or thylacine, it was Australia's largest marsupial predator. About the size of a German shepherd dog, it is classified among the Australian Pleistocene megafauna, but unlike *Diprotodon* or *Zygomaturus*, which died off 46,000 years ago, *Thylacinus* managed to hold on until 1936, when the last known surviving specimen died in the Hobart Zoo. "The Thylacine," wrote David Bowman (2001), "is of great scientific interest because it is a classic example of convergent evolution giving rise

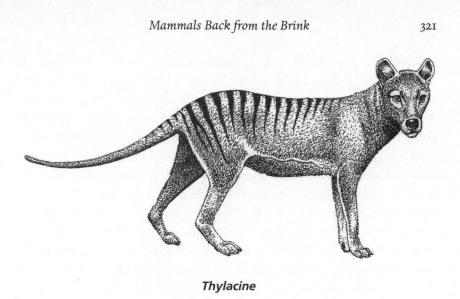

Thylacine

to a wolf-like marsupial." According to Robert Paddle (2000), there were thylacines on the Australian mainland until around 1840, but the mainland population was hunted to extinction by Aborigines because the thylacines were in "direct competition for a finite food resource." (Paddle also argues that the extinction of the thylacine was a continuation of the trend of the extinction of the Pleistocene megafauna by aboriginal over-hunting.)

Because it was purported to be a sheep killer, the Australian government initiated a bounty program on Tasmania that was so successful that the thylacine was hunted to extinction. On the Australian continent, they probably fared badly in competition with the dingo, but dingoes were never introduced to Tasmania. Thylacines almost certainly killed an occasional sheep, but they also hunted kangaroos, wallabies, and other small mammals and birds. (They were getting along nicely before the settlers introduced sheep.) The Tasmanian tiger inhabited dense forests. Its wild and inaccessible habitat has produced numerous "sightings," but none have been verified. Is the Thylacine gone forever? Maybe not, say Australian scientists.

In the Australian Museum in Sydney, there is a small thylacine pup that was preserved in alcohol in 1866. Michael Archer, the director of the museum, announced in early 2002 that he and his staff would attempt to clone the thylacine, thereby bringing it back to life after sixty-six years of nonexistence. By chance, this specimen had been stored in a jar of ethanol rather than formalin, which would have destroyed the DNA. In May 2002, the Evolutionary Biology Unit at the Australian Museum announced that they had successfully replicated individual Tasmanian Tiger genes using a process known as polymerase chain reaction (PCR). The DNA extracted from the pup's cells (and that of two other pups) is undamaged, and the scientists believe that there is no reason that it could not be transplanted into a living cell. Conservationists argue that trying to clone animals is pointless if the reasons behind the animals' disappearance, such as loss of habitat, have not been addressed first. There is no question that much of the thylacine's habitat is still there, as impenetrable as ever, so that would not be a problem. Aussie sheep herders will probably still shoot them, though.

The cloning of the thylacine is not like the far-fetched plot device used in *Jurassic Park,* where fragmentary DNA from dinosaur blood cells was collected from a mosquito trapped in 85-million-year-old amber, and then combined with frog DNA to make a living dinosaur. There are any number of reasons why that would not be practical—or even possible—and in 1997, four years after the movie appeared, Rob DeSalle and David Lindley wrote *The Science of Jurassic Park and the Lost World,* in which they identified the myriad problems that Michael Crichton glossed over in his novel. But the thylacine is not a dinosaur; it is an animal that people alive today might have seen, either in the wild or in the zoo. On its face, the process of bringing the thylacine back from extinction sounds wonderful; it would give us the rare opportunity to apologize to an animal that we made extinct. But after the jubilation and self-congratulation end, we might want to think about the implications of reintroducing a recently extinct wild animal to the world. Where would we put it? In a zoo? Would we try to reintroduce it into the Tasmanian forests? Then who (or what) would teach it to hunt? Should we clone more than one? An entire population? "Dolly" the sheep was cloned in 1997, and since then, other animals have been successfully cloned, but they have been animals that can be easily maintained in captivity—like the gaur.

The gaur (*Bos gaurus*) is a large, oxlike creature native to Southeast Asia. In January 2001, a gaur calf named "Noah" was born to a domestic cow named Bessie, thus becoming the first endangered animal successfully cloned by scientists. Genetic material from the skin cells of a male gaur, who died eight years ago, was fused with the emptied egg cells of common cows. A total of 692 eggs were used in the experiment, which resulted in one live clone.* Noah weighed eighty pounds at birth, but died of dysentery within forty-eight hours. The gaur population has dwindled to about 36,000 because of hunting and because of degradation in the forests, bamboo jungles, and grasslands where they live in India and Southeast Asia.

Like the gaur, the banteng (*B. javanicus*) is a large wild ox that is found in Southeast Asia from Myanmar to Borneo. The animals resemble large, dark-colored domestic cattle, and indeed belong to the same genus. (All domestic cattle, including milk cows, beef cattle, and zebus are *B. taurus*; the kouprey, found only in Cambodia, is *B. sauveli*; and the yak is *B. grunniens*.) In April 2003, it was announced that two banteng calves were born in the San Diego Zoo from cells taken from an animal that died more than twenty years earlier. Robert Lanza, chief scientist for the Massachusetts company Advanced Cell Technology, said that "the bantengs were cloned by transferring the [previously frozen] cells into empty eggs from ordinary domestic cows. We implanted the cloned embryos into beef cattle which served as surrogate mothers. Although we started with sixteen pregnancies, only two of them went to term" (Fox 2003). One of the two was oversized, and had to be euthanized.

Ross MacPhee's quest for signs of hyperdisease in Siberian mammoths (pp. 117–24) was only part of a larger program. As its title suggests, Richard Stone's book, *Mammoth: The Resurrection of an Ice Age Giant* (2001), details the search for DNA by scientists for the purpose of cloning or re-creating a living mammoth, and by others to see if they could learn more about the vanished lifestyle of the Pleistocene pachyderms. In late 2002, Japanese scientists announced that they intended to clone a

---

* A much faster, more efficient method of cloning has been developed, according to an article in *New Scientist* (2002) by Sylvia Westphal. In place of the "micromanipulator," which allows a technician to grab an egg cell, insert a very fine needle, suck out the nucleus, and insert a nucleus of the animal to be cloned, researchers at a dairy facility in Australia developed a method whereby egg cells are split in half, and the half containing the nucleus is discarded. By zapping the cells with an electric current, the empty cytoplast is fused with a cell from the animal to be cloned, which in many cases results in a healthy blastocyst that can be implanted in the womb of the surrogate mother. This method, known as "handmade cloning," has already been successfully tested in endangered African antelopes, such as the bontebok, the giant eland, and the black impala.

mammoth, and establish an Ice Age wildlife park in Siberia. If they are unable to find sufficient DNA for the job, they will obtain a sperm cell (which would contain half of the mammoth DNA) and impregnate an Indian elephant, considered the closest living relative of the mammoth. But the DNA in cells repeatedly frozen and thawed will be so badly damaged that it is very unlikely to be usable for cloning. (In *Jurassic Park,* the dinosaur DNA obtained from a mosquito that had bitten a dinosaur and was preserved in amber had to be supplemented with frog DNA, and even if this were possible—which it isn't—the resulting animal would probably have been more frog than dinosaur.)* The Japanese have countered this argument by successfully cloning two cows using skin cells from a cow embryo that were kept frozen at the same temperature ($-35°C$) as the Siberian permafrost. In this instance, of course, the cow's DNA was complete and undamaged, and was not 10,000 years old. As of this writing, none of the cloning projects have succeeded, but Stone believes that a mammoth will eventually be cloned. He concludes his book: "Whether it is five years, five decades, or five centuries from now, wooly mammoths will once again walk the earth."

In 2003, Danish scientists announced that they had recovered DNA from twenty-eight species of trees, shrubs, mosses, herbs, and eight species of mammals, including the steppe bison, horse, and mammoth (Adler 2003). Eske Willerslev and his colleagues at the University of Copenhagen sampled the permafrost along 1,200 miles of the Arctic coast of Siberia, drilling to depths of a hundred feet, and removing soil cores which they then tested for DNA. According to Willerslev *et al.* (2003):

> Genetic analyses of permafrost and temperate sediments reveal that plant and animal DNA may be preserved for considerable

---

* DeSalle and Lindley's *The Science of Jurassic Park* should be required reading for all those who believe, with Michael Crichton, that resurrecting extinct animals from ancient DNA is possible. The authors explain why retrieving DNA from a mosquito trapped in amber would be nigh impossible; why the genetics, mathematical analyses, chaos theory, dinosaur breeding, and dinosaur behavior (not to mention the peculiar behavior of the paleontologists) in Crichton's book and the ensuing movies are all wrong; and finally why re-creating long-extinct animals is fraught with ethical considerations. In other words, even if we *could* do it, should we?

time periods, even in the absence of obvious macrofossils. In Siberia, five permafrost cores ranging from 400 to 10,000 years (kyr) contained at least 19 different plant taxa, including the oldest authenticated ancient DNA sequences known, and megafaunal sequences, including mammoth, bison, and horse.

It now remains to be seen if the mammoth DNA contains the evidence of disease that Ross MacPhee is looking for, and if the DNA can be used to bring the extinct mammoths back to life.

The Oceans

It is within the realm of human endeavor to collect wild horses, black-footed ferrets, macaws, and even whooping cranes, transfer them to enclosures where they can breed, and then relocate them to the habitat where they once lived. This becomes much more difficult if that habitat is gone, as in the case of Spix's Macaw, where the particular environment in which the blue parrot once thrived has now been almost completely eradicated. But what of those animals that cannot be transported and bred in captivity? What of the great whales whose numbers were so depleted by nineteenth- and twentieth-century whalers that they are now listed as endangered? What of the gray whales, hunted nearly to extinction in the Pacific, who once swam in the Atlantic Ocean as well?

Gray whales once fed in the cold waters off Iceland and Greenland, and came south—perhaps to the Bay of Biscay, or the English Channel—to breed. Morphologically, they were the same as the California gray whales (now known as *Eschrichtius robustus*), which today confine their migratory meanderings to the Pacific coast of North America, annually swimming from the Bering Sea south to Baja California and back again. There used to be a sizable *western* Pacific population of gray whales, summering off Siberia and wintering in breeding grounds off Korea and Japan, but in the twentieth century this population was mostly eliminated by Japanese and Korean whalers. Only a few western gray whales still

Gray Whale

come to the lagoons and bays of Sakhalin Island. But no living person has ever seen an *Atlantic* gray whale. Fortunately, we do have conclusive pale-ontological and historical evidence to verify the existence of this popula-tion of the animal the whalers used to call the "devil-fish."

The earliest mention of what might have been the Atlantic gray whale can be found in an Icelandic bestiary from about A.D. 1200, which describes different kinds of whales accurately enough for modern cetolo-gists to identify some of them by species. Then, in the *Konungsskuggsjá* ("king's mirror"), a thirteenth-century document written in Norwegian, probably as a set of instructions for a king's son, there is a list of twenty-one sea creatures, some of which appear to be living whales, dolphins and pinnipeds (and some of which—mermaids and mermen, for example—are clearly mythological). Although it is not identified very clearly, the gray whale seems to be one of the listed whales.

A seventeenth-century Icelandic work by Jon Gudmundsson (quoted in Hermannsson, 1924) lists various whales that might be found in Ice-landic waters. One of these is *sandlaegja*, which has been translated as "sand-lier"—one who lies in the sand. The description of the sandlaegja* is accompanied by a picture of a whale that is admittedly inconclusive, but is obviously not one of the verified Icelandic species; it is likely that this is an illustration of an Atlantic gray whale. The description translates

---

* Until the publication of Ole Lindquist's 2000 treatise on Atlantic gray whales, most popular works had spelled the name wrong. Lindquist (an Icelander) wrote that "Fraser (1970) through-out erroneously spells the name *sandloegja*, a mistake which, for example, Gaskin (1982:270), Mead and Mitchell (1984:37, 50), R. Ellis (1992:44), and P. J. Bryant (1995:859) have upheld."

as: "Sandlaegja. . . . Good eating. It has whiter baleen plates, which project from the upper jaw instead of teeth, as in all other baleen whales, which will be discussed later. It is very tenacious of life and can come to land to lie as a seal to rest the whole day. . . . Sandlaegja, reaches 30 ells [an ell is about thirty inches], has baleen and is well edible." Many of these characteristics, such as the "whiter baleen plates" and the sand-lying behavior that gave the whale its Icelandic name, would appear to refer to the gray whale, which does indeed have short, whitish baleen, and a habit of coming into very shallow water. But other whales share some of these attributes, so they cannot be said to be absolutely diagnostic.

So far, the literature has produced nothing that can be positively identified as an Atlantic gray whale. But before the modern era, gray whales swam in the Atlantic Ocean, as we know from another source. Fossil remains of a species similar to—if not identical with—the Pacific gray whale have been found in western Europe (Sweden, England, and the Netherlands), and on the east coast of North America from New Jersey to South Carolina. The Atlantic gray whale apparently fed in cold northern waters (perhaps Iceland and Greenland), and then moved south (to Spain, France, or England?) to breed and calve. In protected bays, the cows would likely have delivered their calves and become impregnated, prior to their northward journey in the spring.

With the exception of the fossil evidence, the only strong clues to the identity of this whale are found in the work of Gudmundsson, and in a debatable reference in a New England work of 1725, where Paul Dudley describes a "scrag whale" with characteristics that are not applicable to any other species but the gray whale. Dudley's entire citation, as quoted in the *Philosophical Transactions of the Royal Society of London* for 1725, reads as follows: "The Scrag whale is near a-kin to the Fin-back, but instead of a Fin on his Back, the Ridge of the After part of his Back is scragged with a half Dozen Knobs or Knuckles; he is nearest the right Whale in Figure and for Quantity of Oil; his bone is white but won't split."

In their 1984 study of the Atlantic gray whale, Mead and Mitchell recognize as authoritative only Fraser's sandlaegja, the 1725 description of the "scrag whale" by Paul Dudley, and the 1611 instructions given by the directors of the Muscovy Company to one Thomas Edge. Edge's instructions include descriptions of all sorts of whales he might look for, including the "otta sotta," which was described as being "the same colour as the Trumpa [sperm whale] having finnes in his mouth all white but not above a yard long, being thicker than the Trumpa but not so long. He yeeldes the best oyle but not above 30 hogs' heads."

The most thorough account of the Atlantic gray whale in the historical record is Ole Lindquist's treatise (2000), "The North Atlantic Gray Whale (*Eschrichtius robustus*): An Historical Outline Based on Icelandic, Danish-Icelandic, English and Swedish Sources Dating from ca 1000 A.D. to 1792." Unlike Fraser, Gaskin, Mead and Mitchell, P. J. Bryant, or me, Lindquist reads Icelandic, Danish, and Swedish, and therefore found many more sources than those of us who relied upon earlier, mostly English-language authors. One such source was Thomas Bartholin, a University of Copenhagen professor, who produced a work in 1657 called "Record of the Fishes of Iceland," which contained this description:

> The fifteenth type is the *sandlaegja*. It is twenty or nearly thirty ells long and lies quietly in the sand. It takes the greatest possible pleasure in sand and greedily seeks out the tiny little fish which are abundant there. It is equipped with horny plates, and although it is eaten by humans, it does not have a pleasant taste, nor is it particularly fat. It is difficult to kill and dies slowly as seals do. It is happy to rest on land. If one comes upon it in the sand, one cannot get near it because it throws up the surrounding sand and moves vigorously in an extraordinary way. But once the force of the waves had driven it into the shallows and it has been run through in several places by spears, it lies dead.

Especially in the lagoons of Baja California, Pacific gray whales inhabit fairly shallow waters, and have been known to strand on the beach or on sandbars, but no living whale habitually comes ashore. To do so would mean almost certain death, for whales are ill equipped to move on land, and a whale on the beach in the sun is a whale that cooks in its own blubber insulation. It is therefore curious to read the Icelanders' description of the habits of the sandlaegja, almost every one of which alludes to the whales' habit of lying in the sun like a seal. Lindquist mentions the Swedish clergyman Olaus Magnus (1490–1557), who wrote of a whale, "clearly distinguishable from the walrus, which comes on to the beach in sunshine where it sleeps soundly like the seal and which people frequently manage to capture by tying it with ropes." Lindquist wrote that "the only cetacean that has a habit like that is the gray whale," but while the Atlantic version may have regularly come ashore as the Icelanders said it did, its Pacific counterpart does not engage in such self-destructive behavior.

Lindquist concludes "that the North Atlantic gray whale was hunted primarily by coastal inhabitants (a) around the North Sea and the English Channel, from prehistoric times at least into the high Middle Ages; (b) in Iceland, from about 900 A.D. until about 1730; and (c) in New England by European settlers from the mid-seventeenth century until about the same time, possibly also by Indians there; secondly that it was caught by Basques in the latter half of the sixteenth century and in the early seventeenth century."

We have no way of knowing if the Basques and the Icelanders hunted the Atlantic gray whale to extinction. Their numbers might have been low before the first Basque *chaloup* (a small swing boat, or dory) was ever launched. We do know that these very Basque whalers wreaked havoc on the right whale populations of the Bay of Biscay, and then headed across the North Atlantic to do the same thing. But for all this killing, the right whale is not extinct. During the past two centuries, it appeared for all the world as if the idea was to kill all the whales, but despite our massive, concentrated efforts, we failed to eliminate a single great whale species. If industrial whaling could not eliminate any species of whale, how could seventeenth-century open boat whalers, armed with hand-thrown harpoons, have accomplished what diesel-powered catcher boats armed with exploding harpoons could not? It would have been an extraordinary accomplishment for the early hunters to kill *all* the Atlantic gray whales. They probably contributed to its demise, but other factors, such as disease, climate change, or the mysterious force we call "extinction" also played a part. Extinction is one of the most powerful forces on earth, and one of the most enigmatic. It affects every species that has ever lived, and has eliminated most of them. Nobody is quite sure what extinction is or how it works, except that killing every last member of a given species will make it extinct. The Atlantic gray whale joins Steller's sea cow as the only large marine vertebrates to become extinct in historical times.

The first great whale that was hunted "commercially"—that is, its products were sold as well as consumed by the whalers—was probably

Right Whale

the right whale, known as *sarda* to the Basques. Vague records indicate that the Basques began killing whales about a thousand years ago. Spotting the whales from the shores of the Bay of Biscay, they launched their chalupas and threw wrought iron spears, which they called *arpón*, whose name has come down to us as *harpoon*. The right whales of Biscay were hunted for their oil and the baleen plates that hung from the roof of their cavernous mouths. Sometimes twelve feet long, these keratinous plates (made of the same stuff as your hair and fingernails) were used wherever a strong, flexible material was required: wagon springs, corset stays, buggy whips, and, because its edges were fringed, it was used for the crests of knight's helmets. Most of the sardas were killed for their oil, which was used for heating, lighting, lubrication, tanning, and hundreds of other uses. In *Moby-Dick*, Melville asked: "What kind of oil is used at coronations? Certainly it cannot be olive oil, nor macassar oil, nor castor oil, nor bear's oil, nor train oil, nor cod-liver oil. What then can it possibly be, but sperm oil in its unmanufactured, unpolluted state, the sweetest of all oils?" As far as we know, the Biscay right whales were never very plentiful, and it didn't take long for the Basques to hunt them out of existence.

But there were more right whales across the Atlantic Ocean, and the Basques set out to find them. Some five centuries after Leif Ericsson island-hopped from Norway to Iceland to Greenland and Labrador, the Basques followed Breton codfishermen to Labrador, but they were on the trail of the whale. They fetched up at a place called Red Bay, and in 1603 they built a little village, dedicated to the processing of whales. Canadian archaeological historian Selma Barkham identified the ruins and wrote:

Vast quantities of curved red tiles are to be found in all the places used by Basques as establishments for processing and boiling down whale blubber. They were used to roof both the ovens or furnaces where heavy copper cauldrons were apparently kept boiling day and night, and the cooperages or cabins where the coopers often made up well over 1,000 barrels per voyage for storing whale oil . . . anywhere from 6,000 to 9,000 barrels of whale oil would have been sent back to Europe every year from Red Bay during the peak period of exploitation.

Basque whalers hunted the whales off Labrador, and fished for codfish as well. The right whales ran out, but when Dutchmen like Willem Barents and British explorers such as Henry Hudson and Martin Frobisher sailed north from England seeking a northeast route to Asia, they found the waters around Spitsbergen thick with whales. They were even more "right" than the right whales; they were fatter, slower, and had even longer baleen plates. We now know them as bowheads, but in the eighteenth and nineteenth centuries, they were known as Greenland whales, polar whales, mysticetus, or simply, "the whale." (In Dutch they were called *Groenlandse walvisch*.) Just as Charles Scammon chronicled the history and natural history of the gray whale, and Herman Melville the sperm whale, the historian of the bowhead fishery was William Scoresby, Jr., who, like Scammon and Melville, was himself a whaleman. Scoresby preceded Scammon and Melville—he was born in 1789 and died in 1857; in that sense he set the tone for the whaling histories that followed. (*Moby-Dick*, of course, is much more than a "whaling history," but its descriptions of whales and the process of Yankee whaling are unparalleled and remain the best and most accurate account of square-rigged whaling that we have.) In Scoresby's 1820 *Account of the Arctic Regions with a History and Description of the Northern Whale Fishery*, we learn that "This valuable and interesting animal, generally called *The Whale* by way of eminence, is the object of our most important commerce to the Polar Seas, is productive of more oil than any of the other Cetacea, and being less active, slower in motion, and more timid than any other of the kind, of similar or nearly similar magnitude, is more easily captured."

The bowhead whale, whose common name is derived from the great arc of its jaws, can reach a length of sixty feet and a weight of at least as many tons. It is black in color, often with a white patch on the leading edge of the lower jaw. Like the right whale, the bowhead has no dorsal fin, and shows a smooth back as it rolls through the water. Also like the right whale, the bowhead is a "skimmer," swimming through schools of plankton and trapping them in the fringes of its baleen plates. (Other baleen whales take in a mouthful of water and small organisms, force the water out through the sievelike baleen, and then swallow the organisms.) Bowheads today are usually found in small groups of three or four, but in earlier times they were commonly seen in groups of up to a hundred.

So eager were they to avoid the annual voyage to Spitsbergen, the Dutch established a permanent colony on the little island of Amsterdamoya in 1633, which they called *Smeerenburg*, or "blubber-town." Smeerenburg failed—the overwintering crews died—but British and Dutch whale ships sailed north every year and killed the polar whales with avaricious abandon. By this time, baleen was the object of the hunt; after the plates had been removed, the sixty-ton carcasses were set adrift in the icy northern waters of Greenland and Spitsbergen. In the nineteenth century, ladies had tiny waists and billowing skirts, both of which required slices of baleen. Tens of thousands of whales died to fill the bottomless maw of Dame Fashion, and by the turn of the next century, the whales had become so scarce that the hunt dragged to a close. It is not clear if the scarcity of baleen caused the change in styles, or if styles changed—as styles will—rendering baleen redundant, but whatever the fashion cause-and-effect, hunting the whales of the Davis Strait and Baffin Bay was over by 1860.

Thomas Welcome Roys (1816–1877), a whaling captain out of Sag Harbor, Long Island, had heard rumors of stocks of whales north of the Bering Strait, and in 1848 he sailed the little whaleship *Superior* through that narrow passage and encountered "the fattest whales ever seen and soon the deck space was filled with casks of oil" (Schmitt, de Jong and Winter, 1980). As with Scammon's discovery of the breeding grounds of the California gray whales, Roys's find brought an armada of whaleships, mostly out of San Francisco, to the ice-choked waters north of Alaska, where the bowheads came to breed. The last remaining population of these great whales was on the way to an extirpation not unlike that of their eastern Arctic cousins, when Mother Nature, perhaps recognizing the inequality of the battle, stepped in to save the whales. She arranged for ice conditions so brutal that in 1871, the entire whaling fleet was locked in the ice. Although there was no loss of human life, the fleet was crushed, and the whaling industry suffered a huge setback. Human technology once again found a way to dominate nature; steam-powered whaleships replaced the sailing ships that had been lost in 1871. Now there was a fleet that stood less chance of being trapped in the closing ice—steam-powered ships had a reverse gear, sailing ships did not—and American whalers returned year after year to hunt the bowheads off the North Slope of Alaska. They even established settlements on islands such as Herschel so that the ships would not have to make the difficult and time-consuming journey every year from Seattle or San Francisco. As with almost all whaling enterprises, the whalers of the western Arctic killed every whale in sight, with no regard whatever for the diminishing stock. Another problem—for the whalers as well as the whales—was that a wounded whale often dived under the ice, where it died and could not be retrieved.

Fate, chance, and coincidence played important parts in the ongoing drama that was commercial whaling. The first Yankee sperm whaler seems to have been one Tristam Coffin out of Nantucket, whose ship was blown out to sea during a right whale hunt, and who killed the first *cachalot* (another name for sperm whale), effectively inaugurating the industry that would define the economy of nineteenth-century New England. When the right whales seemed to be running out, the bowheads were discovered, and when the eastern bowheads were eliminated, Captain Roys found a previously unexploited population off Alaska. Captain Scammon more or less accidentally found the Baja breeding grounds of the California gray whales at Laguna Ojo de Liebre (later named Scammon's Lagoon), leading to an unprecedented slaughter. And though it appeared that the last of the bowheads would be safe in their icy fortresses after petroleum was discovered in Pennsylvania in 1859, the discovery of oil deposits off Alaska in 1969 would, among other things, spell disaster for the bowheads again.

The building of the pipeline to carry the oil across Alaska from Prudhoe Bay required many hands, and they often belonged to previously unemployed Inuit men. To entice workers to this harsh environment, wages were high, and many young Inuits quickly became rich. While tradition had previously assigned the role of whaling captain to the village elders, newly prosperous young men could now afford the boat and the equipment, and many became instant captains. As the number of hunting boats went up, the killing also increased, and whale numbers fell accordingly. In 1973, only thirty-seven whales were killed by Inuits, but in 1977, 108 whales were killed or "struck and lost" at the hands of the Inuit whalers. (The problem of "struck and lost" whales—wounded whales that sought refuge under the ice and died—was an important factor in the negotiations that were to follow.) As quotas for other species of great whales were falling (the moratorium on commercial whaling was passed in 1983), the bowheads took center stage at the IWC meetings of the early 1980s, because the commission had to deal with "aboriginal whaling," an extremely contentious subject.* It was one thing to eliminate (or suspend) "commercial" whaling, but depriving Native Americans of their "traditional" rights smacked of discriminatory racism. No U.S. IWC commissioner was prepared to suffer the wrath of an American populace that remembered the slaughter of the plains buffalo to deprive the Indians of their primary food source; or the smallpox-infected blankets given to the Mandans to reduce their resistance to "manifest destiny" (a self-proclaimed justification for taking the entire continent from its original inhabitants); or any number of other transgressions against the peoples we now refer to as "Native Americans."

In 1978 the bowhead population of the North Slope was estimated at 2,000. Analysts decided that such a small population could not survive the pressure of even minimal hunting, so they advocated the complete shutdown of Eskimo whaling before the species became extinct. The Eskimos and their advocates responded by asking for an *increase* in the number of whales they would be allowed to take. The public arguments were bitter; the private ones more so. "Struck and lost" whales were a major part of

* I was a member of the U.S. Delegation to the IWC from 1980 to 1990, and attended the five annual meetings in England, and one each in California, Argentina, Sweden, and New Zealand, and I participated in numerous discussions on the subject of aboriginal whaling quotas. I have discussed my own position on this subject in *Men and Whales* (1991), but for the record, I believe that the preservation of a species—the bowhead whale—must take precedence over the preservation of Inuit traditions, especially when the traditional methods have been replaced by outboard-powered boats, rifle-powered harpoons, and tractors to haul the whales onto the ice for butchering.

the problem: the Eskimos did not want them to count in their totals, while those who were lobbying for low (or no) figures believed that every whale that was shot should count against the Eskimos' total. New studies were commissioned by both sides, and not surprisingly, the results confirmed the position of each side: anti-whaling scientists somehow managed to produce data that showed that the population was declining, while advocates of Eskimo whaling showed that the previous estimates were far too low. By 1988 the estimated number of bowheads had risen to 7,000, and multiyear quotas were being recommended. (For example, the bowhead numbers for 1980 to 1983 were "a three-year quota of 45 whales landed and 65 struck, no more than 17 to be taken in any one year" [Ellis 1991]).

Despite numerous efforts by commercial and aboriginal whalers at various locations in the high northern latitudes to eliminate them, *Balaena mysticetus* still lives. Like many species of great whales, the bowhead was driven perilously close to extinction by rapacious whalers. The Atlantic gray whales are gone, but the Pacific population is alive and thriving. Humpbacks and rights, always the first whales hunted because they breed close to shore, have been reduced to scattered small populations around the world. The rorquals ("groove-throated whales"), the last of the great whales to be hunted because they were too fast and too strong to be caught by harpoon-throwing whalers in rowing boats, are reduced to small populations in remote oceans. The huge blue and fin whales, taken relentlessly in the Antarctic, have not recovered to their former numbers. The sei whale, which lives at polar latitudes in both hemispheres, was reduced in the south by Norwegian and British whalers, and in the north by Norwegians, Icelanders, and Canadians. Of the five species of grooved whales, only the little (by whale standards) minke has survived in substantial numbers. In fact, the minke is probably increasing. When its larger cousins, the blue, fin, and sei whales, were removed from the Antarctic food chain, minke whales, which feed on the same stocks of krill, proliferated to fill the blue and fin whale-sized holes in the ocean.

And what of the sperm whale, hunted around the world and glorified in *Moby-Dick*? Yankee whalers, who killed whales the hard way—whale spotted, boats lowered, whale chased, whale harpooned, whale killed, whale towed back to ship, whale processed—probably didn't kill more

Sperm Whale

than 20,000 whales during their entire history. But modern technology enabled the whalers to kill whales much more efficiently; they chased them in diesel catcher boats, shot them with exploding grenade harpoons, and then shot some more before bringing their catch back to the huge floating factories for processing. Indeed, the heyday of sperm whaling was not the mid-nineteenth century, but rather the mid-twentieth: in the 1960s, Japanese and Soviet whalers, working in the North Pacific, killed 50,000 sperm whales per year.

For years, whaling historians (including me) accepted the conventional estimate of a million-plus sperm whales still occupying their remote offshore habitats, but in 2001 Hal Whitehead of St. John's University, Newfoundland, recalculated the numbers, and came up with only 350,000 sperm whales worldwide, which means that twentieth-century sperm whalers were actually hunting an endangered species—one that they had endangered themselves.

Also known as the cochito, the vaquita is a close relative of the harbor porpoise; it is found only in the Sea of Cortés (Gulf of California), between mainland Mexico and Baja California. Mexican fishermen had been seeing them for years (often in their nets), but *Phocoena sinus* was only described for science in 1958 (Norris and MacFarland). It is even smaller than the harbor porpoise, and may be the smallest of all the cetaceans, at a maximum length of four and a half feet. It is similar in coloration to the harbor porpoise, although the mouth-to-flipper stripe is not as distinct, and the dorsal fin is more sharply curved. Directed

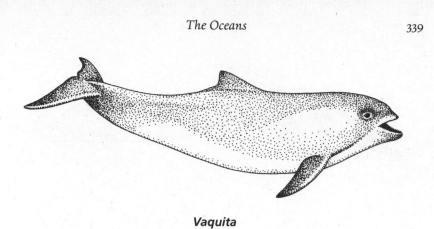

Vaquita

hunting and entanglement in the gillnets of the totoaba (a kind of large sea bass) fishery have seriously reduced the population of vaquitas. With only a few hundred remaining, they are listed among the most endangered of all marine mammals, and they are still being caught in fishing nets in their isolated habitat.

By 1997, when Barlow, Gerrodette, and Silber investigated vaquita numbers in the Gulf of California, things had only gotten worse. They quoted several authorities who agreed that it is "the most endangered marine cetacean in the world." Their surveys indicated that the vaquita population, which had been in decline since they began counting in 1986, was plummeting. For the 1986–88 surveys, they had counted a total of 503 animals, but by 1993, they were able to find only 224. They wrote, "All of these estimates . . . indicate that the species is at a critically low level." Known for less than half a century, the vaquita appears destined to vanish like its vague, almost invisible spout.

The Chinese river dolphin (*Lipotes vexillifer*) now vies with the vaquita and the Indus River dolphin for the title of most endangered cetacean in the world. Known locally as the *baiji*, it lives only in lakes and rivers of central China. It is grayish blue above and white below, with a long, upturned beak and eyes placed high on its head. Although efforts are being made to protect this animal, its propensity for becoming entangled in fishermen's nets has greatly reduced its numbers, and the total population probably does not number more than a hundred animals. For twenty-three years, a single male baiji lived at the Wuhan Institute of Hydrobiology after having been rescued from a fishing net in the Yangtze River. All attempts to find a mate for "Qiqi" failed, and the lonely dolphin died in July 2002.

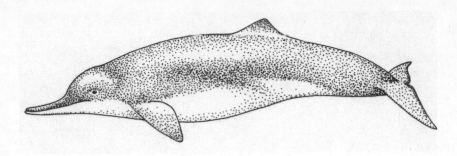

Chinese River Dolphin (baiji)

The Three Gorges dam, the largest hydroelectric project in history, and the largest construction project in China since the building of the great wall, started to go on line in 2003. A mile and a half long and 600 feet high, it will create a reservoir hundreds of feet deep and nearly 400 miles long, and flood the Chang Jiang (Yangtse) River in Sandouping, Yichang, and Hubei provinces. The reservoir, the engineers say, will enable 10,000-ton ocean-going freighters to sail directly into the nation's interior for six months of each year, opening a region burgeoning with agricultural and manufactured products. The dam's hydropower turbines are expected to create as much electricity as eighteen nuclear power plants. At a cost variously estimated at anywhere from 25 billion to 75 billion dollars, it will force the resettlement of almost two million people. Although the relocation of all those people has raised public outcries in China and outside, few voices have been raised in defense of the baiji, which lives only in the middle and lower reaches of the Chang Jiang. The dam is scheduled for completion in 2009, but well before then, the riverbed will be scoured by rushing waters; the river and the reservoir will be polluted by industrial wastes; and the little dolphin's food sources will be wiped out.

In India and Bangladesh there are two very similar species of freshwater dolphins, the Ganges River dolphin (*Platanista gangetica*), and the Indus River dolphin (*P. minor*), separated by minor skeletal differences and by the geographical intrusion of the entire Indian subcontinent. Both species have a long, narrow snout, wide, squared-off flippers, and eyes that have degenerated to the degree that they are almost nonfunctional. Early observers believed that the absence of a visual apparatus meant that these dolphins grubbed in the mud for food, but they were later shown to be excellent echolocators, able to find and catch swimming fishes in the almost opaque waters in which they live. They managed quite well until humans began hunting them and messing with their habitat. Both Indian dolphins have suffered major population losses as a result of hunting and the building of dams in their native rivers, and the numbers of *P. minor* have been reduced to less than 300 animals, making it one of the most endangered cetaceans in the world. In a 1998 study, Reeves and Chaudry wrote:

> The freshwater dolphins in the Indus River system of Pakistan, *Platanista minor*, have been considered endangered sine the 1970s. Measures taken to protect them from deliberate capture seem to have stopped a rapid decline, and combined counts in Sindh and Punjab provinces since the early 1980s suggest a population of at least a few hundred animals. . . . In addition to the risks inherent to any species with an effective population size in the few hundreds (at most), these dolphins are subject to long-term threats associated with living in an artificially controlled waterway used intensively by humans. Irrigation barges partition the aggregate population into discrete populations for much of the year. Dolphins that "escape" during the flood season into irrigation canals or into the reaches downstream of barrages where water levels are low have little chance of survival.

In April 2002, The Dolphin Conservation Society (DCS) of India reported an alarming decline in the population of dolphins in the Brahmaputra River, which runs into the Ganges in Bangladesh. The recent

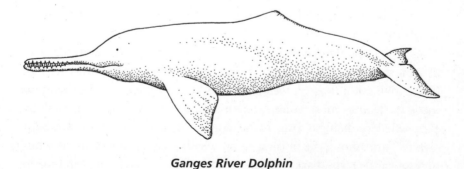

Ganges River Dolphin

census found only 198 dolphins in the river as compared to 218 in 1997. The fall in the population is mainly due to large-scale poaching for dolphin oil, which is used as fish bait and is also believed to have medicinal properties. Strangulation by gill nets, loss of habitat due to siltation, and a depleting prey base due to overfishing are other threats. Riverbed sand extraction (permitted by government lease in 1996) caused dolphin numbers in the Kulsi River to decline from twenty-five in 1993 to only eleven in 2002. With total numbers in the low hundreds, the Indus River dolphin finds itself in a three-way race with the baiji and the vaquita for the dubious title of first cetacean species to become extinct in modern times.

In 1861, showman P. T. Barnum led an expedition to the St. Lawrence River in Canada to capture a white whale for exhibit in his museum at Ann Street and Broadway in lower Manhattan. It was the first cetacean of any kind to be exhibited in an aquarium. In his 1870 autobiography, Barnum wrote:

> Of this whole enterprise, I confess I was very proud that I had originated it and brought it to such a successful conclusion. It was a very great sensation, and it added thousands of dollars to my treasury. The whales, however, soon died—their sudden and immense popularity was too much for them—and I then despatched agents to the coast of Labrador, and not many weeks thereafter I had two more live whales disporting themselves in my monster aquarium.

The beluga (*Delphinapterus leaucas*) is a chubby, sixteen-foot-long snow-white whale with a rounded, prominent forehead (called the "melon"), a low ridge instead of a dorsal fin, and a permanent grin. Unlike most other cetaceans, the beluga's seven neck vertebrae are not fused, giving it a flexible, well-defined neck, and therefore the ability to move its head as most other cetaceans cannot. Called "sea canaries" by early whalers, belugas (the name comes from the Russian word for "white") are among the noisiest of all whales, with a repertoire of whistles, squeaks, chirps, barks, whinnies, and snores. They have been bred in

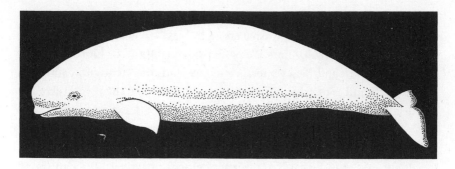

Beluga

captivity, and although they don't perform the acrobatic stunts of bottle-nose dolphins or killer whales, their gentle antics and expressive faces make them the darlings of oceanarium shows.

Belugas occur throughout the arctic and subarctic waters of North America, Greenland, Europe, and Asia. They are often found in ice-covered regions in winter and spring and in coastal waters in summer and autumn. Belugas of the Bering Sea population range throughout the Bering, Chukchi, and Beaufort Seas. They winter in the drifting ice of the Bering Sea, moving in summer to concentration areas scattered along the coast from Bristol Bay to the Mackenzie River Delta in Canada. In Alaska, major concentrations occur in Bristol Bay, Norton Sound, Kotzebue Sound, and Kasegaluk Lagoon. There is a large population in Hudson Bay, seen most often in the vicinity of Churchill, and another in the Gulf of St. Lawrence. There is a small population in Cook Inlet, Alaska, in the vicinity of Anchorage, that has recently been shown to be heavily preyed upon by killer whales (Shelden *et al.* 2003). The world population of belugas has been estimated at around 60,000, and although they have been traditionally hunted by native peoples for centuries, most of the circumpolar belugas are not considered endangered. It is mostly in areas where they come into contact with "advanced" civilizations that they become vulnerable.

The Saguenay River, which feeds the St. Lawrence, once held as many as 5,000 belugas. Fishermen complaining that the white whales were eating their fish induced the Canadian government to introduce a bounty system. A hunt was organized in which sportsmen could shoot a whale from a boat

(the large-scale version of shooting fish in a barrel); the population was reduced to about 500. If this wasn't enough, high levels of toxic chemicals
such as PCBs (polychlorinated biphenyls) and DDT (dichlorodiphenyl-
trichloroethane), and heavy metals such as lead, mercury, and cadmium
have been found in the river. These toxins are thought to be responsible for
the deaths and strandings of many belugas in the St. Lawrence River. Levels of these toxins in St. Lawrence belugas were found to be significantly
higher than in Arctic belugas (Smith, St. Aubin, and Geraci 1990).

It appears that we can only impact belugas negatively. We can shoot
them, harpoon them, and poison them, but while we have had some
success in breeding them in captivity, we cannot offset the tremendous
losses we have inflicted upon them through hunting, habitat degradation,
and chemical poisoning. The beluga population of the St. Lawrence is
dropping, and there is very little we can do about it except to reverse the
trend of human intrusion into their lives and their habitat, and this seems
a forlorn hope indeed.

Probably the most noteworthy of all extinctions are those of the great
predators. The carnivorous dinosaurs, such as *T. rex*—arguably the
most popular of all dinosaurs, if its appearance in movies is any indication—
are gone, and so are the great cats of the Pleistocene. But the animal whose
extinction impresses and puzzles us most is probably *Carcharodon megalodon*,
a fifty-foot-long monstrous shark that resembled today's great white—only
it was three times as long, ten times heavier, and had a mouthful of razor-
sharp, serrated teeth as large as your hand. In the 1996 collection of studies
of the great white shark, Michael Gottfried, Leonard Compagno, and Cur-
tis Bowman introduced the giant "megatooth" shark this way:

> The extinct lamnid shark, *Carcharodon megalodon*, variously re-
> ferred to as the "megatooth shark," "great-tooth shark," or "mega-
> lodon," is the largest macropredatory shark ever to live and is
> among the largest known fishes. *Carcharodon megalodon* is ar-
> guably the most spectacular marine predator of the Cenozoic,
> and has accordingly received much attention from the popular
> and scientific media.

Whether or not it was an ancestor of the great white (*C. carcharias*)—
or even a close relative—is still a subject of debate among elasmobranch
(sharks and rays) taxonomists. Before these debates began, the similarities
in the triangular, serrated teeth seemed more than sufficient for scientists
and nonscientists to lump them together in the genus *Carcharodon,* and indeed, many still do. For convenience, this giant fish is often called simply
by its genus name *Megalodon,* which translates as "great tooth." Its teeth,
our primary souvenirs of its existence, closely resemble those of its
smaller relative. The differences are nonetheless sufficient to justify separating the extinct and recent sharks as distinct species: besides their obviously greater size, megalodon teeth have more and relatively smaller
serrations and, above the root, a scar or "chevron" that is lacking on the
teeth of adult *C. carcharias*. That they are conspicuously different in color,
as well, is simply a consequence of fossilization: the teeth of all living
sharks are white, whereas the teeth of extinct or fossil species range from
light brown to black.

After dinosaur bones (and maybe trilobites), the teeth of this huge extinct shark are probably the most popular of all fossils. These teeth—always
fossilized, despite some claims to the contrary—are heavy and triangular,
with serrations along the blade. The largest known megalodon tooth is six
and a half inches in height, measured vertically. These teeth are shaped very
much like those of today's great white shark (*C. carcharias*), the star of *Jaws*
and other Hollywood movies, so megalodon comes to us with its reputation already in place. Abundant fossil evidence shows that megalodon lived
during the middle and late Tertiary, as far back as 50 million years ago, and

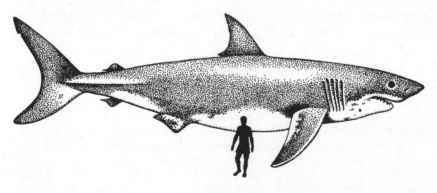

Megalodon

probably wandered the same seas as the modern white shark, plying pretty much the same trade—but on a substantially larger scale. In 1909, when a gigantic megalodon jaw was placed on exhibit in the American Museum of Natural History, curator Bashford Dean wrote, "At the entrance of the Hall of Fossil Fishes, there is now exhibited a restoration of the jaws of a shark (*Carcharodon megalodon*), which lived along the coast of South Carolina in Tertiary time. There can be no doubt that this was the largest and most formidable fish living or extinct of which we have any record. The jaws of a fully grown specimen measured about nine feet across, and must have had a gape of five or six feet."*

We do not actually know what megalodon looked like, since all we have to go on are its teeth, but we assume (from good paleontological practice) that it looked a lot like the living white shark, although it was considerably larger and bulkier. In Gottfried *et al.*, a reconstruction drawing shows a fish with a massive head and jaws, and a body that is much more robust than the puny little white shark's. A very large shark swam in the sea as recently as 2 million years ago, and fed on large prey, probably whales, and is now—mercifully for swimmers, anyway—extinct. But which seas did megalodon inhabit? Robert Purdy (1996) segregates white sharks into "small-toothed species," which includes the current great white, and the "giant-toothed," which includes *C. orientalis, C. auriculata, C. angustidens, C. subauriculatus,* and the giant of the giant-teeth, *C. megalodon*. From the fossil evidence, Purdy deduces that the various species of white sharks "first appeared in the Paleocene seas of southern Russia, Morocco, Angola, and the United States." The fossil teeth of giant-toothed sharks do not necessarily coincide with the fossils of cetacean prey, because these sharks evolved in the Paleocene, before the appearance of whales. Indeed, the eventual appearance of larger and larger whales may have been a factor in the exponential increase in the size of the giant-toothed sharks. The areas most productive in fossils of the

* A photograph of this jaw appears in almost every book about sharks published up to 1996, usually with one or more gents in lab coats sitting in it. The formidable jaw has been removed from exhibit, and a downsized version installed in the Hall of Vertebrate Origins at the museum. The exaggeration was caused by the curators erroneously assuming that all the teeth in a shark's jaw are the same size, and therefore placing six-inch teeth at the rear of the jaw, where, in fact, the teeth would have been much smaller (Randall 1973).

giant-tooths are the southeastern coast of the United States from Florida to Maryland, and, to a much lesser extent, southern England and the shore of what is now Belgium.

Megalodon probably reached a length of around fifty feet, but everywhere there are those who want to increase its length. In a 1964 discussion of fossil teeth found at Uloa, South Africa, David Davies, the Director of the Oceanographic Research Institute in Durban, wrote: "Estimates of size made from the fossil teeth of this wide-ranging shark obtained in various parts of the world indicate that it may have reached sixty to eighty ft. in total length."

People seem more than a little reluctant to concede the disappearance of some prehistoric animals (think of the "plesiosaur" of Loch Ness) and of these, the one that surfaces most regularly is the giant shark. Since the 1975 publication of *Jaws*, and the subsequent blockbuster movies, authors have tried to cash in on Peter Benchley's formula for success, figuring that if a twenty-five-foot shark could generate all that moola, think of what a one-hundred-footer could do. Or a 200-footer!

In 1981, Robin Brown wrote a novel he called *Megalodon,* which was about a 200-foot-long ancestor of the great white shark that was blind, covered with a coat of crustaceans, and living at great depths. When it decides to come to the surface, it eats every person and boat in sight; before we are finished, we have talking dolphins, edible submarines and diving bells, interservice rivalries, sexual politics, disquisitions on the similarity of dolphins and man (thanks to John Lilly), and, for good measure, a Soviet counterplot. Have I mentioned the airlifted and trained sperm whales? In 1987, George Edward Noe self-published a little number he called *Carcharodon,* in which the giant shark has been imprisoned in an iceberg for a couple of million years, but because it is *really* hungry when it thaws out, all hell breaks loose. It goes on a rampage like its predecessor, and before we are finished here, we have the marine biologist hero renting a Norwegian whaling ship and shooting the shark with a grenade harpoon.

Then in 1997 came *Meg* by Steve Alten. The book opens in the late Cretaceous as a *Tyrannosaurus rex* is grabbed by a sixty-foot-long shark and dragged down to be eaten. Take that, *Jurassic Park*! One of the more imaginative inventions in the megalodon canon is Alten's explanation for why the giant sharks have remained unnoticed for so long. It seems that they live in the very deepest part of the ocean, the bottom of the Marianas Trench, which has somehow become a hydrothermal vent area, bubbling with superheated water. "The water temperature above the warm layer," he writes, "is near freezing. The *Meg* could never survive the transition through the cold in order to surface." Whoa! What happened to physics as we know it? Only in Alten's topsy-turvy world can there be a situation where warmer water remains below colder water. "Listen, man," you can hear the author say, "this is *fiction*—I get to make stuff up."

In the previous megalodon books, the shark is dispatched by an intrepid marine biologist in one way or another, but there is nothing to compare with Alten's unbelievable conclusion. The hero, named Jonas Taylor (*Jonas*—I ask you!) is in his one-man submersible when he is swallowed by the shark. He climbs out of the submarine, reaches into his backpack where he always carries a fossil megalodon tooth, and because he has retained an image of the interior anatomy of the shark, *he carves it up and kills it from the inside.* Then he climbs back into the submarine (which he locates by shining his flashlight around in the belly of the shark) and ejects himself from the shark's mouth. Jonas Taylor gets the girl (and also the bends), but you know he will survive. The shark, of course, dies.

Despite the thousands of megalodon teeth that have been dredged from the ocean floor or found embedded in the chalky cliffs of California, Maryland, Florida, North Carolina, Belgium, and Morocco, *not a single white one has ever been found.* All the teeth that have been unearthed or dredged up are brown or black, but the fact that they are not white does not mean they are not bone. A fossil bone is still bone, but one that sometimes contains a hard and heavy infilling of other minerals as well. In the same way that the "bones" of the dinosaurs in various museums are not exactly bones, but bones composed of compacted minerals that have gradually replaced the inorganic material (apatite) in the bones over time. Should someone dredge up a *white* megalodon tooth, we would know that the giant shark became extinct quite recently—or is flourishing somewhere in the vastness of the oceans and has simply lost a tooth.

Sharks of all species have multiple rows of teeth; those currently doing the biting are replaced regularly in a process that has been likened to the action of a moving escalator. Behind the functional front rows of teeth, there are other rows, waiting to move forward as those in the front rank fall out or are otherwise dislodged. A shark will therefore have many more teeth in its mouth during its lifetime than would any other verte-brate, the total number from a given animal across a lifetime numbering perhaps in the thousands. One might then suppose that it would not have required many individual megalodons to have scattered all the teeth ever collected, but not so: we can be certain that hundreds of thousands or even millions more teeth remain in the seafloors or in the ground, most of them never to be seen.

There are many references to gigantic teeth dredged from the oceans depths, but not one of these described a tooth as white. The *Challenger*, the first vessel dedicated to oceanographic research, dredged up several of these manganese-encrusted teeth in 1873, but only later did their origin become clear. (Manganese nodules form very slowly, but they exist on the ocean floor in uncountable millions. Not all of them are formed around sharks' teeth, of course. According to Heezen and Hollister's *The Face of the Deep*, "ice-rafted debris, pumice, the earbones of whales, the teeth of sharks, and other nearly indestructible projectiles become the nuclei of the golf-ball to grapefruit-size manganese nodules which litter much of the deep-sea floor.")

We are probably fortunate that megalodon is extinct. Consider what our attitude toward a day at the beach would be if there were fifty-foot carnivorous sharks in our offshore waters—sharks large enough to swal-low a cow. One of the more interesting questions of shark evolution con-cerns the disappearance of this fifty-foot-long monster that is known to have fed on whales. The whales continued, but megalodon, one of the most fearsome hunters that ever lived, became extinct. Why? Did the whales develop more speed and maneuverability? Did the oceans cool off? Did killer whales replace them as the oceans' apex predators?

Sea creatures should be at least partially protected by the dense medium in which they live, but this has not been the case. People catch fish from the beach, from the banks, from the dock. Water also provides sup-port for fishing boats, and from these platforms of opportunity, human be-ings have been hauling in fish and other marine creatures with their nets, lines, and trawls since the dawn of history. In their 2001 article about his-torical overfishing, Jeremy Jackson and eighteen other authors wrote:

> Ecological extinction caused by overfishing precedes all other pervasive human disturbance to coastal ecosystems, including

pollution, degradation of water quality, and anthropogenic climate change. Historical abundances of large consumer species were fantastically large in comparison with recent observations. Paleoecological, archaeological, and historical data show that time lags of decades to centuries occurred between the onset of overfishing and consequent changes in ecological communities, because unfished species of similar trophic level [diet] assumed the ecological roles of overfished species until they too were overfished, or died of epidemic diseases related to overcrowding. . . . Whales, manatees, dugongs, sea cows, monk seals, crocodiles, codfish, jewfish, swordfish, sharks, and rays are other large marine vertebrates that are now functionally or entirely extinct in most coastal ecosystems.

In May 2003, the journal *Nature* published an article by Ransom Myers and Boris Worm, in which the preindustrial level of large predatory fishes—tuna, swordfish, marlin, and groupers—and groundfish such as cod, halibut, skates and flounder—was shown to have been reduced to 10 percent of their former numbers. This is not a typo: *Ninety percent of all these fish are gone.* Myers and Worm, both fisheries biologists at Dalhousie University in Halifax, Nova Scotia, compiled their data from an analysis of worldwide commercial fisheries, and concluded that "declines of large predators in coastal regions have extended throughout the global ocean, with potentially serious consequences for ecosystems." In a comment on their study, Daniel Pauly of the University of British Columbia said, "It shows how Japanese longlining has expended globally. It is like a hole burning through paper. As the hole expands, the edge is where the fisheries concentrate until there is nowhere else to go. Because longlining technology has improved, the author's estimates are conservative. If the catch rate has dropped by a factor of ten *and* the technology has improved, the declines are even greater than they are saying."

Many of these species are now considered to be hovering on the brink of extinction, and will be gone if something isn't done to slow the fall. Probably the most surprising and unexpected near-extinction in recent years has been that of the barndoor skate (*Raja laevis*) that nobody was fishing for at all. For generations, cod fishermen hauled in these unwanted elasmobranchs, which, at a total of sixteen square feet, approached their namesakes in size. Like many elasmobranchs, *Raja laevis* is "K-selected":

slow to mature, reproduces slowly, and has few offspring at a time. Indeed, newborn barndoor skates are already ten inches across, sizeable enough to get caught in trawls from their day of birth, and therefore never having a hope of reproducing. "Forty-five years ago," Jill Casey and Ransom Myers noted in a 1998 *Science* article, "research surveys on the St. Pierre Bank (off southern Newfoundland) recorded barndoor skates in 10 percent of their tows; in the last 20 years, none has been caught and this pattern of decline is similar throughout the range of the species." What happened? When the "distant water" fleets were scooping up codfish, redfish, and everything else that swam in eastern Canadian and New England waters in the 1970s, a large part of the bycatch was barndoor skates. Fisheries biologists, lately studying the decline of more valuable food fishes, didn't notice the disappearance of barndoor skates until it was too late. "If current population trends continue," wrote Casey and Myers, "the barndoor skate could become the first well-documented example of extinction in a marine fish species."

In 2003, I wrote *The Empty Ocean*, a study of the depletion of the world's marine resources: fishes, sharks, whales, dolphins, porpoises, seals, sea lions, sea otters, sea birds, coral reefs, etc. The Myers and Worm study appeared just as the book was being published, and therefore could not be included, but also after publication, news stories appeared that will be incorporated into future editions. All these stories are developments in

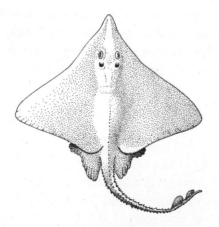

Barndoor Skate

topics covered in *The Empty Ocean*: Canada completely closed all its cod fisheries; more than 700 tons of cod froze to death in the unusually cold waters off Newfoundland; an unusually high number of sea otters have died along the California coast; the Asian trade in shark fins may be much larger than expected, and so on, ad nauseam. In *The Empty Ocean,* I wrote:

> Humans have taken advantage of the ocean's bounty for virtually all of recorded history, probably starting when a prehistoric beachcomber found a freshly dead fish washed ashore. From that innocuous beginning, humans became whalers, sealers, aquaculturists, netters, trollers, purse seiners, longliners, bottom trawlers, rod-and-reelers, dynamiters, poisoners, and a myriad of other professions dedicated to removing living things from the ocean. Sometimes the animals were killed for oil, sometimes for baleen, and sometimes for their fur coats, but for the most part, they were used for food, and this seemed more than enough justification for the continuing slaughter of the oceans' wildlife. People had to eat, didn't they? Besides, the ocean was so big and so deep and so filled with edible items that there seemed no end to its productivity. If one population of whales (or seals, or fishes, or sharks) was depleted, the fishers simply moved to another area and attacked another population, or changed the object of the fishery. A number of fish species, previously regarded as so plentiful as to be unaffected by human enterprise, have instead shown themselves to be vulnerable to fishing to such a degree that they are now considered endangered. The idea that Mother Ocean would continue to provide for her dependants forever has shown itself to be another gross misjudgement on the part of those dependants.

Everybody Off the Train

In the past half century, we have seen hundreds of species go extinct, their like never to be seen again on earth. During the same period, however, some people, having recognized the ignominy and enormity of anthropogenic extinctions, have made valiant attempts to pull some animals back from the brink. Efforts to save the California condor, whooping crane, Père David's deer, Przewalski's horse, and the giant panda have raised the hopes of those who once thought that taking the off-ramp to extinction was irreversible. Some species, like the California gray whale, rebounded simply because people stopped killing them. The science of cloning, still in its infancy, might be a useful tool to forestall extinction, or it may be another wrongheaded attempt to meddle with nature. And even though we did nothing but find them, the appearance of unexpected animals, such as the okapi, the coelacanth, the bonobo, and the Indopacific beaked whale has made people everywhere aware that all the news is not bad. These newly or rediscovered animals cannot compensate for the vast depredations of the past; they are but a handful compared to a multitude. Even as new titi monkeys and lemurs appear to the delight of primatologists, more and more animals are going under, and their numbers are on the increase.

*   *   *

"The world," wrote Purvis, Jones, and Mace, in their 2000 "Extinction" paper, "is entering a major extinction spasm. Present rates of species extinction are reckoned to be between 1000 and 10,000 times the rates seen through much of geological history." They attribute this to people, writing, "Unlike the mass extinctions of the past, the driving processes are not attributable to abiotic geological changes or extraterrestrial impacts. Instead, the direct and indirect causes are almost entirely due to the activities of a vast, growing, and global human population competing for global resources with all the other species on earth." Another threat to biodiversity is human encroachment, where rain forests, old growth woodlands, wetlands, plains, pastures, and almost all other locations on earth have been chopped, channeled, replanted, paved over, developed, or turned into dump sites at the expense of the creatures that once called those habitats home. As Edward O. Wilson (2001) wrote, "Around the world, biodiversity—defined as the full variety of life, from genes to species to ecosystems—is in trouble. Not a week goes by without reports of the imminent end of one species or another. For every celebrity animal that vanishes, biologists can point to thousands of species of plants and smaller animals either recently extinct or on the brink." Peter Raven, writing in the same publication (the book that accompanied the 1998 opening of the American Museum of Natural History's Hall of Biodiversity), said,

> Today, Earth may be undergoing a new mass extinction. Sadly, up to a quarter of the species on Earth may become extinct, or be on the way to extinction by the end of the first quarter of the next century, and up to three-quarters of the species may be extinct or endangered by the year 2200. Unlike earlier mass extinctions, however, the current one appears to result from the actions of a single biological species, *Homo sapiens,* the human species. . . . Finally, Earth's human population is increasing by an estimated 80 million people annually. Not only does this rapid population growth accelerate the depletion of crucial natural resources, but it also increases the risk of famine and social and political instability around the world.

Once upon a time, the sea was considered threatening. Its unknown tides, currents, waves, and weather often spelled doom for early seafarers, and until the sixteenth century, few adventurers were willing to sail out of sight of land. Sailors feared the monsters and serpents that might be lurking in the depths; even today, many people are afraid to swim in deep water. We know little about the oceans, and for the most part, what we do know is restricted to the surface layers. As the oceanographer said to the young student who was marveling at the amount of water he could see from shore, "and that's only the top of it." We are beginning to explore the depths, but of the two-thirds of the planet's surface that is covered by water, two-thirds of *that* is more than two miles deep, so it can be said that the predominant habitat for living things on Planet Earth is deep ocean, most of which remains totally unexplored. The immensity of the sea that thwarted so many explorers has also served as a haven for its wildlife. Yes, we fished some species too heavily—anchovies off Peru, for example, and now codfish in the North Atlantic—but isn't the ocean wide and deep and protective of its own? Can't prey escape the predations of fishermen by roaming farther afield or descending to the depths? In a word, no. Humans have never called the sea home, but it has served and protected them, and provided sustenance for eons. They have demonstrated their gratitude by raping, pillaging, and poisoning it.

In an essay published in *Science* in September 2001 ("Humans as the World's Greatest Evolutionary Force"), Stephen Palumbi of Harvard University's Department of Organismic and Evolutionary Biology wrote:

> In addition to altering global ecology, technology and human population growth also affect evolutionary trajectories, dramatically accelerating evolutionary change in other species, especially in commercially important, pest, and disease organisms. Such changes are apparent in antibiotic and human immunodeficiency virus (HIV) resistance to drugs, plant and insect resistance to pesticides, rapid changes in invasive species, life-history change in commercial fisheries, and pest adaptation to biological engineering products. . . . Human impact on the global biosphere now controls many major facets of ecosystem function. Currently, a large fraction of the world's available fresh water, arable land, fisheries production, nitrogen budget, CO_2 balance, and biotic

turnover are dominated by human effects. Human ecological impact has enormous evolutionary consequences as well and can greatly accelerate evolutionary change in the species around us, especially disease organisms, agricultural pests, commensals, and species hunted commercially.

In the very next issue of *Science,* an article appeared that was signed by thirty-three practicing conservationists, entitled, "Can We Defy Nature's End?" (The lead author is Stuart Pimm of Columbia University.) It begins, "In the catalog of global environmental insults, extinction stands out as irreversible." Of the oceans, they wrote, "Fishing contributes only 5 percent of the global protein supply, yet it is the major threat to the oceans' biodiversity. The multitude of fish species caught on coral reefs constitutes only a small, though poorly known, fraction of the total catch, but fisheries severely damage these most diverse marine ecosystems. Most major fish stocks are overfished; thus, mismanagement diminishes our welfare and biodiversity simultaneously. Conversely, protected areas enhance biodiversity and fish stocks." To forestall catastrophe, the authors recommend "greater research efforts into the links between biodiversity, ecosystems, their services, and people. Infectious diseases are entering human populations as our numbers increase and as we encroach upon tropical forests and other pathogen reservoirs. Global climate change will have major impacts on human health through changes in food production, access to fresh water, exposure to vector- and water-borne disease, sea-level rise and coastal flooding, and extreme weather events."

In *The Ghost with Trembling Wings* (2002), a book about the (occasional) rediscovery of species thought to be extinct, Scott Weidensaul wrote,

> It has become clear that the Cretaceous extinction was caused, or at least exacerbated, by the impact of a meteor, whose crater was finally located off the Yucatán Peninsula in 1991, and there is some evidence that a cosmic collision also caused the great Permian death. But we are witness to the gathering pace of what ecologists have labeled the Sixth Extinction, and this time we are the meteor, blistering the world with our appetites, our habitat

destruction and climate change, all driven by the crushing weight of our population. By conservative reckoning, the planet loses three or four species an hour, eighty or more a day, thirty thousand a year—the highest extinction rate in 65 million years.

"We have met the enemy," said Pogo, "and he is us." We have now reached the point where we must decide what to do about what Oxford professor Robert May (quoted in Gibbs, 2001) calls "the breaking edge of the sixth great wave of extinction in the history of life on Earth." Any number of books and articles have been written and symposia convened to try to comprehend the nature and magnitude of the problem, and to determine what—if anything—can be done about it. In March 2000, the National Academy of Sciences held a colloquium on "The Future of Evolution." In the introduction to his presentation ("Declines of biomes and biotas and the future of evolution"), David Woodruff wrote:

> Although panel discussants disagreed whether the biodiversity crisis constitutes a mass extinction event, all agreed that current extinction rates are 50 to 500 times background and are increasing and that the consequences for the future evolution of life are serious. In response to the on-going rapid decline of biomes and homogenization of biotas, the panelists predicted changes in species geographic ranges, genetic risks of extinction, genetic assimilation, natural selection, mutation rates, the shortening of food chains, the increase in nutrient-enriched niches permitting the ascendancy of microbes, and the differential survival of ecological generalists. Rates of evolutionary processes will change in different groups, and speciation in the larger vertebrates is essentially over. Action taken over the next few decades will determine how impoverished the biosphere will be in 1,000 years when many species will suffer reduced evolvability and require interventionist genetic and ecological management. . . . The ultimate test of evolutionary biology as a science is not whether it solves the riddles of the past but rather whether it enables us to manage the future of the biosphere. Our inability to make clearer predictions about the future of evolution has serious consequences for both biodiversity and humanity.

We know that humans are changing their home planet in ways that will affect us—and every other living thing—for eons to come. We have slashed and burned rain forests, polluted our air and waterways, dammed the rivers and drained the lakes, warmed the atmosphere and punched holes in it, removed some species and introduced others. For all this befouling of our nest, we thought that at least evolution, that inexorable force that controls the rise and fall of species (including us), was somehow beyond human modification. Oh yes, we can breed dogs, cattle, pigs, or pigeons selectively, and create animals that look slightly different from their parents and very different from their distant ancestors, but affecting evolution? That idea went out with Lamarck, didn't it? Not exactly. In a paper published in 2001, Norman Myers of Oxford and Andrew Knoll of Harvard wrote that the extinction crisis now in progress will have serious and profound effects on evolution:

> The biotic crisis overtaking our planet is likely to precipitate a major extinction of species. That much is well known. Not so well known but probably more significant in the long term is that the crisis will surely disrupt and deplete certain basic processes of evolution, with consequences likely to persist for millions of years. Distinctive features of future evolution could include a homogenization of biotas, a proliferation of opportunistic species, a pest-and-weed ecology, an outburst of speciation among taxa that prosper in human-dominated ecosystems, a decline of biodisparity, an end to the speciation of large vertebrates, the depletion of "evolutionary powerhouses" in the tropics, and unpredictable emergent novelties. Despite this likelihood, we have only a rudimentary understanding of how we are altering the evolutionary future. As a result of our ignorance, conservation policies fail to reflect long-term evolutionary aspects of biodiversity loss.

In a book called *Extinction: Evolution and the End of Man* (2002), Michael Boulter, a paleobiologist at the University of East London, writes that we have brought the world to the brink of an extinction event that is comparable in magnitude to the previous "big five" mass extinctions. Boulter argues that the first "man made extinction event" was our elimination of the Neanderthals, our only rivals for dominion over the earth. The balance that existed between the two upright hominids was destroyed when we learned to talk: "This supremely important event in the history of planet Earth marked the end of 10,000 years of relative harmony." In addition to speech, humans possess many other skills that differentiate them from all other animals: "a bipedal gait, dextrous thumbs, an ability to plan ahead, and a certain selfishness." To this list Boulter adds one more: the ability to change the environment." By our worldwide indifference to the effects of our own activities, we have drastically upset what more innocent observers used to refer to as "the balance of nature." Boulter writes:

> Since Nevil Shute's novel about a nuclear holocaust, *On the Beach*, written in the 1950s, there have been countless tales of the end of man on this planet. Now, it seems, the joke is that we are doing very well on our own, just with our use of fossil fuels. There is no need for nuclear weapons or the inventions of science fiction writers. It is our own aggressive selfishness that has led to our lifestyle, and this has evolved its own political system to maintain the status quo. Now it's too late to change and we cannot organize ourselves to stop. I speculate that our system is in free fall, out of control.

In recent years, there has appeared a spate of "The End of " books. Bill McKibben's *The End of Nature* and Francis Fukuyama's *The End of History* were followed by Peter Ward's *The End of Evolution* and John Horgan's *The End of Science*. (Horgan's book contains a chapter called "The End of Evolutionary Biology.") But the subtitle of Michael Boulter's book hits closest to home: "Evolution and the End of Man." He says, "The extinction event we are experiencing now could be leading to a second diversification starting when the large mammals, especially we humans, are gone. Already a good proportion of the families of the large mammals has become, or is becoming, extinct." The world has managed to withstand major extinction events, and will probably survive the current one—unless of course, in our great wisdom, we manage to blow up the planet. The history of life has been punctuated by a vast number of extinctions; some carried off large numbers of creatures, others simply saw the last individual of a race of beings expire. For the most part the earth's inhabitants had no choice but to sit back and watch the whirlwind forces of nature swirl around them, as the various creatures migrated, reproduced, evolved, or became extinct. No dinosaur ever studied evolutionary theory; no passenger pigeon knew there was such a thing as extinction. Our unique self-awareness has changed that: we are the only species that knows there are such things as evolution and extinction. We have invented something that Boulter calls "the self-organized mass extinction from within," and with exquisite irony, that invention will prove the end of mankind.

References

Abbot, D. H. and A. E. Isley. 2002. Extraterrestrial influences on mantle plume activity. *Earth Planet. Sci. Let.* 205 (1):52–62.

Ackerman, D. 1995. *The Rarest of the Rare: Vanishing Animals, Timeless Worlds.* New York: Vintage.

Adam, P. J. and G. G. Garcia. 2003. New information on the natural history, distribution, and skull size of the extinct West Indian monk seal, *Monachus tropicalis. Mar. Mam. Sci.* 19 (2):297–310.

Adler, R. 2003. Genes reveal signs of Ice Age life. *New Scientist* 178 (2392):16–17.

Alber, W. L. 1999. The teeth of the tyrannosaurs. *Scientific American* 281 (3):50–52.

Alexander, R. McN. 1989. *Dynamics of Dinosaurs and Other Extinct Giants.* New York: Columbia University Press.

Alford, J. J. 1973. The American bison: an Ice Age survivor. *Proc. Ass. Am. Geog.* 5:1–6.

Allee, W. C. 1931. *Animal Aggregations. A Study in General Sociology.* Chicago: University of Chicago Press.

Allen, A. A. 1937. Hunting with a microphone the voices of vanishing birds. *National Geographic* 71 (6):697–723.

Allen, A. A. and P. P. Kellogg. 1937. Recent observations on the ivory-billed woodpecker. *Auk* 54:164–84.

Allen, G. M. 1942. *Extinct and Vanishing Mammals of the Western Hemisphere.* American Committee for Wildlife Protection.

Allen, J. A. 1880. *History of North American Pinnipeds: A Monograph of the Walrus, Sea-Lions, Sea-Bears and Seals of North America.* Washington, D.C.: Government Printing Office.

Allen, R. P. 1957. *On the Trail of Vanishing Birds*. New York: McGraw-Hill.

Alroy, J. 1999. Putting North America's End-Pleistocene megafaunal extinction in context: large-scale analyses of spatial patterns, extinction rates and size distributions. In R. D. E. MacPhee, ed., *Extinctions in Near Time: Causes, Contexts and Consequences*, pp. 105–43. San Diego: Academic Press.

———. 2001. A multispecies overkill simulation of the End-Pleistocene Megafaunal Mass Extinction. *Science* 292:1893–96.

Altangerel, P., M. Norell, L. Chiappe, and J. Clark. 1993. Flightless bird from the Cretaceous of Mongolia. *Nature* 362:623–26.

Alten, S. 1997. *Meg*. New York: Doubleday.

Alvarez, L. W. 1983. Experimental evidence that an asteroid impact led to the extinction of many species 65 million years ago. *Proc. Nat. Acad. Sci.* 80:627–42.

Alvarez, L., W. Alvarez, F. Asaro, and H. V. Michel. 1980. Extraterrestrial cause for the Cretaceous-Tertiary Extinction. *Science* 208:1095–1108.

Alvarez, W. 1997. *T. rex and the Crater of Doom*. Princeton, N.J.: Princeton University Press.

Alvarez, W. and F. Asaro. 1990. What caused the great extinctions? An extraterrestrial impact. *Scientific American* 263 (4):78–84.

Alvarez, W., P. Claeys, and S. W. Kieffer. 1995. Emplacement of Cretaceous-Tertiary boundary shocked quartz from Chicxulub Crater. *Science* 216:249–56.

Alvarez, W., L. Alvarez, F. Asaro, and H. V. Michel. 1984. The end of the Cretaceous: sharp boundary or gradual transition? *Science* 223:1183–1186.

Alvarez, W., E. G. Kaufmann, F. Surlyk, L. W. Alvarez, F. Asaro, and H. V. Michel. 1984. Impact theory of mass extinctions and the invertebrate fossil record. *Science* 223 (4641):1135–1141.

Anderson, A. 1995. The extinction of moa in southern New Zealand. In P. S. Martin and R. G. Klein, eds., *Quaternary Extinctions: A Prehistoric Revolution*, pp. 728–40. Tucson: University of Arizona Press.

Anderson, E. 1995. Who's who in the Pleistocene: a mammalian bestiary. In P. S. Martin and R. G. Klein, eds., *Quaternary Extinctions: A Prehistoric Revolution*. pp. 40–89. Tucson: University of Arizona Press.

Anderson, P. K. 1995. Competition, predation, and the evolution and extinction of Steller's sea cow, *Hydrodamalis gigas*. *Mar. Mam. Sci.* 11 (3):391–94.

Anderson, P. K. and D. P. Domning. 2002. Steller's sea cow. In W. F. Perrin, B. Würsig, and J. G. M. Thewissen, eds., *Encyclopedia of Marine Mammals*. pp. 1178–1181. San Diego: Academic Press.

Anon. 2001. Discovery of a remarkably preserved fossil dinosaur on display at AMNH. *Rotunda* 26 (6):2–3.

Anon. 2001b. 200 tigers killed last year, says noted conservationist. *Cat News* 34:5–6.

Anon. 2003. Charged in condor killing. *New York Times* April 30:A24.

Archibald, J. D. 1996. *Dinosaur Extinction and the End of an Era*. New York: Columbia University Press.

Archibald, J. D. and W. A. Clemens. 1982. Late Cretaceous extinctions. *American Scientist* 70:377–85.

Aristotle. n.d. *Historia animalium*. Translated by A. L. Peck. 1965. Loeb Classical Library, Cambridge, Mass.: Harvard University Press.

Aronson, R. 1990. Rise and fall of life at sea. *New Scientist* 128 (1741):34–37.

Austin, O. L. 1949. The status of Steller's albatross. *Pacific Science* 3 (4):283–95.

———. 1961. *Birds of the World*. New York: Golden Press.

Bailey, A. M. 1956. *Birds of Midway and Laysan Islands*. Denver, Colo.: Denver Museum of Natural History.

Bains, S., R. M. Corfield, and R. D. Norris. 1999. Mechanisms of climate warming at the end of the Paleocene. *Science* 285:724–27.

Bains, S., R. D. Norris, R. M. Corfield, and K. L. Faul. 2000. Termination of global warmth at the Paleocene/Eocene boundary through productivity feedback. *Nature* 407:171–74.

Bakker, R. T. 1971. Dinosaur physiology and the origin of mammals. *Evolution* 25:636–58.

———. 1972. Anatomical and physiological evidence of endothermy in dinosaurs. *Nature* 238:81–85.

———. 1975. Dinosaur renaissance. *Scientific American* 232 (4):58–78.

———. 1986. *The Dinosaur Heresies*. New York: William Morrow.

Bakker, R. T. and P. M. Galton. 1974. Dinosaur monophyly and a new class of vertebrates. *Nature* 248:168–72.

Barkham, S. H. 1984. The Basque whaling establishment in Labrador 1536–1632—A summary. *Arctic* 37:515–19.

Barlow, G. W. 2000. *The Cichlid Fishes: Nature's Grand Experiment in Evolution*. New York: Perseus.

Barlow, J., T. Gerrodette, and G. Silber. 1997. First estimates of vaquita abundance. *Mar. Mam. Sci.* 13 (1):44–58.

Barnes, L. G., D. P. Domning, and C. E. Ray. 1985. Status of studies on fossil marine mammals. *Mar. Mam. Sci.* 1 (1):15–53.

Barnes, R. F. W. 2002. The bushmeat boom and bust in West and Central Africa. *Oryx* 36 (3):236–42.

Barnett, A. 2001. Safety in numbers. *New Scientist* 169:38–41.

Barnum, P. T. 1870. *Struggles and Triumphs, or, Forty Years' Recollections*. J. B. Burr.

Basu, A. R., M. I. Petaev, R. J. Poreda, S. B. Jacobsen, and L. Becker. 2003. Chondritic meteorite fragments associated with the Permian-Triassic boundary in Antarctica. *Science* 302:1388–92.

Beard, P. H. 1965. *The End of the Game*. New York: Viking.

Becker, L., R. J. Poreda, A. G. Hunt, T. E. Bunch, and M. Rampino. 2001. Impact event at the Permian-Triassic Boundary: Evidence from extraterrestrial noble gases in fullerines. *Science* 291:1530–33.

Beebe, W. 1915. A tetrapleryx stage in the ancestry of birds. *Zoological* 2(2): 38–52.

———. 1932. *Nonsuch: Land of Water*. New York: Harcourt Brace.

———. 1935. Bermuda cahow rediscovered. *Bull. N.Y. Zool. Soc.* 38 (6):187–90.

Belon, P. 1996. *Beluga: A Farewell to Whales*. New York: Lyons and Burford.

Benchley, P. 1974. *Jaws*. Garden City, N.Y.: Doubleday.

Bennett, J., M. Cooper, M. Hunter, and L. Jardine. 2003. *London's Leonardo: The Life and Work of Robert Hooke*. New York: Oxford University Press.

Benton, M. J. 1990a. Scientific methodologies in collision: the history of the study of the extinction of the dinosaurs. In M. K. Hecht, B. Wallace, and R. J. McIntyre, eds., *Evolutionary Biology*, pp. 371–400. New York: Plenum.

———. 1990b. *The Reign of the Reptiles*. Crescent.

———. 1990c. Origin and interrelationships of dinosaurs. In D. B. Weishampel, P. Dodson, and H. Osmólska, eds., *The Dinosauria*, pp. 11–30. Berkeley: University of California Press.

———. 1993. Late Triassic extinctions and the origin of the dinosaurs. *Science* 260:769–70.

———. 1996. *The Penguin Historical Atlas of the Dinosaurs*. New York: Penguin.

———. 1997. *Vertebrate Paleontology*. New York: Chapman & Hall.

———. 2003a. Wipeout. *New Scientist* 178 (2392):38–41.

———. 2003b. *When Life Nearly Died: The Greatest Mass Extinction of All Time*. London: Thames and Hudson.

Benton, M. J., M. A. Wills, and R. Hitchin. 2000. Quality of the fossil record through time. *Nature* 403:534–37.

Berger, J., J. E. Swenson, and I.-L. Persson. 2001. Recolonizing carnivores and naïve prey: conservation lessons from Pleistocene extinctions. 2001. *Science* 291:1036–1039.

Berger, L., R. Spear, P. Daszak, D. E. Green, A. A. Cunningham, C. L. Goggin, R. Slocombe, M. A. Ragan, A. D. Hyatt, K. R. McDonald, H. B. Hines, K. R. Lips, G. Marantelli, and H. Parkes. 1998. Chytridiomycosis causes amphibian mortality associated with population declines in the rain forests of Australia and Central America. *Proc. Nat. Acad. Sci.* 95 (15):9031–9036.

Beston, H. 1928. *The Outermost House*. New York: Holt, Rinehart and Winston.

Birkeland, C., ed. 1997. *Life and Death of Coral Reefs*. New York: Chapman & Hall.

Blakeslee, S. 1997. Protozoan attacks frogs. *New York Times* September 16:F1.

———. 2003. Straighten up and fly right. *New York Times* June 3:F1–4.

Bodsworth, F. 1955. *Last of the Curlews*. New York: Dodd, Mead.

Bohannon, J. 2002. An 11th-hour rescue for the great apes? *Science* 297:2203.

Bohor, B. F., P. J. Modreski, and E. E. Foord. 1987. Shocked quartz in the Cretaceous-Tertiary boundary clays: evidence for a global distribution. *Science* 236:705–9.

Bohor, B. F., E. E. Foord, P. J. Modreski, and D. M. Triplehorn. 1984. Mineralogic evidence for an impact event at the Cretaceous-Tertiary boundary. *Science* 224:867–69.

Bostwick, J. A. and F. T. Kyte. 1996. The size and abundance of shocked quartz in Cretaceous-Tertiary boundary sediments from the Pacific basin. In G. Ryder, D. Fastovsky, and S. Gartner, eds., *The Cretaceous-Tertiary Event and Other Catastrophes in Earth History*, pp. 403–15. Geological Society of America Special Paper 307.

Boulter, M. 2002. *Extinction: Evolution and the End of Man*. London: Fourth Estate.

Boulva, J. 1979a. Mediterranean monk seal. In *Mammals in the Seas II: Pinniped Species Summaries and Report on Sirenians*, pp. 95–100. Rome: FAO.

———. 1979b. Caribbean monk seal. In *Mammals in the Seas II: Pinniped Species Summaries and Report on Sirenians*, pp. 101–3. Rome: FAO.

Bouman, I. and J. Bouman. 1994. The history of Przewalski's horse. In L. Boyd and K. A. Houpt, eds., *Przewalski's Horse: The History and Biology of an Endangered Species*, pp. 5–38. Albany: State University of New York Press.

Bouman, I., J. Bouman, and L. Boyd. 1994. Reintroduction. In L. Boyd and K. A. Houpt, eds., *Przewalski's Horse: The History and Biology of an Endangered Species*, pp. 255–63. Albany: State University of New York Press.

Bourgeois, J., T. A. Hansen, P. L. Wiberg, and E. G. Kauffman. 1988. A tsunami deposit at the Cretaceous-Tertiary boundary in Texas. *Science* 241:567–70.

Bowman, D. M. J. S. 2001. Future eating and country keeping: what role has environmental history in the management of biodiversity? *Jour. Biogeog.* 28:549–64.

Bowring, S. A., D. H. Erwin, Y. G. Jin, M. W. Martin, K. Davidek, and W. Wang. 1998. U/Pb zircon geochronology and tempo of the End-Permian Mass Extinction. *Science* 80:1039–45.

Boyd, L. 2002. Reborn free [Przewalski's Horse]. *Natural History* 111 (6):56–61.

Boyd. L. and K. A. Houpt, eds. 1994. *Przewalski's Horse: The History and Biology of an Endangered Species*. Albany: State University of New York Press.

Boynton, G. 2003. Gorillas in our midst. *Condé Nast Traveler* May: 180–86, 265–68.

Broad, W. J. 1985. Authenticity of bird fossil is challenged. *New York Times* May 7:C1.

———. 2002. New theory on dinosaurs: multiple meteorites did them in. *New York Times* November 5: F1, 4.

Brown, R. 1981. *Megalodon*. New York: Coward, McCann & Geoghegan.

Browne, M. W. 1985. Dinosaur experts resist meteor extinction idea: Paleontologists say dissenters risk harm to their careers. *New York Times* October 29:21–22.

————. 1988. Debate over dinosaur extinction takes an unusually rancorous turn. *New York Times* January 19:19, 23.

Brunet, M., F. Guy, D. Pilbeam, H. T. Mackaye, A. Likius, D. Ahounta, A. Beauvilain, C. Blondel, H. Bocherens, J.-R. Boisserie, L. De Bonis, Y. Coppens, J. Dejax, C. Denys, P. Duringer, V. Eisenmann, G. Fanone, P. Fronty, D. Geraads, T. Lehmann, F. Lihoreau, A. Louchart, A. Mahamat, G. Merceron, G. Mouchelin, O. Otero, P. P. Campomanes, M. P. De Leon, J.-C. Rage, M. Sapanet, M. Schuster, J. Sudre, P. Tassy, X. Valentin, P. Vignaud, L. Viriot, A. Zazzo, and C. Zollikofer. 2002. A new hominid from the Upper Miocene of Chad, Central Africa. *Nature* 418:145–51.

Buckland, W. 1823. *Reliquiae Diluviunae, or, Observations on the Organic Remains Contained in Caves, Fissures, and Diluvial Gravel, and on other Geological Phenomena, Attesting the Action of an Universal Deluge.* London: John Murray.

————. 1824. Notice on the *Megalosaurus* or great fossil lizard of Stonesfield. *Trans. Geol. Soc. London* 1:390–96.

Buffetaut, E. 1984. Selective extinctions and terminal Cretaceous events. *Nature* 310:276.

————. 1990. Vertebrate extinctions and survival across the Cretaceous-Tertiary boundary. *Tectonophysics* 171:337–45.

Bunce, M., T. H. Worthy, T. Ford, W. Hoppitt, E. Willerslev, A. Drummond, and A. Cooper. 2003. Extreme reversed sexual size dimorphism in the extinct New Zealand moa, *Dinornis. Nature* 425:172–75.

Busby, C. J., G. Yip, L. Blikra, and P. Renne. 2002. Coastal landsliding and catastrophic sedimentation triggered by Cretaceous-Tertiary bolide impact: a Pacific margin example? *Geology* 30 (8):687–90.

Buscemi, D. 1982. The Labrador duck. *Oceans* 15 (5):3.

Cadbury, D. 2000. *The Dinosaur Hunters: A True Story of Scientific Rivalry and the Discovery of the Prehistoric World.* London: Fourth Estate.

Caesar. J. n.d. *The Gallic Wars.* Loeb Classical Library, Cambridge, Mass.: Harvard University Press.

Calzada, N., C. H. Lockyer, and A. Aguilar. 1994. Age and sex composition of the striped dolphin die-off in the western Mediterranean. *Mar. Mam. Sci.* 10 (3):299–310.

Cardillo, M. and A. Lister. 2002. Death in the slow lane. *Nature* 419:440–41.

Carlton, J. T., J. B. Geller, M. L. Reaka-Kudla, and E. A. Norse. 1999. Historical extinctions in the sea. *Ann. Rev. Ecol. Syst.* 30:515–38.

Carpenter, K. 1997. Tyrannosauridae. In P. J. Currie and K. Padian, eds., *Encyclopedia of Dinosaurs*, pp. 766–68. San Diego: Academic Press.

Carrington, R. 1957. *Mermaids and Mastodons.* New York: Rinehart.

Casey, J. M. and R. A. Myers. 1998. Near extinction of a large widely distributed fish. *Science* 281:690–92.

Chadwick, D. 2002. Down to a handful [Atwater's prairie chicken]. *National Geographic* 201 (3):49–61.

Chaloner, W. G. and A. Hallam, eds. 1989. *Evolution and Extinction*. New York: Cambridge University Press.

Chan, S., A. V. Maksimuk, L.V. Zhirnov, and S. V. Nash. 1995. *From Steppe to Store: The Trade in Saiga Antelope Horn*. Traffic International.

Chang, K. 2001. Clues to a meteor that aided dinosaurs. *New York Times* May 12: A11.

Charig, A . 1989. The Cretaceous-Tertiary boundary and the last of the dinosaurs. *Phil. Trans. Roy. Soc. London* B325:387–400.

————. 1993. Disaster theories of dinosaur extinction. *Modern Geology* 18:299–318.

Charig, A. J., F. Greenaway, A. C. Milner, C. A. Walker, and P. J. Whybrow. 1986. *Aracheopteryx* is not a forgery. *Science* 232:623–25.

Chatterjee, S. 1997. *The Rise of Birds*. Baltimore: Johns Hopkins University Press.

Chauvet, J.-M., E. B. Deschamps, and C. Hillaire. 1996. *Dawn of Art: The Chauvet Cave*. New York: Abrams.

Chen, P.-J., Z.-M. Dong, and S.-N. Zhen. 1998. An exceptionally well-preserved theropod dinosaur from the Yixian Formation of China. *Nature* 391:147–52.

Chiappe, L. M. 1995a. The first 85 million years of avian evolution. *Nature* 378:349–55.

————. 1995b. A diversity of early birds. *Natural History* 104 (6):52–55.

Chiappe, L. M., M. A. Norell, and J. L. Clark. 1998. The skull of a relative of the stem-group bird *Mononykus*. *Nature* 392:275–78.

Chiappe, L. M., J. Shu-an, J. Qiang, and M. A. Norell. 1999. Anatomy and systematics of the Confuciusornithidae (Theropoda: Aves) from the late Mesozoic of Northeastern China. *Bull. Amer. Mus. Nat. Hist.* 242:3–89.

Chorn. J. and R. S. Hoffman. 1978. *Auluopoda melanoleuca* [Giant Panda]. *Mammalian Species* 110:1–6. American Society of Mammalogists.

Chundawat, R. S., N. Gogate, and A. J. T. Johnsingh. 1999. Tigers in Panna: preliminary results from an Indian tropical dry forest. In J. Seidensticker, S. Christie and P. Jackson, eds., *Riding the Tiger: Tiger Conservation in Human-Dominated Landscapes*. pp. 123–29. New York: Cambridge University Press.

Clemens, E. S. 1986. Of asteroids and dinosaurs: The role of the press in the shaping of scientifc debate. *Soc. Stud. Sci.* 16:421–56.

Clemens, W. A., J. D. Archibald, and L. J. Hickey. 1981. Out with a whimper, not a bang. *Paleobiology* 7:293–98.

Climo, G. and A. Balance. 1997. *Hoki: The Story of a Kakapo*. Godwit.

Clutton-Brock, J. 1999. *A Natural History of Domesticated Mammals*. New York: Cambridge University Press.

Coghlan, A. 2002. Not such close cousins after all. *New Scientist* 175:20.

Cokinos, C. 2000. *Hope Is the Thing with Feathers*. New York: Warner Books.

Colbert, E. H. 1965. *The Age of Reptiles*. New York: W. W. Norton and Co.

———. 1973. *Wandering Lands and Animals*. New York: Dutton.

———. 1983. *An Illustrated History of Dinosaurs*. New York: Hammond.

Collar, N. J. and P. Andrew. 1988. *Birds to Watch: The ICBP World Checklist of Threatened Birds*. Washington, D. C.: Smithsonian Institution Press.

Conniff, R. 1999. Africa's wild dogs. *National Geographic* 195 (5):36–63.

Conrad, J. R. 1957. *The Horn and the Sword: The History of the Bull as a Symbol of Power and Fertility*. New York: Dutton.

Coolidge, H. J. 1933. *Pan paniscus*, pygmy chimpanzee from south of the Congo River. *Amer. Jour. Phys. Anthro.* 18 (1):1–59.

———. 1984. Historical remarks bearing on the discovery of *Pan paniscus*. In R. L. Susman, ed., *The Pygmy Chimpanzee*. pp. ix-xiii. New York: Plenum Press.

Coolidge, H. J. and B. Shea. 1982. External body dimensions of *Pan paniscus* and *Pan troglodytes* chimpanzees. *Primates* 23:245–51.

Coria, R. A. and L. Salgado. 1995. A new giant carnivorous dinosaur from the Cretaceous of Patagonia. *Nature* 377:224–26.

Cosivi, O., J. M. Grange, C. J. Daborn, M. C. Raviglione. T. Fujikura. D. Cousins, R. A. Robinson, H. F. A. K. Huchzermeyer, I. de Kantor, and F.-X. Meslin. 1998. Zoonotic tuberculosis due to *Myobacterium bovis* in developing countries. *Emerging Infectious Diseases* 4 (1):59–70.

Cottrell, L. 1962. *The Bull of Minos*. New York: Grosset and Dunlap.

Courchamp, F., T. Clutton-Brock, and B. Grenfell. 1999. Inverse density dependence and the Allee effect. *Trends in Ecology and Evolution* 14:405–10.

———. 2000. Multipack dynamics and the Allee effect in the African wild dog *Lycaon pictus*. *Animal Conservation* 3:277–85.

Courchamp, F., B. Grenfell, and T. Clutton-Brock. 1999. Population dynamics of obligate cooperators. *Proc. Royal Soc. London* 266:557–63.

———. 2000. Impact of natural enemies on obligately cooperative breeders. *Oikos* 91:311–22.

Courtillot, V. 1990. What caused the great extinctions? A volcanic eruption. *Scientific American* 263 (4):85–92

———. 1999. *Evolutionary Catastrophes: The Science of Mass Extinction*. New York: Cambridge University Press.

Courtillot, V., G. Féraud, H. Maluski, D. Vandamme, M. G. Moreau, and J. Besse.

1988. Deccan flood basalts and the Cretaceous/Tertiary boundary. *Nature* 333:843–46.

Cowen, R. 2000. *History of Life*. Oxford: Blackwell Science.

Cracraft, J. 2001a. Avian evolution, Gondwana biogeography and the Cretaceous-Tertiary mass extinction event. *Proc. Roy. Soc. London* 268 (1466):459–69.

———. 2001b. Gondwana genesis. *Natural History* 110 (10):64–73.

Cuppy, W. 1941. *How to Become Extinct*. Chicago: University of Chicago Press.

Currie, P. J. 1996. Out of Africa: Meat-eating dinosaurs that challenge *Tyrannosaurus rex*. *Science* 272:971–91.

———. 2000. Feathered dinosaurs. In G. S. Paul, ed., *The Scientific American Book of Dinosaurs*. pp. 183–89. New York: St. Martins Press.

Currie, P. J. and K. Padian, eds. 1997. *Encyclopedia of Dinosaurs*. San Diego: Academic Press.

Cuvier, G. 1800–1805. *Leçons d'anatomie comparée*. Paris.

Czerkas, S. M. and D. F. Glut. 1982. *Dinosaurs, Mammoths and Cavemen: The Art of Charles R. Knight*. New York: Dutton.

Dalebout, M. L., G. J. B. Ross, C. S. Baker, R. C. Anderson, P. B. Best, V. G. Cockcroft, H. L. Hinsz, V. Peddemors, and R. L. Pitman. 2003. Appearance, distribution, and genetic distinctiveness of Longman's beaked whale, *Indopacetus pacificus*. *Mar. Mam. Sci.* 19 (3):421–61.

Dalton, R. 2000a. Feathers fly over Chinese fossil bird's legality and authenticity. *Nature* 403:689–90.

———. 2000b. Fake bird fossil highlights the problems of illegal trading. *Nature* 404:696.

———. 2000c. Chasing the dragons. *Nature* 406:930–32.

Dampier, W. 1699. *A New Voyage Round the World*. Knapton (Argonaut Press edition, 1927).

Darwin, C. 1859. *The Origin of Species by Means of Natural Selection*. London.

Daszak, P., A. A. Cunningham, and A. D. Hyatt. 2000. Emerging infectious diseases of wildlife—threats to biodiversity and human health. *Science* 287:443–49.

Daszak, P., L. Berger, A. A. Cunningham, A. D. Hyatt, D. E. Green,and R. Speare. 1999. Emerging infectious diseases and amphibian population declines. *Emerging Infectious Diseases* 5 (6):735–48.

Datta, A., J. Pasnsa, M. D. Madhusudan, and C. Mishra. 2003. Discovery of the leaf deer *Muntiacus putaoensis* in Arunachal Pradesh: an addition to the large mammals of India. *Current Science* 84 (3):454–58.

———. 1999 *The Fifth Miracle: The Search for the Origin and Meaning of Life*. New York: Simon and Schuster.

———. 2001. The living dead [viruses]. *New Scientist* 172 [Inside Science 144]:1–4.

Day, D. 1989. *Vanished Species*. Gallery.

Dayton, L. 2001. Mass extinction pinned on Ice Age hunters. *Science* 292:1819.

De Camp, L. S. and C. C. De Camp. 1968. *The Day of the Dinosaur*. Garden City, N.Y.: Doubleday.

De Laubenfels, M. W. 1956. Dinosaur extinction: one more hypothesis. *Jour. Paleo.* 30 (1):207–18.

DeSalle, R. and D. Lindley. 1997. *The Science of Jurassic Park and the Lost World*. New York: HarperCollins.

Desmond, A. J. 1976. *The Hot-Blooded Dinosaurs*. New York: Dial Press.

De Waal, F. B. M. 1998. *Bonobo: The Forgotten Ape*. Berkeley: University of California Press.

———. 1999. Cultural primatology comes of age. *Nature* 399:635–36.

Diamond, J. 1984. "Normal" extinctions of isolated populations. In M. H. Nitecki, ed., *Extinctions*, pp. 191–246. Chicago: University of Chicago Press.

———. 1989. The present, past and future of human-caused extinctions. *Phil. Trans. Roy. Soc. London* B325:469:77.

———. 1995. Historic extinction: A Rosetta Stone for understanding prehistoric extinctions. In P. S. Martin and R. G. Klein, eds. *Quaternary Extinctions: A Prehistoric Revolution*. pp. 824–66. Tucson: University of Arizona Press.

———. 1997. *Guns, Germs, and Steel*. New York: W. W. Norton and Co.

———. 2000. Blitzkrieg against the moas. *Science* 287:2170–71.

Dietz, R. S. 1961. Astroblemes. *Scientific American* 205 (2):50–58.

Dingus, L. and T. Rowe. 1998. *The Mistaken Extinction: Dinosaur Evolution and the Origin of Birds*. New York: Freeman.

Dingus, L., E. S. Gaffney, M. A. Norell, and S. D. Sampson. 1995. *The Halls of Dinosaurs: A Guide to Saurischians and Ornithischians*. New York: American Museum of Natural History.

Dodson, P., and L. P. Tatarinov. 1990. Dinosaur extinction. In D. B. Weishampel, P. Dodson, and H. Osmólska, eds., *The Dinosauria*, pp. 55–62. Berkeley: University of California Press.

Dott, R. H. 1983. Itching eyes and dinosaur demise. *Geology* 11:126.

Doyle, A. C. 1912. *The Lost World*. New York: Puffin edition, 1981.

Dudley, P. 1725. An essay upon the natural history of whales. *Phil. Trans. Roy. Soc. London* 33:256–69.

Dumont, H. J. 1995. Ecocide in the Caspian. *Nature* 377:673–74.

Dung, V. V., P. M. Giao, N. N. Chinh, D. Tuoc, P. Arctander, and J. MacKinnon. 1993. A new species of living bovid from Vietnam. *Nature* 363:443–45.

Eckert, A. W. 1963. *The Great Auk*. New York: Signet.

Ehrlich, P. R., and A. H. Ehrlich. 1981. *Extinction: The Causes and Consequences of the Disappearance of Species*. New York: Random House.

Ehrlich, P. R., D. S. Dobkin, and D. Wheye. 1992. *Birds in Jeopardy: The Imperiled and Extinct Birds of the United States and Canada, Including Hawaii and Puerto Rico.* Palo Alto, Calif.: Stanford University Press.

Eldredge, N. 1975. Survivors from the good old, old, old days. *Natural History* 81 (10):52–59.

———. 1976. Collecting trilobites in North America. Part One: the East. *Fossils* 1 (1):58–67.

———. 1984. Simpson's Inverse: bradytely and the phenomenon of living fossils. In N. Eldredge and S. M. Stanley, eds., *Living Fossils,* pp. 272–77. New York: Springer-Verlag.

———. 1991a. *Fossils: The Evolution and Extinction of Species.* Princeton, N.J.: Princeton University Press.

———. 1991b. *The Miner's Canary: Unraveling the Mysteries of Evolution.* Englewood Cliffs, N.J.: Prentice-Hall.

———. 1998. *The Pattern of Evolution.* New York: Freeman.

———. 2000. *The Triumph of Evolution and the Failure of Creationism.* New York: Freeman.

———. 2001. Evolution, extinction, and humanity's place in nature. In M. J. Novacek, ed., *The Biodiversity Crisis: Losing What Counts,* pp. 76–80. New York: American Museum of Natural History.

Eldredge, N. and S. M. Stanley, eds. 1984. *Living Fossils.* New York: Springer-Verlag.

Ellis, R. 1991. *Men and Whales.* New York: Alfred A. Knopf.

———. 2003. *The Empty Ocean.* Washington, D. C.: Island Press.

Elzanowski, A. 2000. The flying dinosaurs. In G. S. Paul, ed., *The Scientific American Book of Dinosaurs,* pp. 169–82. New York: St. Martins Press.

Emiliani, C. 1993a. Viral extinctions in deep-sea species. *Nature* 366:217–18.

———. 1993b. Extinction and viruses. *BioSystems* 31:155–59.

Emiliani, C., E. B. Kraus, and E. M. Shoemaker. 1981. Sudden death at the end of the Mesozoic. *Earth Plan. Sci. Letters* 55:317–34.

Erickson, G. M. 1999. Breathing life into *Tyrannosaurus rex. Scientific American* 281 (3):42–49.

Erwin, D. H. 1993. *The Great Paleozoic Crisis: Life and Death in the Permian.* New York: Columbia University Press.

———. 1994. The Permo-Triassic extinction. *Nature* 367:231–36.

———. 1996. The mother of mass extinctions. *Scientific American* 273 (1):72–78.

Erwin, D. H., J. W. Valentine, and D. Jablonski.1997. The origin of animal body plans. *American Scientist* 85 (2):126–37.

Estes, R. D. and R. K. Estes. 1974. The biology and conservation of the giant sable antelope, *Hippotragus niger variani,* Thomas, 1916. *Proc. Acad. Nat. Sci. Phila.* 126:73–104.

————. 1976. Behaviour and ecology of the giant sable. *Nat. Geo. Reps.* (1968 Projects): 115–29.

Fanshawe, J. H. 1989. Serengeti's painted wolves. *Natural History* 98 (3):56–67.

Fanshawe, J. H. and C. D. FitzGibbon. 1993. Factors influencing the hunting success of an African wild dog pack. *Animal Behaviour* 45:479–90.

Fanshawe, J. H., L. H. Frame, and J. R. Ginsberg. 1991. The wild dog—Africa's vanishing carnivore. *Oryx* 25:137–46.

Fassett. J., R. A. Zelinsky, and J. R. Budahn. 2002. Dinosaurs that did not die; evidence for Paleocene dinosaurs in the Ojo Alamo Sandstone, San Juan Basin, New Mexico. In C. Koeberl and K. MacLeod, eds, *Catastrophic and Mass Extinctions: Impacts and Beyond*. pp. 307–36. Geological Society of America.

Fastovsky, D. E. and D. B. Weishampel. 1996. *The Evolution and Extinction of the Dinosaurs*. New York: Cambridge University Press.

Ferrari, S. F. and M. A. Lopes. 1992. A new species of marmoset, genus *Callithrix* Erxleben 1777 (Callitrichidae, Primates) from Western Brazilian Amazonia. *Goeldiana, Zoologia* 11:1–13.

Field, J. and R. Fullagar. 2001. Archaeology and Australian megafauna. *Science* 284:7a.

Fiffer, S. 2000. *Tyrannosaurus Sue: The Extraordinary Saga of the Largest, Most Fought Over T. Rex Ever Found*. New York: Freeman.

Fisher, J., N. Simon, and J. Vincent. 1969. *Wildlife in Danger*. New York: Viking.

Fisher, P. E., D. A. Russell, M. K. Stoskopf, R. E. Barrick, M. Hammer, and A. A. Kuzmitz. 2000. Cardiovascular evidence for an intermediate or higher metabolic rate in an ornithischian dinosaur. *Science* 288:503–5.

Fisher, W. K. 1903. Notes on the birds peculiar to Laysan Island, Hawaiian Group. *Auk* 20:384–97.

Fitter, R. 1968. *Vanishing Wild Animals of the World*. New York: Franklin Watts.

Flannery, T. 1994. *The Future Eaters*. New York: Grove Press.

————. 2001. *The Eternal Frontier*. New York: Atlantic Monthly Press.

Flannery, T. and R. G. Roberts. 1999. Late Quaternary extinctions in Australia: an overview. In R. D. E. MacPhee, ed., *Extinctions in Near Time: Causes, Contexts and Consequences*. pp. 239–55. San Diego: Academic Press.

Flannery, T. and P. Schouten. 2001. *A Gap in Nature: Discovering the World's Extinct Animals*. New York: Atlantic Monthly Press.

Flynn, J. J., M. A. Nedbal, J. W. Dragoo, and R. L. Honeycutt. 2000. Whence the red panda? *Molec. Phylogen. Evol.* 17 (2):190–99.

Foote, M. and J. Sepkoski. 1999. Absolute measures of the completeness of the fossil record. *Nature* 398:415–17.

Forcada, J., P. S. Hammond, and A. Aguilar. 1999. Status of the Mediterranean monk seal *Monachus monachus* in the western Sahara and the implications of a mass mortality event. *Mar. Ecol. Prog. Ser.* 188:246–61.

Ford, C. 1966. *Where the Sea Breaks Its Back*. Boston: Little, Brown.

Foreman, D. 2002a. Early awareness of extinction. *Wild Earth* 11 (3/4):2–4.

———. 2002b. The causes and processes of extinction. *Wild Earth* 12 (1):2–4.

Forshaw, J. M. 1973. *Parrots of the World*. Garden City, N.Y.: Doubleday.

Forster, C. A., S. D. Sampson, L. M. Chiappe, and D. W. Krause. 1998. The theropod ancestry of birds: new evidence from the late Cretaceous of Madagascar. *Science* 279:1915–1919.

Fortey, R. 1989. There are extinctions and there are extinctions: examples from the Lower Palaeozoic. *Phil. Trans. Roy. Soc. London* B325:307–26.

———. 1998. *Life: A Natural History of the First Four Billion Years of Life on Earth*. New York: Alfred A. Knopf.

Fossey, D. 1983. *Gorillas in the Mist*. Boston: Houghton-Mifflin.

Fox, D. 2002. Wallaby nations. *New Scientist* 175 (2354):32–35.

Fox, M. 2003. Scientists clone endangered Asian banteng. *Yahoo! News* http://news.yahoo.com/news?tmpl=story2&cid=570&u=/nm200304

Frankel, C. 1999. *The End of the Dinosaurs*. New York: Cambridge University Press.

Fraser, F. C. 1970. An early 17th century record of the California gray whale in Icelandic waters. *Invest. on Cetacea* 2:13–20.

Fuller, E. 1999. *The Great Auk*. Self-published.

———. 2001. *Extinct Birds*. Ithaca, N.Y.: Cornell University Press.

———. 2002. *Dodo*. New York: Universe Publishing.

Gaban-Lima, R., M. A. Raposo, and E. Höfling. 2002. Description of a new species of *Pionospitta* (Aves; Psittacidae) endemic to Brazil. *Auk* 119 (3):815–19.

Gaffney, E. S., L. Dingus, and M. K. Smith. 1995. Why cladistics? *Natural History* 104 (6):33–35.

Gallenkamp, C. 2001. *Dragon Hunter: Roy Chapman Andrews and the Central Asiatic Expeditions*. New York: Viking.

Ganapathy, R. 1980. A major meteorite impact on the Earth 65 million years ago: evidence from the Cretaceous-Tertiary boundary clay. *Science* 209:921–23.

Garrett, L. 1994. *The Coming Plague*. New York: Farrar Straus and Giroux.

Garretson, M. S. 1938. *The American Bison*. New York: New York Zoological Society.

Garshelis, D. 1997. Sea otter mortality estimated from carcasses collected after the *Exxon Valdez* oil spill. *Cons. Bio.* 11:905–16.

Gartner, S. and J. Keany. 1978. The terminal Cretaceous event: A geologic problem with an oceanographic solution. *Geology* 6:708–12.

Gartner, S. and J. P. McGuirk. 1979. Terminal Cretaceous extinction: Scenario for a catastrophe. *Science* 206:1272–1276.

Gazo, M., F. Aparicio, M. A. Cedenilla, J. F. Layna, and L. M. González. 2000. Pup survival in the Mediterranean monk seal (*Monachus monachus*) colony at Cabo Blanco Peninsula (Western Sahara-Mauritania). *Mar. Mam. Sci.* 16 (1):158–68.

Gee, H. 2000. O, for the wings of a dinosaur. *Nature* Science Update 11 April 2000. (http://www.nature.com/nsu/nsu-pf/000413/000413-6.html)

Gee, H. and L. Rey. 2003. *A Field Guide to Dinosaurs*. New York: Barron's.

Geist, V. 1998. *Deer of the World*. Stackpole.

Geraci, J. R. and D. J. St. Aubin, eds. 1979. *Biology of Marine Mammals: Insights Through Strandings*. Marine Mammal Commission, Washington, D.C.

Gesner, C. 1551–87. *Historia Animalium*. Zurich.

Gibbs, W. W. 2001. On the termination of species. *Scientific American* 285 (5):40–49.

Gillette, D. D. 1994. *Seismosaurus: The Earth Shaker*. New York: Columbia University Press.

Gilliard. E. T. 1958. *Living Birds of the World*. New York: Doubleday.

Gilmore, R. M. 1960. A census of the California gray whale. *U.S. Fish and Wildlife Service Scient. Rept: Fisheries*. No. 342.

Gingerich, P. G. 1995. Pleistocene extinctions in the context of origination-extinction equilibria in Cenozoic mammals. In P. S. Martin and R. G. Klein, eds., *Quaternary Extinctions: A Prehistoric Revolution*. pp. 223–49. Tucson: University of Arizona Press.

Gittleman, J. L. and M. E. Gompper. 2001. The risk of extinction—what you don't know will hurt you. *Science* 291:997–99.

Gonzalez, S., A. C. Kitchener, and A. M. Lister. 2000. Survival of the Irish elk into the Pleistocene. *Nature* 405:753–54.

Goodman, S. 2001. More funding needed to wipe out rinderpest. *Nature* 411:403.

Goodwin, G. G. 1946. The end of the great northern sea cow. *Natural History* 55 (2):56–61.

Gore, R. 1989. Extinctions: What caused the earth's great dyings? *National Geographic* 175 (6):662–69.

———. 1993. The Cambrian Period: explosion of life. *National Geographic* 184 (4): 120–36.

Gottfried, M. D., L. J. V. Compangno, and S. C. Bowman. 1996. Size and skeletal anatomy of the giant "Megatooth" shark *Carcharodon megalodon*. In A. P. Klimley and D. G. Ainley, eds., *Great White Sharks*. pp. 55–66. San Diego: Academic Press.

Gould, S. J. 1973. The misnamed, mistreated, and misunderstood Irish elk. *Natural History* 82 (3):10–19.

———. 1978. Nature's odd couples. *Natural History* 87 (1):38–41.

———. 1983. Nature's great era of experiments. *Natural History* 7/83:12–20.

———. 1987. The fossil fraud that never was. *New Scientist* 113 (1551):32–36.

———. 1994. The coherence of history. In S. Bengston, ed., *Early Life on Earth. Nobel Symposium No. 84*, pp. 1–8. New York: Columbia University Press.

———. 1996a. Up against a wall. *Natural History* 105 (7):16–22, 70–73.

———. 1996b. The dodo in the caucus race. *Natural History* 105 (11):22–33.

Gould, S. J. and N. Eldredge. 1977. Punctuated equilibria: The tempo and mode of evolution reconsidered. *Paleobiology* 3:115–51.

Grayson, D. G. 1995a. Nineteenth-century explanations of Pleistocene extinctions: A review and analysis. In P. S. Martin and R. G. Klein, eds., *Quaternary Extinctions: A Prehistoric Revolution*, pp. 5–39. Tucson: University of Arizona Press.

———. 1995b. Explaining Pleistocene extinctions: thoughts on the structure of a debate. In P. S. Martin and R. G. Klein, eds., *Quaternary Extinctions: A Prehistoric Revolution*, pp. 807–23. Tucson: University of Arizona Press.

Greenway, J. C. 1958. *Extinct and Vanishing Birds of the World*. American Committee for International Wildlife Protection.

Greenwood, A. D., C. Capelli, G. Possnert, and S. Pääbo. 1999. Nuclear DNA sequences from late Pleistocene megafauna. *Molec. Biol. Evol.* 16:1466–73.

Grieve, R. A. F. 1987. Terrestrial impact structures. *Ann. Rev. Earth Plan. Sci.* 15:245–70.

Gross, A. 1931. The last heath hen. *Bull. Mass. Audubon Soc.* 15 (5):12.

Grossman, M. L. and J. Hamlet. 1964. *Birds of Prey of the World*. New York: Clarkson N. Potter.

Groves, C. P. 1972. *Ceratotherium simum* [white rhinoceros]. *Mammalian Species* 8:1–6. American Society of Mammalogists.

Groves, C. P. and F. Kurt. 1972. *Dicerorhinus sumatrensis* [Sumatran rhinoceros]. *Mammalian Species* 21:1–6. American Society of Mammalogists.

Groves, C. P. and G. B. Schaller. 2000. The phylogeny and biogeography of the newly discovered Annamite artiodactyls. In E. S. Vrba and G. B. Schaller, eds., *Antelopes, Deer, and Relatives*. pp. 261–82. New Haven, Conn.: Yale University Press.

Grubb, P. 1981. *Equus burchelli* [Burchell's Zebra]. *Mammalian Species* 157:1–9. American Society of Mammalogists.

Grzelewski, D. 2002. Going to extremes [kakapo]. *Smithsonian* 33 (7):90–95.

Guggisberg, C. A. W. 1966. *S. O. S. Rhino*. October House.

Guthrie, R. D. 1984. Mosaics, allelochemics, and nutrients: an ecological theory of late Pleistocene megafaunal extinctions. In P. S. Martin and R. G. Klein, eds., *Quaternary Extinctions: A Prehistoric Revolution*. pp. 259–98. Tucson: University of Arizona Press.

————. 2000. Paleolithic art as a resource in artiodactyl paleobiology. In E. S. Vrba and G. B. Schaller, eds., *Antelopes, Deer, and Relatives.* pp. 96–127. New Haven, Conn.: Yale University Press.

Hahn, E. 1967. *Animal Gardens.* Garden City, N.Y.: Doubleday.

Haile-Selassie, Y. 2001. Late Miocene hominids from the Middle Awash, Ethiopia. *Nature* 412:178–81.

Haines, F. 1970. *The Buffalo.* New York: Columbia University Press.

Haley, D. 1978a. Saga of Steller's sea cow. *Natural History* 87 (9):9–17.

————. 1978b. Steller sea cow. In D. Haley, ed., *Marine Mammals of Eastern North Pacific and Arctic Waters.* pp. 236–41. Pacific Search Press.

————. 1980. The great northern sea cow: Steller's gentle siren. *Oceans* 13 (5):7–11.

Hallam, A. 1972. Continental drift and the fossil record. *Scientific American* 227 (11):56–66.

————. 1986. End-Cretaceous mass extinction event: argument for terrestrial causation. *Science* 238:1237–42.

————. 1989. The case for sea-level change as a dominant causal factor in mass extinction of marine invertebrates. *Phil. Trans. Roy. Soc. London* B325:437–55.

Hallam, A. and P. B. Wignall. 1997. *Mass Extinctions and Their Aftermath.* New York: Oxford University Press.

Halliday, T. 1979. The great auk: one that got away. *Oceans* 12 (5):27–31.

Halstead, L. B. 1968. *The Pattern of Vertebrate Evolution.* New York: Freeman.

Halstead, L. B. and J. Halstead. 1987. *Dinosaurs.* New York: Sterling.

Hare, J. 1997. The wild Bactrian camel *Camelus bactrianus ferus* in China: the need for urgent action. *Oryx* 31 (1):45–48.

————. 1998. *The Lost Camels of Tartary.* London: Abacus.

Harwood, J. 1989. Lessons from the seal epidemic. *New Scientist* 121 (1652): 38–42.

Haubitz, B., M. Prokop, W. Döhring, J. H. Ostrom, and P. Wellnhofer. 1988. Computer tomography of *Archaeopteryx. Paleobiology* 14 (2):206–13.

Hecht, J. 1988. Evolving theories for old extinctions. *New Scientist* 118:16.

————. 2000. Feathers fly. *New Scientist* 167:18.

————. 2001. Telltale bones. *New Scientist* 172:14.

————. 2003. Great apes plunge towards extinction. *New Scientist* 178:11.

Heck, H. 1951. The breeding-back of the aurochs. *Oryx* 1 (3):119.

Heezen, B. C., and C. D. Hollister. 1971. *The Face of the Deep.* New York: Oxford University Press.

Hemingway, E. 1935. *Green Hills of Africa.* New York: Scribner's (Touchstone edition, 1996).

Henderson, D. 2000. *Asteroid Impact.* New York: Dial.

Hermansson, H. 1924. Jon Gudmundsson and his natural history of Iceland. *Islandica* 15, I-XXVIII, 1–40.

Heuvelmans, B. 1995. *On the Track of Unknown Animals.* London: Kegan Paul.

Hildebrand, A. R. and W. V. Boynton. 1990. Proximal Cretaceous-Tertiary boundary impact deposits in the Caribbean. *Science* 248:843–47.

———. 1991. Cretaceous ground zero. *Natural History* 100:47–53.

Hillman, C. N. and T. W. Clark. 1980. *Mustela nigripes* [Black-footed ferret]. *Mammalian Species* 126:1–3. American Society of Mammalogists.

Hillman-Smith, A. K. K. 1998. The current status of the northern white rhino in Garamba. *Pachyderm* 25:104–5.

Hillman-Smith, A. K. K. and C. P. Groves. 1994. *Diceros bicornis* [black rhinoceros]. *Mammalian Species* 455:1–8. American Society of Mammalogists.

Hilton-Taylor, C., comp. 2000. *2000 IUCN Red List of Threatened Species.* World Conservation Union (IUCN).

Hoffman, A. 1989. What, if anything, are mass extinctions? *Phil. Trans. Roy. Soc. London* B325:13–21.

Hoffman, H. J. 2000. When life nearly came to an end: the Permian Extinction. *National Geographic* 198 (3):100–13.

Hofman, R. J. and W. N. Bonner, 1985. Conservation and protection of marine mammals: Past, present and future. *Mar. Mam. Sci.* 1 (2):109–27.

Hogan, J. 2003. Cave art date splits critics. *New Scientist* 178 (2391):8.

Holdaway, R. N. and C. Jacomb. 2000. Rapid extinction of the moas (Aves: Dinornithiformes): model, test, and implications. *Science* 287:2250–2254.

Holmes, B. and J. Hecht. 2003. Will overcrowding sink Noah's Ark? *New Scientist* 180 (2422):6–7.

Holtz, T. R. 1998. Theropod paleobiology: More than just bird origins. *Gaia* 15:1–3.

Holyoak, D. T. 1971. Comments on the extinct parrot *Lophopsittacus mauritianus*. *Ardea* 59:50–51.

Horton, D. R. 1995. Red kangaroos: last of the Australian megafauna. In P. S. Martin and R. G. Klein, eds., *Quaternary Extinctions: A Prehistoric Revolution.* pp. 639–80. University of Arizona Press.

Hotton, N. 1968. *The Evidence of Evolution.* Washington, D.C.: Smithsonian Institution Press.

Hou, L. 2001. *Mesozoic Birds of China.* Translated by W. Downs. Phoenix Valley Provincial Aviary of Taiwan.

Hou, L.-H., L. D. Martin, Z. Zhou, and A. Feduccia. 1996. Early adaptive radiation of birds: evidence from fossils of northeastern China. *Science* 274:1164–1167.

Hou, L.-H., Z. Zhou, L. D. Martin, and A. Feduccia. 1995. A beaked bird from the Jurassic of China. *Nature* 337:616–18.

Hoyle, F. and C. Wickramasinghe. 1978. *Lifecloud: The Origin of Life in the Universe.* New York: Harper & Row.

Hsü, K. J. 1980. Terrestrial catastrophe caused by cometary impact at the end of the Cretaceous. *Nature* 285:201–3.

———. 1986. *The Great Dying: Cosmic Catastrophe, Dinosaurs, and the Theory of Evolution.* San Diego: Harcourt Brace Jovanovich.

———. 1989. Rare events, mass extinction and evolution. *Historical Biology* 2:1–4.

Hunter, J. A. 1952. *Hunter.* New York: Harper & Brothers.

Hut, P., W. Alvarez, W. P. Elder, T. Hansen, E. G. Kauffman, G. Keller, E. M. Shoemaker, and P. R. Weissman. 1987. Comet showers as a cause of mass extinctions. *Nature* 329:118–26.

Hutchison, R., and A. Graham. 1993. *Meteorites.* New York: Sterling.

Huxley, J. 1963. *Wild Lives of Africa.* New York: Harper & Row.

Huynen, L., C. D. Millar, R. P. Schofield, and D. M. Lambert. 2003. Nuclear DNA sequences detect species limits in ancient moa. *Nature* 425:175–78.

Hwang, S. H., M. A. Norell, J. Quiang, and G. Kequin. 2002. New specimens of *Microraptor zhaoianus* (Theropoda: Dromaeosauridae) from northeastern China. *Amer. Mus. Novitates* 381:1–44.

Ichihara, T. 1963. Identification of the pygmy blue whale in the Antarctic. *Norsk Hvalfangst-tidende* 52 (5):128–30.

Ives, R. 1996. *Of Tigers and Men.* New York: Doubleday.

Jablonski, D. 1989. The biology of mass extinction: a palaeontological view. *Phil. Trans. Roy. Soc. London* B325:357–68

———. 1999. The future of the fossil record. *Science* 284:2114–16.

Jackson, J. B. C., M. X. Kirby, W. H. Berger, K. A. Bjorndal, L. W. Botsford, B. J. Bourque, R. H. Bradbury, R. Cooke, J. Erlandson, J. A. Estes, T. P. Hughes, S. Kidwell, C. B. Lange, H. S. Lenihan, J. M. Pandolfi, C. H. Peterson, R. S. Steneck, M. J. Tegner, and R. R. Warner. 2001. Historical overfishing and the recent collapse of coastal ecosystems. *Science* 293:629–38.

Janis, C. M. 1986. Evolution of horns and related structures in hoofed mammals. *Discovery* 19 (2):8–17.

Jauniaux, T., G. Charlier, M. Desmecht, M. Haelters, T. Jaques, B. Losson, J. Van Gompel, J. Tavernier, and F. Coignoul. 2000. Pathological findings in two fin whales (*Balaenoptera physalus*) with evidence of morbillivirus infection. *Jour. Comp. Pathol.* 123(2-3):198–201.

Jensen, T., M. van de Bildt, H. H. Dietz, T. H. Andersen, A. D. Hammer, T. Kuiken, and A. D. Osterhaus. 2002. Another phocine distemper outbreak in Europe. *Science* 297:209.

Jepsen, G. L. 1964. Riddle of the terrible lizards. *American Scientist* 52:227–46.

Ji, Q., L. M. Chiappe, and S.-A. Ji. 1999. A new late Mesozoic confuciusornithid bird from China. *Jour. Paleo.* 19 (1):1–7.

Ji, Q., P. J. Currie, M. A. Norell, and S.-A Ji. 1998. Two feathered dinosaurs from northeastern China. *Nature* 393:753–61.

Ji, Q., M. A. Norell, K.-Q Gao, S-A Ji, and D. Ren. 2001. The distribution of integumentary structures in a feathered dinosaur. *Nature* 410:1084–1088.

Jin, Y. G., Y. Wang, W. Wang, Q. H. Shang, C. Q. Cao, and D. H. Erwin. 2000. Pattern of marine mass extinction near the Permian-Triassic boundary in South China. *Science* 289:432–36.

Johnson, C. N. 1998. Species extinction and the relationship between distribution and abundance. *Nature* 394:272–74.

———. 2002. Determinants of loss of mammal species during the Late Quaternary "megafauna" extinctions: life history and ecology, but not body size. *Proc. Roy. Soc. London* DOI: 10.1098 / rspb.2002.2130.

Jones A. P., G. D. Price, N. J. Price, P. S. DeCarli, and R. A. Clegg. 2002. Impact induced melting and the development of large igneous provinces. *Earth Planet. Sci. Let.* 202 (3):551–61.

Jones, C. G., W. Heck, R. E. Lewis, Y. Mungroo, G. Slade, and T. Cade. 1995. The restoration of the Mauritius kestrel (*Falco punctatus*) population. *Ibis* 137:173–80.

Jones, T. D., J. A. Ruben, L. D. Martin, E. N. Kurochkin, and A. Feduccia. 2000. Nonavian feathers in a Late Triassic archosaur. *Science* 288:2202–2225.

Juniper, T. 2002. *Spix's Macaw: the Race to Save the World's Rarest Bird.* London: Fourth Estate.

Kaiho, K., Y. Kajiwara, T. Nakano, Y. Miura, H. Kawahata, K. Tazaki, M. Ueshima, Z. Chen and G.-R. Shi. 2001. End-Permian catastrophe by a bolide impact: evidence of a gigantic release of sulfur from the mantle. *Geology* 29:815–18.

Kaiser, J. 2003. Ebola, hunting push ape populations to the brink. *Science* 300:232.

Kano, T. 1980. Social behavior of wild pygmy chimpanzees (*Pan paniscus*). *Jour. Human Evol.* 11:35–39.

———. 1992. *The Last Ape: Pygmy Chimpanzee Behavior and Ecology.* Palo Alto, Calif.: Stanford University Press.

Kaplan, M. 2002. Plight of the condor. *New Scientist* 176:34–36.

Keller, G. 1994. K-T boundary issues. *Science* 264:641.

———. 2001. The end-Cretaceous mass extinction in the marine realm: year 2000 assessment. *Plan. Space Sci.* 49:817–30.

Keller, G., T. Adatte, W. Stinnesbeck, M. Affolter, L. Schilli, and J. G. Lopez-Oliva. 2002. Multiple spherule layers in the late Maatrichtian of northeastern Mexico. *Geol. Soc. Amer. Sp. Paper* 356:145–61.

Kelley, S. P. and E. Gurov. 2002. The Boltysh, another end-Cretaceous impact. *Meteor. Plan. Sci.* 37 (8):1031–1044.

Kellner, A. W. A. and D. de A. Campos. 2002. The function of the cranial crest and jaws of a unique pterosaur from the early Cretaceous of Brazil. *Science* 297:389.

Kemp, N., M. Dilger, N. Burgess, and C. V. Dung. 1997. The saola *Pseudoryx nghetinhensis* in Vietnam—new information on distribution and habitat preferences, and conservation needs. *Oryx* 31 (1):37–44.

Kenyon, K. W. 1977. Caribbean monk seal extinct. *Jour. Mammal.* 58:979–98.

———. 1978. Hawaiian monk seal. In D. Haley, ed., *Marine Mammals of Eastern North Pacific and Arctic Waters.* pp. 212–16. Seattle: Pacific Search Press.

———. 1980. No man is benign: the endangered monk seal. *Oceans* 13 (3):48–54.

———. 1981. Monk seals—*Monachus* Fleming, 1822. In S. H. Ridgway and R. J. Harrison, eds., *Handbook of Marine Mammals, Vol. II: Seals.* pp. 195–220. San Diego: Academic Press.

Kerr, R. 1987. Asteroid impact gets more support. *Science* 236:666–68.

———. 1992. Huge impact tied to mass extinction. *Science* 257:878–80.

———. 2000. Did volcanoes drive ancient extinctions? *Science* 289:1130–1131.

———. 2001. Whiff of gas points to impact mass extinction. *Science* 291:1469–1470.

———. 2003. Megafauna died from big kill, not big chill. *Science* 300:885.

Kiesecker, J. M., A. R. Blaustein, and L. K. Belden. 2001. Complex causes of amphibian population declines. *Nature* 410:681–84.

Kiltie, R. A. 1995. Seasonality, gestation time, and large mammal extinctions. In P. S. Martin and R. G. Klein, eds., *Quaternary Extinctions: A Prehistoric Revolution.* pp. 299–314. Tucson: University of Arizona Press.

Klein, R. G. 1974. On the taxonomic status, distribution and ecology of the blue antelope, *Hippotragus leucophaeus* (Pallas, 1766). *Ann. S. Afr. Mus.* 65:99–143.

———. 2000. Human evolution and large mammal extinctions. In E. S. Vrba and G. B. Schaller, eds., *Antelopes, Deer, and Relatives,* pp. 128–39. New Haven: Yale University Press.

Knoll, A. H., R. K. Bambach, D. E. Canfield, and J. P. Grotzinger. 1996. Comparative Earth history and Late Permian Mass Extinction. *Science* 273:452–57.

Knudtson, P. M. 1977. The case of the missing monk seal. *Natural History* 86 (8):78–83.

Kohl, L. 1982. Pére David's deer saved from extinction. *National Geographic* 162 (4):478–85.

Kring, D. A. 2000. Impact events and their effect on the origin, evolution, and distribution of life. *GSA Today* 10 (8):1–7.

Kring, D. A. and D. D. Durda. 2003. The day the world burned. *Scientific American* 289 (6):98–105.

Kurtén, B. 1991. *The Innocent Assassins.* New York: Columbia University Press.

Kurtén, B. and E. Anderson. 1980. *Pleistocene Mammals of North America.* New York: Columbia University Press.

Laurie, W. A., E. M. Lang, and C. P. Groves. 1983. *Rhinoceros unicornis* [Indian rhinoceros]. *Mammalian Species* 211:1–6. American Society of Mammalogists.

LeBoeuf, B. J., K. W. Kenyon, and B. Villa-Ramirez. 1986. The Caribbean monk seal is extinct. *Mar. Mam. Sci.* 2 (1):70–72.

Lemonick, M. D. 1994. Too few fish in the sea. *Time* 143 (14):70–71.

Leroy, E. M., P. Rouquet, P. Formenty. S. Souquière, A. Kilbourne, J.-M. Froment, M. Barmejo, S. Smit, W. Karesh, R. Swanepoel, S. R. Zaki, and P. E. Rollin. 2004. Multiple Ebola virus transmission events and rapid decline of central African wildlife. *Science* 303:387–90.

Leslie, M. 2001. Tales of the sea. *New Scientist.* 169 (2275):32–35.

Ley, W. 1948. *The Lungfish, the Dodo, and the Unicorn.* New York: Viking.

———. 1951. *Dragons in Amber.* New York: Viking.

———. 1962. *Exotic Zoology.* New York: Viking.

———. 1968. *Dawn of Zoology.* Englewood Cliffs, N.J.: Prentice-Hall.

Li, H.-C., T. Funiyoshi, H. Lou, S. Yashiki, S. Sunoda, L. Cartier, L. Nunez, I. Munoz, S. Horai, and K. Tajima. 1999. The presence of ancient human T-cell lymphotrophic virus type 1 provirus DNA in an Andean mummy. *Nature Medicine* 5 (12):1428–32.

Lillegraven, J. A. and J. J. Eberle. 1999. Vertebrate faunal changes through Lancian and Puercan time in southern Wyoming. *Jour. Paleo.* 73 (4):691–710.

Linden, E. 1992a. Chimpanzees with a difference: bonobos. *National Geographic* 181 (3):46–53.

Linden, E. 1992b. A curious kinship: apes and humans. *National Geographic.* 181 (3):3–45.

Lindique, M. and K. P. Erb. 1996. Research on effects of temporary horn removal on black rhinos in Namibia. *Pachyderm* 21:27–30.

Lindquist, O. 2000. The North Atlantic gray whale (*Eschrichtius robustus*): an historical outline based on Icelandic, Danish-Icelandic, English and Swedish sources dating from ca 1000 AD to 1792. *Occasional Papers, Centre for Environmental History and Policy* 1:1–53.

Long, J. A. 1995. *The Rise of Fishes: 500 Million Years of Evolution.* Baltimore: Johns Hopkins University Press.

Long, M. E. 1998. The vanishing prairie dog. *National Geographic* 193 (4):116–31.

Lopes, M. A. and S. F. Ferrari. 1996. Preliminary observations on the ka'apor capuchin *Cebus kaapori* Queiroz 1992 from Eastern Brazilian Amazonia. *Biological Conservation* 76:321–24.

Lundelius, E. and R. Graham. 1999. The weather changed. *Discovering Archaeology* 1 (5):48–53.

Luschekina, A. A., S. Dulamtseren, L. Amgalan, and V. M. Neronov. 1999. The status and prospects for conservation of the Mongolian saiga, *Saiga tatarica mongolica*. *Oryx* 33 (1):21–30.

Lyell, C. 1832. *Principles of Geology*. London: Murray.

MacDonald, D., ed. 2001. *The Encyclopedia of Mammals*. Abingdon, England: Andromeda.

McDonald, J. N. 1995. The recorded North American selection regime and Late Quaternary megafaunal extinctions. In P. S. Martin and R. G. Klein, eds., *Quaternary Extinctions: A Prehistoric Revolution*, pp. 404–39. Tucson: University of Arizona Press.

McGowan, C. 1991. *Dinosaurs, Spitfires, and Sea Dragons*. Cambridge, Mass.: Harvard University Press.

MacKenzie, D. 2000a. Sick to death. *New Scientist* 167 (2250):32–35.

———. 2000b. Vultures face oblivion. *New Scientist* 167 (2258):5.

———. 2001. Will Viagra save the seals? *New Scientist* 170 (2286):13.

McKinley, D. 1980.The balance of decimating factors and recruitment in extinction of the Carolina parakeet. *Indiana Audubon Quarterly* 59 (1):8–18.

———. 1985. *The Carolina Parakeet in Florida*. Florida Ornithologoical Society.

McLaren, D. J. 1970. Time, life, and boundaries. *Jour. Paleo.* 44:801–15.

———. 1989. Detection and significance of mass killings. *Historical Biology* 2:5–15.

McNamara, K., and J. Long. 1998. *The Evolution Revolution*. New York: John Wiley.

McNeil, D. G. 1999. The great ape massacre. *New York Times Magazine*. May 9:46–48.

McNeill, W. H. 1993. Patterns of disease emergence in history. In S. Morse, ed., *Emerging Viruses*, pp. 29–36. New York: Oxford University Press.

McNulty, F. 1970. *Must They Die? The Strange Case of the Prairie Dog and the Black-Footed Ferret*. New York: Doubleday.

MacPhee, R. D. E. 1999 (ed.) *Extinctions in Near Time: Causes, Contexts and Consequences*. San Diego: Academic Press.

MacPhee, R. D. E and C. Flemming. 2001. Brown-eyed, milk-giving . . . and extinct: Losing mammals since A. D. 1500. In M. J. Novacek, ed., *The Biodiversity Crisis: Losing What Counts*. pp. 94–99. New York: American Museum of Natural History.

MacPhee, R. D. E. and P. A. Marx. 1997. The 40,000-year plague: humans, hyperdisease, and first-contact extinctions. In S. M. Goodman and B. D. Patterson, eds., *Natural Change and Human Impact in Madagascar*. pp. 169–217. Washington, D.C.: Smithsonian Institution Press.

McRae, M. 2000. Central Africa's orphan gorillas: will they survive in the wild? *National Geographic* 197 (2):84–97.

Magnus, O. 1555. *Historia de gentibus septentrionalibus*. Antwerp.

Maisey, J. G. 1996. *Discovering Fossil Fishes*. New York: Henry Holt.

Malakoff, D. 2001. Scientists use strandings to bring species to life. *Science* 293:1754–1757.

Marshall, C. R. 1999. Missing links in the history of life. In J. W. Schopf, ed., *Evolution: Facts and Fallacies*, pp. 37–69. San Diego: Academic Press.

Marshall, C. R. and P. D. Ward. 1996. Sudden and gradual molluscan extinctions in the Latest Cretaceous of western European Tethys. *Science* 274:1360–1363.

Marshall, L. G. 1995. Who killed Cock Robin? An investigation of the extinction controversy. In P. S. Martin and R. G. Klein, eds., *Quaternary Extinctions: A Prehistoric Revolution*, pp. 785–806. Tucson: University of Arizona Press.

Martill, D. and D. Naish. 2000. *Walking with Dinosaurs: The Evidence*. London: BBC.

Martin, E. B. 1984. They're killing off the rhino. *National Geographic* 405–22.

———. 1989. Report on the trade in rhino products in Eastern Asia and India. *Pachyderm* 11:13–22.

———. 1990. Medicines from Chinese treasures. *Pachyderm* 13:12–13.

Martin, E. B. and C. Martin. 1982. *Run Rhino Run*. London: Chatto & Windus.

Martin, E. B. and T. C. I. Ryan. 1990. How much rhino horn has come into international markets since 1970? *Pachyderm* 13:20–25.

Martin, E. B. and L. Vigne. 1996. Nepal's rhinos: one of the greatest conservation success stories. *Pachyderm* 21:10–26.

———. 1997. An historical perspective of the Yemeni rhino horn trade. *Pachyderm* 23:29–40.

Martin, P. S. 1984. Catastrophic extinctions and late Pleistocene blitzkrieg. In M. H. Nitecki, ed., *Extinctions*, pp. 153–89. Chicago: University of Chicago Press.

———. 1985. Prehistoric overkill: the global model. In P. S. Martin and R. G. Klein, eds., *Quaternary Extinctions: A Prehistoric Revolution*, pp. 354–403. Tucson: University of Arizona Press.

Martin, P. S. and R. G. Klein, eds. 1995. *Quaternary Extinctions: A Prehistoric Revolution*. Tucson: University of Arizona Press.

Martin, P. S. and D. W. Steadman. 1999. Preshistoric extinctions on islands and continents. In R. D. E. MacPhee, ed., *Extinctions in Near Time: Causes, Contexts and Consequences*, pp. 257–69. San Diego: Academic Press.

Martindale, D. 2000. No mercy. *New Scientist* 168 (2260):28–32.

Marzuola, C. 2003. Camelid comeback. *Science News* 163 (2):26–28.

Masland, T. 2003. Gorillas in the midst of an outbreak? *Newsweek* 141 (3):10.

Matthiessen, P. 1959. *Wildlife in America*. New York: Viking.

———. 2002. *The Birds of Heaven: Travels with Cranes*. New York: North Point.

Mayor, A. 2000. *The First Fossil Hunters: Paleontology in Greek and Roman Times*. Princeton, N.J.: Princeton University Press.

Mead, J. G. and E. D. Mitchell. 1984. Atlantic gray whales. In M. L. Jones, S. L. Swartz, and S. Leatherwood, eds., *The Gray Whale, Eschrichtius robustus,* pp. 33–53. San Diego: Academic Press.

Mead, J. I. and D. J. Meltzer. 1995. North American Late Quaternary extinctions and the radiocarbon record. In P. S. Martin and R. G. Klein, eds., *Quaternary Extinctions: A Prehistoric Revolution,* pp. 440–50. Tucson: University of Arizona Press.

Meagher, M. 1986. *Bison bison* [*North American bison*]. *Mammalian Species* 226:1–8. American Society of Mammalogists.

Meegaskumbura, M., F. Bossuyt, R. Pethiyagoda, K. Manamendra-Arachchi, M. Bahir, M. C. Milinkovitch, and C. J. Schneider. 2002. Sri Lanka: an amphibian hot spot. *Science* 298:379.

Melosh, H. J., N. M. Schneider, K. J. Zahnle, and D. Latham. 1990. Ignition of global wildfires at the Cretaceous/Tertiary boundary. *Nature* 343:251–54.

Melville, Herman. 1996. *Moby-Dick*. Quality Paperback Book Club: New York.

Menkhorst, P. 2001. *A Field Guide to the Mammals of Australia*. New York: Oxford University Press.

Menon, V. 1996. *Under Siege: Poaching and Protection of Greater One-Horned Rhinoceros in India*. Traffic International.

Merton, D. and M. Clout. 1999. Kakapo: back from the brink. *Wingspan* 9 (2):14–17.

Merton, D. V., R. D. Morris, and I. A. E. Atkinson. 1984. Lek behaviour in a parrot: the kakapo *Strigops habroptilus* of New Zealand. *Ibis* 126:277–83.

Meyer, A. Evolutionary Celebrities. Review of *The Cichlid Fishes: Nature's Grand Experiment in Evolution*, by George W. Barlow. *Nature* 410:17–18.

Miceler, J. 2002. Chasing the chiru. *Explorers Journal* 80 (3):26–33.

Milius, S. 2003. After West Nile virus. *Science News* 163 (13):203–5.

Miller, B., R. P. Reading, and S. Forrest. 1996. *Prairie Night: Black-Footed Ferrets and the Recovery of an Endangered Species*. Washington, D.C.: Smithsonian Institution Press.

Miller, G. H., J. W. Magee, B. J. Johnson, M. L. Fogel, N. A. Spooner, M. T. McCulloch, and L. K. Ayliffe. 1999. Pleistocene extinction of *Genyornis newtoni*: human impact on Australian megafauna. *Science* 283:205–8.

Milliken, T., K. Nowell, and J. B. Thomsen. 1993. *The Decline of the Black Rhino in Zimbabwe*. Cambridge, England: Traffic International.

Mills, C. 1999. The wild, wild pest. *Sciences* 37 (2):10–13.

Mills, J. A. 1993. *Market Under Cover: The Rhinoceros Horn Trade in South Korea*. Cambridge, England: Traffic International.

Mills, J. A. 1997. *Rhinoceros Horn and Tiger Bone in China: An Investigation of Trade Since the 1993 Ban*. Cambridge, England: Traffic International.

Milner-Gulland, E. J., O. M. Bukreeva, T. Coulson, A. A. Lushchekina, M. V. Kholodova, A. Bekenov, and Y. A. Grachev. 2003. Reproductive collapse in saiga antelope harems. *Nature* 422:135.

Milner-Gulland, E. J., M. V. Kholodova, A. Bekenov, O. M. Bukreeva, Y. A. Grachev, L. Amgalan, and A. A. Lushchekina. 2001. Dramatic decline in saiga antelope populations. *Oryx* 35 (4):340–45.

Mittermeier, R. A., M. Schwarz, and J. M. Ayres 1992. A new species of marmoset, genus *Callithrix* Erxleben, 1777 (Callithrichidae, Primates), from the Rio Maues Region, State of Amazonas, Central Brazilian Amazonia. *Goeldiana, Zoologia.* 14:1–17.

Mochi, U. and T. D. Carter. 1971. *Hoofed Mammals of the World.* New York: Scribners.

Moen, R. A., J. Pastor, and Y. Cohen. 1999. Antler growth and extinction of Irish elk. *Evol. Ecol. Res.* 1:235–49.

Molyneux, T. 1697. A discourse concerning the large horns frequently found under ground in Ireland. *Phil. Trans. Roy. Soc.* 9:485–52.

Monastersky, R. 1999a. Eruptions cleared path for dinosaurs. *Science News* 155(17):260.

———. 1999b. Waking up to the dawn of vertebrates. *Science News* 156 (20):292.

Mooney, H. A. and E. E. Cleland. 2001. The evolutionary impact of invasive species. *Proc. Natl. Acad. Sci.* 98 (10):5446–5451.

Moore, J. C. 1968. Relationships among the living genera of beaked whales with classifications, diagnoses and keys. *Fieldiana: Zoology* 53:209–98.

———. 1972. More skull characteristics of the beaked whale *Indopacetus pacificus.* *Fieldiana: Zoology* 62:1–19.

Moorehead, A. 1957. *No Room in the Ark.* New York: Harper & Brothers.

Morell, V. 1999. The sixth extinction. *National Geographic* 195 (2):43–59.

Mörzer Bruyns, W. F. J. 1971. *Field Guide of Whales and Dolphins.* Amsterdam: Tor.

Mowat, F. 1984. *Sea of Slaughter.* New York: Atlantic Monthly Press.

Murphy, N. L., D. Trexler, and M. Thompson. 2002. Exceptional soft-tissue preservation in a mummified ornithopod dinosaur from the Campanian Lower Judith River Formation. *Jour. Vert. Paleo.* 22 (3) Supp. 91A.

Murphy, R. C. and L. S. Mowbray. 1951. New light on the cahow. *Auk* 68 (3):266–80.

Murray, P. M. 1995. Extinctions down under: a bestiary of extinct Australian Late Pleistocene monotremes and marsupials. In P. S. Martin and R. G. Klein, eds., *Quaternary Extinctions: A Prehistoric Revolution,* pp. 600–628. Tucson: University of Arizona Press.

Murray, P. M. and D. Megirian. 1998. The skull of dromornithid birds: anatomical evidence for their relationship to Anseriformes. *Rec. S. Aust. Mus.* 31 (1):51–97.

Myers, N. 1979. *The Sinking Ark.* New York: Pergamon.

Myers, N. and A. H. Knoll. 2001. The biotic crisis and the future of evolution. *Proc. Nat. Acad. Sci.* 98 (10):5389–5392.

Myers, R. and B. Wurm. 2003. Rapid worldwide depletion of predatory fish communities. *Nature* 423:280–83.

Newell, N. D. 1967. Revolutions in the history of life. *Geol. Soc. America Special Paper* 89:63–91.

Nitecki, M. H., ed. 1984. *Extinctions*. Chicago: University of Chicago Press.

Noe, G. E. 1987. *Carcharodon*. New York: Vantage.

Norell, M. A. and J. A. Clarke. 2001. Fossil that fills a critical gap in avian evolution. *Nature* 409:181–84.

Norell, M. A., L. Chiappe, and J. A. Clarke. 1993. A new limb on the avian family tree. *Natural History* 102 (9):38–43.

Norell, M. A., E. S. Gaffney, and L. Dingus. 1995. *Discovering Dinosaurs in the American Museum of Natural History*. New York: Alfred A. Knopf.

Norell, M. A., Q. Ji, K. Gao, C. Yuan, Y. Zhao, and L. Wang. 2002. "Modern" feathers on a non-avian dinosaur. *Nature* 416:36–37.

Norman, D. 1994. *Prehistoric Life*. New York: Macmillan.

Norris, K. S. and R. N. McFarland. 1958. A new porpoise of the genus *Phocoena* from the Gulf of California. *Jour. Mammal.* 39:22–39.

Norris, R. D., J. Firth, J. S. Blusztajn, and G. Ravissa. 2000. Mass failure of the North Atlantic margin triggered by the Cretaceous-Paleogene bolide impact. *Geology* 28 (12):1119–1122.

Novacek, M. J. 1996. *Dinosaurs of the Flaming Cliffs*. New York: Anchor Doubleday.

———. 1999. 100 million years of land vertebrate evolution: The Cretaceous-Early Tertiary transition. *Ann. Missouri Bot. Gard.* 86:230–58.

———, ed. 2001. *The Biodiversity Crisis: Losing What Counts*. New York: American Museum of Natural History.

Novacek, M. J. and E. E. Cleland. 2001. The current biodiversity extinction event: Scenarios for mitigation and recovery. *Proc. Nat. Acad. Sci.* 98 (10):5466–5470.

Novas, F. E. 1997. South American dinosaurs. In P. J. Currie and K. Padian, eds., *Encyclopedia of Dinosaurs*, pp. 678–89. San Diego: Academic Press.

Nowak, R. M. 1991. *Walker's Mammals of the World (Fifth Edition)*. Baltimore: Johns Hopkins University Press.

Nowell, K. 2003. Revision of the Felidae Red List of threatened species. *Cat News* 37:4–6.

Nowell, K., W.-L. Chyi, and C.-J. Pei. 1992. *The Horns of a Dilemma: The Market for Rhino Horn in Taiwan*. Cambridge, England: Traffic International.

Officer, C. B. and C. L. Drake. 1985a. Terminal Cretaceous environmental events. *Science* 227:1161–66.

———. 1985b. Cretaceous-Tertiary extinctions: Alternative models. *Science* 230: 1292–95.

Officer, C. B. and J. Page. 1996. *The Great Dinosaur Extinction Controversy*. Boston: Addison-Wesley.

Officer, C. B., A. Hallam, C. L. Drake, and J. D. Devine. 1987. Late Cretaceous and paraoxysmal Cretaceous/Tertiary extinctions. *Nature* 326:143–49.

Olsen, P. E. 1999. Giant lava flows, mass extinctions, and mantle plumes. *Science* 284:604–5.

Olson, S. L. and H. F. James. 1995. The role of Polynesians in the extinction of the avifauna of the Hawaiian Islands. In P. S. Martin and R. G. Klein, eds., *Quaternary Extinctions: A Prehistoric Revolution*, pp. 768–80. Tucson: University of Arizona Press.

Omura, H., T. Ichihara, and T. Kasuya. 1970. Osteology of pygmy blue whale with additional information on external and other characteristics. *Sci. Rep, Whales Res. Inst.* 22:1–29.

Osborn, H. F. 1906. The causes of extinction in Mammalia. *Amer. Nat.* 480:769–95, 829–59.

Osterhaus, A. D., G. F. Rimmelzwaan, B. E. E. Martina, T. M. Bestebroer, and R. A. M. Fouchier. 2000. Influenza B virus in seals. *Science* 288:1051–1053.

Ostrom, J. H. 1969. Osteology of *Deinonychus antirrhopus*, an unusual theropod from the Lower Cretaceous of Montana. *Bull. Peabody Mus. Nat. Hist* 30:1–165.

———. 1973. The ancestry of birds. *Nature* 242:136.

———. 1974. *Archaeopteryx* and the origin of flight. *Q. Rev. Biol.* 49:27–47.

———. 1976a. *Archaeopteryx* and the origin of birds. *Biol. Jour. Linn. Soc.* 8:91–182.

———. 1976b. Some hypothetical anatomical stages in the evolution of avian flight. *Smithsonian Contrib. Paleobiol.* 27:1–21.

———. 1976c. On a new specimen of the Lower Cretaceous theropod dinosaur *Deinonychus antirrhopus*. *Berviora*. 439:1–21.

———. 1979. Bird flight: how did it begin? *Amer. Scientist* 67:46–56.

Ovington, D. 1978. *Australian Endangered Species*. Cassell Australia.

Owen, R. 1842. Report on British fossil reptiles, Part II. *Rep. Br. Assoc. Adv. Sci.* 1841:60–294.

Paddle, R. 2000. *The Last Tasmanian Tiger*. New York: Cambridge University Press.

Padian, K. 1983. A functional analysis of flying and walking in pterosaurs. *Paleobiology* 9 (3):218–39.

———. 1988. Triassic-Jurassic extinctions. *Science* 241:1358–1359.

———. 1997. Dinosauria: Definition. In P. J. Currie and K. Padian, eds., *Encyclopedia of Dinosaurs*. pp. 175–78. San Diego: Academic Press.

———. 1997b. Pterosauria. In P. J. Currie and K. Padian, eds., *Encyclopedia of Dinosaurs*. pp. 613–17. San Diego: Academic Press.

———. 1998. When is a bird not a bird? *Nature* 393:729–30.

Padian, K. and L. Chiappe. 1998. The origin of birds and their flight. *Scientific American* 278 (2):38–47.

Pain, S. 2000a. The demon duck of doom. *New Scientist* 166 (2240):37–39.

———. 2000b. Scary monsters, super creeps. *New Scientist* 166 (2241):42–45.

———. 2002. Ready to croak. *New Scientist* 174 (2337):17.

Pálfy, J. and P. L. Smith. 2000. Synchrony between Early Jurassic extinction, oceanic anoxic event, and the Karoo-Ferrar flood basalt volcanism. *Geology* 28 (8):747–50.

Pálfy, J., J. K. Mortensen, E. S. Carter, P. L. Smith, R. M. Friedman, and H. W. Tripper. 2000. Timing the end-Triassic mass extinction: First on land, then in the sea? *Geology* 28 (1):39–42.

Palumbi, S. R. 2001. Humans as the world's greatest evolutionary force. *Science* 293:1786–1790.

Parfit, M. 1995. Diminishing returns: exploiting the ocean's bounty. *National Geographic* 188 (5):2–37.

Parsons, K. M. 2001. *Drawing Out Leviathan: Dinosaurs and the Science Wars.* Bloomington: Indiana University Press.

Paul, G. S. 1988. *Predatory Dinosaurs of the World.* New York: Simon and Schuster.

———. 1991. The many myths, some old, some new, of dinosaurology. *Modern Geology* 16:69–99.

———. 2002. *Dinosaurs of the Air.* Baltimore: John Hopkins University Press.

Pearce, F. 2003. Going the way of the dodo? *New Scientist* 177 (2382):4–5.

Pearson, D. A., T. Schaefer, K. R. Johnson, and D. J. Nichols. 2000. Palynologically calibrated vertebrate record from North Dakota consistent with abrupt dinosaur extinction at the Cretaceous-Tertiary boundary. *Geology* 29 (1):39–42.

Pedersen, A. M. K., A. D. Greenwood, J. L. Jensen, F. Lee, C. Capelli, R. DeSalle, A. Tikhonov, P. A. Marx, and R. D. E. MacPhee. 2001. Evolution of endogenous retrovirus-like elements of the woolly mammoth (*Mammuthus primigenius*) and its relatives. *Mol. Bio. Evol.* 18 (5):840–47.

Pereladova, O. B., V. E. Flint, A. J. Sempéré, N. V. Soldatova, V. U. Dutov, and G. Fisenko. 1999. Przewalski's horse—adaptation to semi-wild life in desert conditions. *Oryx* 33 (1):46–57.

Perkins, S. 2002. Wings Aplenty. *Science News* 163 (4):51.

Perry, R. 1965. *The World of the Tiger.* New York: Atheneum.

Peters, D. S. and J. Qiang. 1999. Had *Confuciusornis* to be a climber? *Jour. Ornith.* 140:41–50.

Peters, S. E. and M. Foote. 2002. Determinants of extinction in the fossil record. *Nature* 416:420–24.

Peterson, D. 2003. *Eating Apes*. Berkeley: University of California Press.

Peterson, K. A. 2003. Human cultural agency in extinction. *Wild Earth* 11 (1):10–13.

Phillips, T. and P. Wilson, eds. 2002. *The Bear Bile Business*. World Society for the Protection of Animals.

Pimm, S. L., M. Ayres, A. Balmford, G. Branch, K. Brandon, T. Brooks, R. Bustamante, R. Costanza, R. Cowling, L. M. Curran, A. Dobson, S. Farber, G. A. B. da Fonseca, C. Gascon, R. Kitching, J. McNeely, T. Lovejoy, R. A. Mittermeier, N. Meyers, J. A. Patz, B. Raffle, D. Rapport, P. Raven, C. Roberts, J. P. Rodríguez, A. B. Rylands, C. Tucker, C. Safina, C. Samper, M. L. J. Stiassny, J. Supriatna, D. H. Wall, and D. Wilcove. 2001. Can we defy nature's end? *Science* 293:2207–2208.

Pitman, R. L. 2002. Alive and whale: a missing cetacean resurfaces in the tropics. *Natural History* 111 (7):32–36.

Pitman, R. L., D. M. Palacios, P. L. R. Brennan, B. J. Brennan, K. C. Balcomb, and T. Miyashita. 1999. Sightings and possible identity of a bottlenose whale in the tropical Indopacific: *Indopacetus pacificus*? *Mar. Mam. Sci.* 15 (2):531–49.

Pliny. Translated by G. P. Goold. 1933. *Naturalis Historia*. Cambridge, Mass.: Loeb Classical Library, Harvard University Press.

Pojeta, J. 2000. Fossils, G-men, money, and museums. *Science* 289:1695–1696.

Pope, K. O., A. C. Ocampo, and C. E. Duller. 1991. Mexican site for K/T impact crater? *Nature* 351:105.

Porter, R. 1997. *The Greatest Benefit to Mankind: The Medical History of Humanity*. New York: W. W. Norton and Co.

Pounds, J. A. 2001. Climate and amphibian declines. *Nature* 410:639–40.

Pounds, J. A., M. P. L. Fogden, and J. H. Campbell. 1999. Biological response to a climate change on a tropical mountain. *Nature* 398:611–15.

Powell, J. L. 1998. *Night Comes to the Cretaceous*. New York: Freeman.

Prothero, D. R. 1985. Mid-Oligocene extinction event in North American land mammals. *Science* 229:550–51.

———. 1986. North American mammalian diversity and Eocene-Oligocene extinctions. *Paleobiology* 11:389–405.

Prum, R. O. 2003. Dinosaurs take to the air. *Nature* 421:323–24.

Prum, R. O. and A. H. Brush. 2003. Which came first, the feather or the bird? *Scientific American* 288 (3):84–93.

Purdy, R. W. 1996. Paleoecology of fossil white sharks. In A. P. Klimley and D. G. Ainley, eds., *Great White Sharks*, pp. 67–78. San Diego: Academic Press.

Purvis, A., K. E. Jones, and G. M. Mace. 2000. Extinction. *BioEssays* 22:1123–1133.

Qiang, J., P. J. Currie, M. A. Norell, and J. Shu-an. 1998. Two feathered dinosaurs from northeastern China. *Nature* 393:753–61.

Quammen, D. 1996. *The Song of the Dodo: Island Biogeography in an Age of Extinctions*. New York: Scribner.

Queiroz, H. L. 1992. A new species of capuchin monkey, genus *Cebus* Erxleben, 1777 (Cebidae Primates) from Eastern Brazilian Amazonia. *Goeldiana, Zoologia* 15:1–13.

Quinn, J. F. 1983. Mass extinctions in the fossil record. *Science* 219:1239–1240.

Rabinowitz, A. 2001. *Beyond the Last Village*. Washington, D.C.: Island Press.

Rampino, M. R. and R. B. Stothers. 1984. Terrestrial mass extinctions, cometary impacts and the Sun's motion perpendicular to the galactic plane. *Nature* 308:709–17.

———. 1988. Flood basalt volcanism during the past 250 million years. *Science* 241:663–67.

Rampino, M. R. and T. Volk. 1988. Mass extinctions, atmospheric sulphur and climatic warming at the K/T boundary. *Nature* 338:247–49.

Rasoloarison, R. M., S. M. Goodman, and J. U. Ganzhorn. 2000. Taxonomic revision of mouse lemurs (*Microcebus*) in the western portions of Madagascar. *Int. Jour. Primat.* 21 (6):963–1019.

Rau, R. E. 1978. Additions to the revised list of preserved material of the extinct Cape Colony quagga and notes on the relationship and distribution of southern plains zebras. *Ann. S. Afr. Mus.* 77:27–45.

———. 1983. The coloration of the extinct Cape Colony quagga. *Afr. Wildl.* 37 (4):136–41.

Raup, D. M. 1984. Death of Species. In M. H. Nitecki, ed., *Extinctions*. pp. 1–19. Chicago: University of Chicago Press.

———. 1989. The case for extraterrestrial causes of extinction. *Phil. Trans. Roy. Soc. London* B325:421–35.

———. 1991a. Extinction: bad genes or bad luck? *New Scientist* 131 (1786):46–49.

———. 1991b. *Extinction: Bad Genes or Bad Luck?* New York: W. W. Norton and Co.

———. 1994. The role of extinction in evolution. *Proc. Nat. Acad. Sci.* 91:6758–6763.

———. 1996a. Extinction models. In D. Jablonski, D. H. Erwin, and J. H. Lipps, eds., *Evolutionary Paleobiology*, pp. 419–33. Chicago: University of Chicago Press.

———. 1996b. *The Nemesis Affair: A Story of the Death of the Dinosaurs and the Ways of Science*. New York: W. W. Norton and Co.

Raup, D. M. and J. J. Sepkoski. 1982. Mass extinctions in the marine fossil record. *Science* 215:1501–1503.

————. 1986. Periodic extinction of families and genera. *Science* 231:833–36.

Raup, D. M. and S. M. Stanley. 1971. *Principles of Paleontology*. New York: Freeman.

Raup, D. M., J. J. Sepkoski, and S. M. Stigler. 1982. Mass extinctions in the fossil record. *Science* 219:1240–1241.

Raven, P. H. 2001. What have we lost, what are we losing? In M. J. Novacek, ed., *The Biodiversity Crisis: Losing What Counts*, pp. 58–62. New York: American Museum of Natural History.

Ray, C. E. and D. P. Domning. 1986. Manatees and genocide. *Mar. Mam. Sci.* 2 (1):77–78.

Reading, R. P., H. Mix, B. Lhagvasuren, and E. S. Blumer. 1999. Status of wild Bactrian camels and other large ungulates in south-western Mongolia. *Oryx* 33 (3):247–55.

Reeves, R. R. and A. A. Chaudhry. 1998. Status of the Indus River dolphin *Platanista minor*. *Oryx* 32 (1):35–44.

Regan, H., R. Lupia, A. N. Drinnan, and M. A. Burgman. 2001. The currency and tempo of extinction. *American Naturalist* 157 (1):1–10.

Retallack, G. J., R. M. H. Smith, and P. D Ward. 2003. Vertebrate extinction across the Permian-Triassic boundary in Karoo Basin, South Africa. *Geol. Soc. Amer. Bull.* 115 (9):1133–1152.

Revkin, A. C. 2000. Extinction turns out to be a slow, slow process. *New York Times* October 24:F1–2.

Reynolds, V. 1967. *The Apes*. New York: Dutton.

Ricciuti, E. R. 1973. *To the Brink of Extinction*. New York: Holt, Rinehart, and Winston.

Rice, D. W. and A. A. Wolman. 1971. *The Life History and Ecology of the California Gray Whale (Eschrichtius robustus)*. Spec. Pub. No. 3. American Society of Mammalogists.

Rich, T. H., P. Vickers-Rich, and R. A. Gangloff. 2002. Polar dinosaurs. *Science* 295:979–80.

Richmond, B. G. and D. S. Strait. 2000. Evidence that humans developed from a knuckle-walking ancestor. *Nature* 404:382–85.

Ridley, M. 2000. *Genome: The Autobiography of a Species in 23 Chapters*. New York: HarperCollins.

Rigby, J. K. 1987. The last of the North American dinosaurs. In S. J. Czerkas and E. C. Olson, eds., *Dinosaurs Past and Present, vol. 2*, pp. 118–35. Los Angeles: Natural History Museum of Los Angeles County and University of Washington Press.

Roberts, R. G., T. F. Flannery, L. K. Ayliffe, H. Yoshida, J. M. Olley, G. J. Prideaux, G. M. Laslett, A. Baynes, M. A. Smith, R. Jones, and B. L. Smith. 2001. New ages for the last Australian megafauna: continent-wide extinction about 46,000 years ago. *Science* 292:1888–1892.

Robichaud, W. G. 1998. Physical and behavioral description of a captive saola, *Pseudoryx nghetinhensis*. *Jour. Mammal.* 79 (2):394–405.

Roelke-Parker, M. E., L. Munson, C. Packer, R. Kock, S. Cleaveland, M. Carpenter, S. I. O'Brien, A. Pospischil, R. Hofmann-Lehmann, H. Lutz, G. L. M. Mwamengele, M. N. Mgasa, G. A. Machange, B. A. Summers, and M. J. G. Appel. 1996. A canine distemper virus epidemic in Serengeti lions (*Panthera leo*). *Nature* 379 (6564):441–45.

Roman, J. 2000. Is the right whale going down? *Wildlife Conservation* 103 (3):26–35.

Romer, A. S. 1962. *Man and the Vertebrates*. New York: Penguin.

———. 1966. *Vertebrate Paleontology*. Chicago: University of Chicago Press.

Rookmaaker, K. 2002. Miscounted population of the southern white rhinoceros in the early 19th century. *Pachyderm* 32:22–27.

Rose, M. R. 1998. *Darwin's Spectre: Evolutionary Biology in the Modern World*. Princeton, N. J.: Princeton University Press.

Ross, P. S., J. G. Vos, L. S. Birnbaum, and A. D. Osterhaus. 2000. PCBs are a health risk for humans and wildlife. *Science* 289:1878–1879.

Rothschild, B. and M. Helbling. 2001.Documentation of hyperdisease in the Late Pleistocene: validation of an early 20th century hypothesis. *Jour. Vert. Paleo.* 21 (3) Supp. 94A.

Rougier, G. W., M. J. Novacek, and D. Dashzeveg. 1997. A new multituberculate from the late Cretaceous locality Ukhaa Tolgod, Mongolia: considerations on multituberculate interrelationships. *Amer. Mus. Novitates* 3191:1–26.

Rowe, T., R. A. Ketcham, C. Denison, M. Colbert, X. Xu, and P. J. Currie. 2001. The *Archaeoraptor* forgery. *Nature* 410:539–40.

Ruderman, D. A. 1974. Possible consequences of nearby supernova explosions for atmospheric ozone and terrestrial life. *Science* 184:1079–1081.

Rudwick, M. J. S. 1972. *The Meaning of Fossils*. Chicago: University of Chicago Press.

———. 1992. *Scenes from Deep Time*. Chicago: University of Chicago Press.

———. 1997. *Georges Cuvier, Fossil Bones, and Geological Catastrophes*. Chicago: University of Chicago Press.

Russell, D. A. 1979. The enigma of the extinction of the dinosaurs. *Ann. Rev. Earth Planet. Sci.* 7:163–82.

———. 1982. The mass extinctions of the Late Mesozoic. *Scientific American* 246 (1):58–65.

———. 1984. The gradual decline of the dinosaurs—fact or fallacy? *Nature* 307:360–61.

———. 1989. *An Odyssey in Time: The Dinosaurs of North America*. Toronto: University of Toronto Press, and National Museum of Natural Sciences.

Russell, D. A. and W. Tucker. 1971. Supernovae and the extinction of the dinosaurs. *Nature* 229:553–54.

Ryskin, G. 2003. Methane-driven oceanic eruptions and mass extinctions. *Geology* 31 (9):741–44.

Sankhala, K. 1977. *Tiger! The Story of the Indian Tiger.* New York: Simon and Schuster.

Scammon, C. M. 1874. *The Marine Mammals of the Northwestern Coast of North America; Together with an Account of the American Whale Fishery.* New York: Carmany, and G. P. Putnam's Sons.

Schaller, G. B. 1964. *The Year of the Gorilla.* Chicago: University of Chicago Press.

———. 1972. *The Mountain Gorilla: Ecology and Behavior.* Chicago: University of Chicago Press.

———. 1981. Pandas in the wild. *National Geographic* 160 (6):735–49.

———. 1993. Tibet's remote Chang Tang. *National Geographic* 184 (2):63–87.

———. 1995. Gentle gorillas, turbulent times. *National Geographic* 188 (4):58–83.

———. 2003. Drop dead gorgeous: why poachers are killing the chiru. *National Geographic* 203 (4):124–25.

Schaller, G. B., and A. Rabinowitz. 1995. The saola or spindlehorn bovid *Pseudoryx nghetinhensis* in Laos. *Oryx* 29 (2):107–15.

Scheffer, V. 1958. *Seals, Sea Lions, and Walruses.* Palo Alto, Calif.: Stanford University Press.

Scheuchzer, J. J. 1708. *Piscium Querelae et Vindiciae.* Zurich.

Schmitt, F. P., C. de Jong, and F. H. Winter. 1980. *Thomas Welcome Roys: America's Pioneer of Modern Whaling.* Charlottesville: University Press of Virginia.

Schubert, C. 2001. Life on the edge: will a mass extinction usher in a world of weeds and pests? *Science News* 160 (11):168–70.

Schwarz, E. 1929. Das Vorkommen des Schimpansen auf den linken Kongo-Ufer. *Rev. zool. botan. Africaines* 16:425–26.

Scoresby, W. 1820. *An Account of the Arctic Regions with a History and Description of the Northern Whale Fishery.* 2 Vols. Archibald Constable (David & Charles edition, 1969).

Scotese, C. R., L. M. Gahagan, and R. L. Larson. 1988. Plate tectonic reconstructions of the Cretaceous and Cenozoic ocean basins. In C. R. Scotese and W. W. Sager, eds., *Mesozoic and Cenozoic Plate Reconstructions, Tectonophysics.* 155:27–48.

Seeley, H. G. 1887. On the classification of fossil animals commonly named Dinosauria. *Proc. Roy. Soc. London* 43 (206):165–71.

———. 1888. The classification of the Dinosauria. *Rep. Br. Assoc. Adv. Sci.* 1887:698–99.

Seidensticker, J. 1996. *Tigers.* Stillwater, Minn.: Voyageur.

Sepkoski, J. J. 1984. A kinetic model of Phanerozoic diversity. III. Post-Paleozoic families and mass extinctions. *Paleobiology* 10 (2):246–67.

Sereno, P. C. 1995. Roots of the family tree. *Natural History* 104 (6):30–32.

———. 1997. The origin and evolution of dinosaurs. *Ann. Rev. Earth Planet. Sci.* 25:435–89.

———. 1999. The evolution of dinosaurs. *Science* 284:2137–47.

Shapiro, B., D. Sibthorpe, A. Rambaut, J. Austin, G. M. Wragg, O. R. P. Bininda-Edmonds, P. L. M. Lee, and A. Cooper. 2002. Flight of the dodo. *Science* 295:1683.

Sharpton, V. L. 1995. Chicxulub impact crater provides clues to Earth's history. *Earth in Space* 8 (4):7.

Sharpton, V. L., G. B. Dalrymple, L. E. Martin, G. Ryder, B. C. Schuraytz, and J. Urrutia-Fucugauchi. 1992. New links between the Chicxulub impact structure and the Cretaceous/Tertiary Boundary. *Nature* 359:819–21.

Shay, D. and J. Duncan. 1993. *The Making of Jurassic Park.* New York: Ballantine.

Sheehan, P. M., D. E. Fastovsky, C. Barreto, and R. G. Hoffman. 2000. Dinosaur abundance was not declining in the "3m gap" at the top of the Hell Creek Formation, Montana and North Dakota. *Geology* 28 (6):523–26.

Sheehan, P. M., D. E. Fastovsky, R. G. Hoffman, C. B. Berghaus, and D. L. Gabriel. 1991. Sudden extinction of the dinosaurs: Latest Cretaceous, Upper Great Plains, USA. *Science* 254:835–39.

Shelden, K. E. W., D. J. Rugh, B. A. Mahoney, and M. E. Dalheim. 2003. Killer whale predation on belugas in Cook Inlet, Alaska: Implications for a depleted population. *Mar. Mam. Sci.* 19 (3):529–44.

Shepherd, A. 1966. *The Flight of the Unicorns.* New York: Abelard-Schuman.

Shuker, K. 1993. *The Lost Ark.* New York: HarperCollins.

Shukolyukov, A. and G. W. Lugmair. 1998. Isotopic evidence for the Cretaceous-Tertiary impactor and its type. *Science* 282:927–29.

Sibley, D. A. 2000. *The Sibley Guide to Birds (National Audubon Society).* New York: Alfred A. Knopf.

Silverberg, R. 1967. *The Auk, the Dodo, and the Oryx: Vanished and Vanishing Creatures.* New York: Crowell.

Simon, N. and P. Géroudet. 1970. *Last Survivors: The Natural History of Animals in Danger of Extinction.* New York: World.

Simons, L. M. 2000. Archaeoraptor fossil trail. *National Geographic* 198 (4):128–32.

Simpson, G. G. 1951. *Horses.* New York: Oxford University Press.

———. 1953. *Life of the Past.* New Haven: Yale University Press.

———. 1984. *Tempo and Mode in Evolution.* New York: Columbia University Press.

Simpson, S. 2001. Fishy Business. *Scientific American* 285 (1):82–88.

Sleep, N. H., K. J. Zahnle, J. F. Kasting, and H. J. Morowitz. 1989. Annihilation of ecosystems by large asteroid impacts on the early Earth. *Nature* 342:139–42.

Sloan, C. 1999. Feathers for T. rex? *National Geographic* 196 (5):98–107.

————. 2003. Lord of the wings. *National Geographic* 203 (5):22–23.

Sloan, R. E. and L. M. Van Valen. 1965. Cretaceous mammals from Montana. *Science* 148:220–27.

Sloan, R. E., J. K. Rigby, L. M. Van Valen and D. Gabriel. 1986. Gradual dinosaur extinction and simultaneous ungulate radiation in the Hell Creek Formation. *Science* 232:629–32.

Smit, J. and J. Hertogen. 1980. An extraterrestrial event at the Cretaceous-Tertiary boundary. *Nature* 285:198–200.

Smith, J. M. 1989. The causes of extinction. *Phil. Trans. Roy. Soc. London* B325:241–52.

Smith, T. G., D. J. St. Aubin, and J. R. Geraci, eds. 1990. *Advances in Research on the Beluga Whale, Delphinapterus leucas.* Department of Fisheries and Oceans, Ottawa.

Sokolov, V. E. 1974. *Saiga tartarica* [Saiga]. *Mammalian Species* 38:1.4. American Society of Mammalogists.

Spalton, J. A., S. A. Brend, and M. W. Lawrence. 1999. Arabian oryx reintroduction in Oman: successes and setbacks. *Oryx* 33 (2):168–75.

Spray, J. G., S. P. Kelley, and D. B. Rowley. 1998. Evidence for a late Triassic multiple impact on Earth. *Nature* 392:171–73.

Stanley, S. M. 1984a. Marine mass extinctions: a dominant role for temperature. In M. H. Nitecki, ed., *Extinctions,* pp. 69–117. Chicago: University of Chicago Press.

————. 1984b. Mass extinctions in the ocean. *Scientific American* 250 (6):64–72.

————. 1984c. Does bradytely exist? In N. Eldredge and S. M. Stanley, eds., *Living Fossils,* pp. 278–80. New York: Springer-Verlag.

————. 1987. *Extinction.* New York: Scientific American Library.

Stejneger, L. 1884. Contributions to the history of the Commander Islands. No. 2. Investigations relating to the date of the extermination of Steller's sea-cow. *Proc. U. S. Natl. Mus.* 8:181–89.

————. 1887. How the great northern sea-cow (*Rytina*) became exterminated. *Amer. Nat.* 21: 1047–54.

————. 1936. *Georg Wilhelm Steller.* Cambridge, Mass.: Harvard University Press.

Steller, G. W. 1781. *Journal of a Voyage with Bering, 1741–1742* (Palo Alto, Calif.: Stanford University Press edition, 1988).

Steno [Niels Stensen]. 1667. *Elementorum Myologiae Specimen . . . Canis Carchariae Dissectum Caput.* Florence.

Stevens, W. K. 2000. Suspects in "blitzkrieg" extinctions: primitive hunters. *New York Times* March 28:F5.

Stokstad, E. 2001. New fossil fills gap in bird evolution. *Science* 291:225.

Stone, R. 2001a. The cold zone: testing a new theory of mammoth extinction. *Discover* 22 (2):58–65.

————. 2001b. *Mammoth: The Resurrection of an Ice Age Giant*. New York: Perseus.

————. 2003. Championing a 17th century underdog. *Science* 301:152.

Stuart, A. J. 1999. Late Pleistocene megafaunal extinctions: a European perspective. In R. D. E. MacPhee, ed., *Extinctions in Near Time: Causes, Contexts and Consequences*, pp. 257–69. San Diego: Academic Press.

Sues, H.-D. 2001. Ruffling feathers. *Nature* 410:1036–1037.

Sutcliffe, A. J. 1985. *On the Track of Ice Age Mammals*. Cambridge, Mass.: Harvard University Press.

Sutherland, W. J. 2002. Science, sex, and the kakapo. *Nature* 419:265–66.

Swisher, C. C., Y.-Q. Wang, X.-L. Wang, X. Xu, and Y. Wang. 1999. Cretaceous age for the feathered dinosaurs of Liaoning, China. *Nature* 400:58–61.

Taldukar, B. K. 2002. Dedication leads to reduced rhino poaching in Assam in recent years. *Pachyderm* 33:58–63.

Tanner, J. T. 1941. Three years with the ivory-billed woodpecker, America's rarest bird. *Audubon* January–February: 5–14.

————. 1942. Present status of the ivory-billed woodpecker. *Wilson Bull.* 54 (1):57–58.

Tarduno, J. A., D. B. Brinkman, P. R. Renne, R. D. Cottrell, H. Scher, and P. Castillo. 1998. Evidence for extreme climatic warmth from Late Cretaceous Arctic vertebrates. *Science* 282 (5397):2241–2243.

Tattersall, I. 2000. Once we were not alone. *Scientific American* 282 (1):56–62.

Taylor, L. R., L. J. V. Compagno, and P. J. Struhsaker. 1983. Megamouth—a new species, genus, and family of lamnoid shark (*Megachasma pelagios*, family Megachasmidae) from the Hawaiian Islands. *Proc. Cal. Acad. Sci.* 43 (8):87–110.

Temple, S. 1977. Plant-animal mutualism: coevolution with dodo leads to near extinction of plant. *Science* 197:885–86.

Terry, D. O., J. A. Chamberlain, P. W. Stoffer, P. Messina, and P. A. Jannett. 2001. Marine Cretaceous-Tertiary boundary section in southwestern South Dakota. *Geology* 29 (11):1055–1088.

Terry, K. D. and W. H. Tucker. 1968. Biologic effects of supernovae. *Science* 159:421– 23.

Teuke, M. R. 2003. Spreading their wings. *Midwest Airlines Magazine* 11 (2):24–29.

Thain, M. and M. Hickman. 1994. *Penguin Dictionary of Biology*. New York: Penguin Reference.

Thorne, E. T. and E. S. Williams. 1988. Disease and endangered species: The black-footed ferret as a recent example. *Conservation Biology* 2:66–74.

Thulborn, T., and S. Turner. 2002. The last dicynodont: an Australian Cretaceous relict. *Proceedings of the Royal Society* B270 (1518):985–93.

Tickell, W. L. N. 1975. Observations on the status of Steller's albatross (*Diomedea albatrus*) 1973. *Bull. Int. Council Bird Preservation*. 12:125–31.

————. 2000. *Albatrosses*. New Haven: Yale University Press.

Townsend, C. H. 1937. *Guide to the New York Aquarium*. New York: New York Zoological Society.

Tratz, E. P. and H. Heck. 1954. Der afrikanische Anthropoide "Bonobo:" Eine neue Menschenaffengattung. *Säugertierkundliche Mitteilungen* 2:97:101.

Trotter, M. M. and B. McCulloch. 1995. Moas, men, and middens. In P. S. Martin and R. G. Klein, eds., *Quaternary Extinctions: A Prehistoric Revolution*. pp. 708–27. Tucson: University of Arizona Press.

Tudge, C. 2000. *The Variety of Life*. New York: Oxford University Press.

Tumlison, R. 1987. *Felis lynx* [Lynx]. *Mammalian Species* 269:1–8. American Society of Mammalogists.

Turner, A. 1997. *The Big Cats and Their Fossil Relatives*. New York: Columbia University Press.

Twigger, R. 2001. *The Extinction Club*. New York: William Morrow.

Vadja, V., J. I. Raine, and C. J. Hollis. 2001. Indication of global deforestation at the Cretaceous-Tertiary boundary by New Zealand fern spike. *Science* 294:1700–1702.

van Lawick, H. and J. Goodall. 1971. *Innocent Killers*. New York: Houghton Mifflin.

van Oosterzie, P. 1997. *Where Worlds Collide: The Wallace Line*. Ithaca, N.Y.: Cornell University Press.

van Roosmalen, M. G. M., T. van Roosmalen, and R. A. Mittermeier. 2002. A taxonomic review of the titi monkeys, genus *Callicebus* Thomas, 1903, with the description of two new species, *Callicebus bernhardi* and *Callicebus stephennashi*, from Brazilian Amazonia. *Neotropical Primates* 10 (Suppl.):1–52.

van Roosmalen, M. G. M., T. van Roosmalen, R. A. Mittermeier, and A. B. Rylands. 2000. Two new species of marmoset, genus *Callithrix* Erxleben, 1777 (Callitrichidae, Primates), from the Tapajós/Madeira interfluvium, south central Amazonia, Brazil. *Neotropical Primates* 8 (1):2–18.

van Schaik, C. P., K. A. Monk, and J. M. Yarrow-Robertson. 2001. Dramatic decline in orang-utan numbers in the Leuser Ecosystem, northern Sumatra. *Oryx* 33 (1)14–25.

van Schaik, C. P., M. Ancrenaz, G. Borgen, B. Galdikas, C. D. Knott, I. Singleton, A. Suzuki, S. S. Utami, and M. Merrill. 2003. Orangutan cultures and the evolution of material culture. *Science* 299:102–5.

Van Valen, L. M. 1984. Catastrophes, expectations, and the evidence. *Paleobiology* 10:121–37.

Van Valen, L. M. and R. E. Sloan. 1977. Ecology and extinction of the dinosaurs. *Evolutionary Theory* 2:37–64.

Vartanyan, S. L., K. A. Arslanov, T. V. Tertychnaya, and S. B. Chernov. 1995. Radiocarbon dating evidence for mammoths on Wrangel Island, Arctic Ocean, until 2000 B.C. *Radiocarbon* 37 (1):1–6.

Veale, S. 2002. They're back! glad to see 'em? *New York Times* October 13:4:2.

Verney, P. 1979. *Animals in Peril: Man's War Against Wildlife.* Provo, Utah: Brigham Young University Press.

Verrill, A. H. 1948. *Strange Prehistoric Animals and Their Stories.* Boston: L. C. Page.

Vickers-Rich, P. and T. H. Rich. 1999. *Wildlife of Gondwana: Dinosaurs and Other Vertebrates from the Ancient Supercontinent.* Bloomington: Indiana University Press.

Vidal, O. and J.-P. Gallo-Reynoso. 1996. Die-offs of marine mammals and sea birds in the Gulf of California, México. *Mar. Mam. Sci.* 12 (4):627–34.

Vigne, L. and E. B. Martin. 1989. Taiwan: the greatest threat to the survival of Africa's rhinos. *Pachyderm* 11:23–25.

Vogel, G. 2000. Conflict in Congo threatens bonobos and rare gorillas. *Science* 287:2386–87.

———. 2003a. Orangutans, like chimps, heed the cultural call of the collective. *Science* 299:27–28.

———. 2003b. Can great apes be saved from Ebola? *Science* 300:1645.

Vrba, E. S. and G. B. Schaller. 2000. *Antelopes, Deer, and Relatives.* New Haven: Yale University Press.

Wada, S., M. Oishi, and T. K. Yamada. 2003. A new discovered species of living baleen whale. *Nature* 426:278–81.

Walker, J. F. 2002. *A Certain Curve of Horn: The Hundred-Year Quest for the Giant Sable Antelope of Angola.* New York: Atlantic Monthly Press.

Wallace, A. R. 1869. *The Malay Archipelago, the land of the Orang-Utan and the Bird of Paradise: A Narrative of Travel with Studies of Man and Nature.* London (Dover edition, 1962).

Walsh, P. D., K. A. Abernathy, M. Bermejo, R. Beyers, P. de Wachter, M. E. Akou, B. Huijbregts, D. I. Mambounga, A. K. Toham, A. M. Kilbourn, S. A. Lahm, S. Latour, F. Maisels, C. Mbina, Y. Mihindou, S. N. Obiang, E. N. Effa, M. P. Starkey, P. Telfer, M. Thibault, C. G. Tutin, L. J. T. White, and D. S. Wilkie. 2003. Catastrophic ape decline in western equatorial Africa. *Nature* 422:611–14.

Ward, G. C. 1997. Making room for wild tigers. *National Geographic* 192 (6):2–35.

Ward, P. D. 1983. *Nautilus pompilius.* In P. R. Boyle, ed., *Cephalopod Life Cycles. Vol. 1, Species Accounts,* pp.11–28. San Diego: Academic Press.

———. 1984. Is *Nautilus* a living fossil? In N. Eldredge and S. M. Stanley, eds., *Living Fossils,* pp. 247–56. New York: Springer-Verlag.

———. 1988. *In Search of Nautilus.* New York: Simon and Schuster.

———. 1992. *On Methuselah's Trail.* New York: Freeman.

———. 1997. *The Call of Distant Mammoths.* Toronto: Copernicus.

———. 1999. *Time Machine: Scientific Explorations in Deep Time.* Toronto: Copernicus.

Ward, P. D. and D. Brownlee. 2000. *Rare Earth.* Toronto: Copernicus.

Ward, P. D. and W. B. Saunders. 1997. *Allonautilus*: a new genus of living nautiloid cephalopod and its bearing on phylogeny of the Nautilida. *Jour. Paleo.* 71 (6):1054–1064.

Ward, P. D., D. R. Montgomery, and R. Smith. 2000. Altered river morphology in South Africa related to the Permian-Triassic extinction. *Science* 289:1740–1743.

Ward, P. D., J. W. Haggart, E. S. Carter, D. Wilbur, H. W. Tipper, and T. Evans. 2001. Sudden productivity collapse associated with the Triassic-Jurassic Boundary mass extinction. *Science* 292:1148–1151.

Webb, J. 1991. Dolphin epidemic spreads to Greece. *New Scientist* 130:14.

Webb, S. D. 1995. Ten million years of mammal extinctions in North America. In P. S. Martin and R. G. Klein, eds., *Quaternary Extinctions: A Prehistoric Revolution*, pp. 189–210. Tucson: University of Arizona Press.

Webster, D. 1996. Dinosaurs of the Gobi: Unearthing a fossil trove. *National Geographic* 190 (1):70–89.

Weidensaul, S. 1999. *Living on the Wind: Across the Hemisphere with Migratory Birds.* New York: North Point Press.

———. 2002. *The Ghost with Trembling Wings.* New York: North Point Press.

Weiner, J. 1994. *The Beak of the Finch.* New York: Alfred A. Knopf.

Weintraub, B. 1995. First look at new Asian mammal. *National Geographic* 187 (1):12.

Weishampel, D. B., P. Dodson, and H. Osmólska, eds. 1990. *The Dinosauria.* Berkeley: University of California Press.

Wellnhofer, P. Archaeopteryx. 1990. *Scientific American* 262 (5):70–77.

Wendt, H. 1959. *Out of Noah's Ark.* Boston: Houghton Mifflin.

———. 1968. *Before the Deluge.* New York: Doubleday.

Wenshi, P. 1995. New hope for China's giant pandas. *National Geographic* 187 (2):100–115.

Wesson, R. 1997. *Beyond Natural Selection.* Cambridge, Mass.: MIT Press.

Western, D. 1989. The undetected trade in rhino horn. *Pachyderm* 11:26–29.

Westphal, S. P. 2002. So simple, almost anyone can do it. *New Scientist* 175 (2356):16–17.

White, T. D., G. Suwa, and B. Asfaw. 1994. *Australopithecus ramidus,* a new species of hominid from Aramis, Ethiopia. *Nature* 371:306–12.

White, T. D., B. Asfaw, D. Degusta, H. Gilbert, G. D. Richards, G. Suwa, and F. C. Howell. 2003. Pleistocene *Homo sapiens* from Middle Awash, Ethiopia. *Nature* 423:742–47.

Whiten, A., J. Goodall, W. C. Mcgrew, T. Nishida, V. Reynolds, Y. Sugiyama, C. E. G. Tutin, R. W. Wrangham, and C. Boesch. 1999. Cultures in chimpanzees. *Nature* 399: 682–85.

Whitfield, J. 2003a. The law of the jungle. *Nature* 421:8–9.

———. 2003b. Ape populations decimated by hunting and Ebola virus. *Nature* 422:551.

Wilkie, D. S. and J. Carpenter. 1999. Bushmeat hunting in the Congo Basin: an assessment of impacts and options for mitigation. *Biodiversity and Conservation* 8:927–55.

Willerslev, E., A. J. Hansen, J. Binladen, T. B. Brand, M. T. P. Gilbert, B. Shapiro, M. Bunce, C. Wiuf, D. A. Gilichinsky, and A. Cooper. 2003. Diverse plant and animal genetic records from Holocene and Pleistocene sediments. *Science* 300:791–95.

Williams, E. S., E. T. Thorne, M. J. G. Appel, and D. W. Belitsky. 1988. Canine distemper in black-footed ferrets (*Mustela nigripes*) from Wyoming. *Journal of Wildlife Diseases* 24 (3):385–98.

Wilson, E. O. 1992. *The Diversity of Life*. Cambridge, Mass.: Harvard University Press.

———. 2001. Biodiversity: wildlife in trouble. In M. J. Novacek, ed., *The Biodiversity Crisis: Losing What Counts*, pp. 18–20. New York: American Museum of Natural History.

———. 2002. *The Future of Life*. New York: Alfred A. Knopf.

Winchester, S. 2001. *The Map that Changed the World: William Smith and The Birth of Modern Geology*. New York: HarperCollins.

Wong, K. 2001. Mammoth kill. *Scientific American* 284 (2):22.

Woodroffe, R. 2001. Assessing the risks of intervention: immobilization, radio-collaring and vaccination of African wild dogs. *Oryx* 35 (3):234–44.

Woodroffe, R. and J. R. Ginsberg. 1999a. Conserving the African wild dog *Lycaon pictus*. I. Diagnosing and treating causes of decline. *Oryx* 33 (2):132–42.

———. 1999b. Conserving the African wild dog *Lycaon pictus*. II. Is there a role for reintoduction? *Oryx* 33 (2):143–51.

Woodroffe, R., J. R. Ginsberg, D. W. R. Macdonald, and the IUCN/SSC Canid Specialist Group. 1997. *The African Wild Dog: Status Survey and Conservation Action Plan*. Gland, Switzerland: IUCN Publications.

Woodruff, D. S. 2001. Declines of biomes and biotas and the future of evolution *Proc. Natl. Acad. Sci.* 98 (10):5471–76.

Woodward, J. 1695. *Essay Toward the Natural History of the Earth: and Terrestrial Bodies, especially Minerals: as also of the Seas, Rivers and Springs. With an Account of the Universal Deluge: and of the Effects that it had upon the Earth*. London.

Wright, K. 2001. The mother of all extinctions. *Discover* 22 (10):28–29.

Wroe, S. 1998. Killer kangaroo. *Australasian Science* 19:25–28.

———. 1999a. Killer kangaroos and other murderous marsupials. *Scientific American* 280 (5):68–74.

———. 1999b. The bird from hell? *Nature Australia* 26:58–64.

————. 2000. Move over sabre-toothed tiger. *Nature Australia* 27:44–51.

————. 2001. The lost kingdoms of Australia. *Newton* 4:98–104.

————. 2002. A review of terrestrial mammalian and reptilian carnivore ecology in Australian fossil faunas, and factors influencing their diversity: the myth of reptilian domination and its broader ramifications. *Aust. Jour. Zool.* 50:1–24.

Wroe, S., and J. Field. 2001a. Mystery of megafaunal extinctions remains. *Australasian Science* 22:21–25.

————. 2001b. Giant wombats and red herrings. *Australasian Science* 22:18.

Wroe, S., J. Field, and R. Fullagar. 2002. Lost giants. *Nature Australia* 27 (5):54–61.

Wroe , S., T. J. Myers, R. T. Wells, and A. Gillespie. 1999. Estimating the weight of the Pleistocene marsupial lion, *Thylacoleo carnifex* (Thylacoleonidae: Marsupialia): implications for the ecomorphology of a marsupial super-predator and hypotheses of impoverishment of Australian marsupial carnivore faunas. *Aust. Jour. Zool.* 47:489–98.

Wynn, J. C., and E. M. Shoemaker. 1998. The day the sands caught fire. *Scientific American* 279 (5):65–71.

Wyss, A. 2001. Digging up fresh clues about the origin of mammals. *Science* 292:1496–1497.

Xu, X. 2000. "Feathers for T. rex?" [Letter to the Editor] *National Geographic* 197 (3):xviii.

Xu, X., X. Wang, and X. Wu. 1999. A dromaeosaurid dinosaur with a filamentous integument from the Xixian Formation of China. *Nature* 401:262–66.

Xu, X., Z. Zhou, and R. O. Prum. 2001. Branched integumental structures in *Sinornithosaurus* and the origin of feathers. *Nature* 408:200–203.

Xu, X. Z. Zhou, and X. Wang. 2000. The smallest known non-avian theropod dinosaur. *Nature* 408:705–8.

Xu, X., Z. Zhou, X. Wang, X. Kuang, F. Zheng, and X. Du. 2003. Four-winged dinosaurs from China. *Nature* 421:335–40.

Yarrow-Robertson, J. M. and C. P. van Schaik. 2001. Causal factors underlying the dramatic decline of the Sumatran orang-utan. *Oryx* 35 (1):26–38.

Zhang, F. and Z. Zhou. 2000. A primitive Enantiornithine bird and the origin of feathers. *Science* 290:1955–1959.

Zhi, L. 1993. Newborn panda in the wild. *National Geographic* 183 (2):60–65.

————. 2002. Kingdom of the panda. *Discover* 23 (11):62–67.

Zhou, Z. and F. Zhang. 2001. Two new genera of ornithurine birds from the Early Cretaceous of Liaoxi related to the origin of modern birds. *Chinese Sci. Bull* 46 (5):371–77.

————. 2002a. A long-tailed, seed eating bird from the Early Cretaceous of China. *Nature* 418:405–9.

———. 2002b. Largest bird from the Early Cretaceous and its implications for the earliest avian ecological diversification. *Naturwissenschaften* 89:34–38.

Zhou, Z., P. M. Barrett, and J. Hilton. 2003. An exceptionally preserved Lower Cretaceous ecosystem. *Nature* 421:807–14.

Zhou, Z., J. A. Clarke, and F. Zhang. 2002. *Archaeoraptor*'s better half. *Nature* 420:285.

Zihlman, A. L., J. E. Cronin, D. L. Kramer, and V. M. Sarich. 1978. Pygmy chimpanzees as a possible prototype for the common ancestor of humans, chimpanzees, and gorillas. *Nature* 275:744–46.

Index